TM 9-775

WAR DEPARTMENT TECHNICAL MANUAL

LANDING VEHICLE TRACKED,

LVT MK. I AND MK. II

TECHNICAL MANUAL

RESTRICTED DISSEMINATION OF RESTRICTED MATTER — The information contained in restricted documents and the essential characteristics of restricted materiel may be given to any person known to be in the service of the United States and to persons of undoubted loyalty and discretion who are cooperating in Government work, but will not be communicated to the public or to the press except by authorized military public relations agencies. (See also paragraph 18 b. AR 380-5, 28 September 1942.)

WAR DEPARTMENT • 5 FEBRUARY 1944

©2013 Periscope Film LLC
All Rights Reserved
ISBN#978-1-937684-36-5
www.PeriscopeFilm.com

DISCLAIMER:

This manual is sold for historic research purposes only, as an entertainment. It contains obsolete information and is not intended to be used as part of an actual operation or maintenance training program. No book can substitute for proper training by an authorized instructor.

©2013 Periscope Film LLC
All Rights Reserved
ISBN#978-1-937684-36-5
www.PeriscopeFilm.com

WAR DEPARTMENT TECHNICAL MANUAL

TM 9-775

Landing Vehicle Tracked, MK. I and MK. II

WAR DEPARTMENT • *5 FEBRUARY 1944*

RESTRICTED *DISSEMINATION OF RESTRICTED MATTER—*
The information contained in restricted documents and the essential characteristics of restricted materiel may be given to any person known to be in the service of the United States and to persons of undoubted loyalty and discretion who are cooperating in Goverment work, but will not be communicated to the public or to the press except by authorized military public relations agencies. (See also paragraph 18 b, AR 380-5, 28 September 1942.)

★All pertinent information from TB 700-10, dated 9 January 1943, and TB 700-11, dated 28 June 1943 insofar as it applies to this materiel, is superseded by TM 9-775.

WAR DEPARTMENT
Washington 25, D.C., 5 February 1944

TM 9-775, Landing Vehicle Tracked Mk. I and Mk. II, is published for the information and guidance of all concerned.

$$\begin{bmatrix} \text{A.G. 300.7 (26 Apr 43)} \\ \text{O.O.M. 461/RA (5 Feb 44) R} \end{bmatrix}$$

BY ORDER OF THE SECRETARY OF WAR:

G. C. MARSHALL,
Chief of Staff.

OFFICIAL:
    J. A. ULIO,
       Major General,
          The Adjutant General.

DISTRIBUTION: IC and H 5 (4); R 9 (2); Bn 9 (1); C 9 (4).

(For explanation of symbols, see FM 21-6.)

*TM 9-775

CONTENTS

PART ONE — VEHICLE OPERATING INSTRUCTIONS

		Paragraphs	Pages
SECTION	I. Introduction	1– 2	5– 8
	II. Description and tabulated data	3– 5	9– 22
	III. Driving controls and operation	6– 8	23– 35
	IV. Turret controls and operation	9– 11	35– 38
	V. Auxiliary equipment controls and operation	12– 14	39– 41
	VI. Operation under unusual conditions	15– 20	41– 44
	VII. First echelon preventive maintenance service	21– 25	44– 56
	VIII. Lubrication	26– 27	56– 64
	IX. Tools and equipment stowage on the vehicle	28– 34	64– 80

PART TWO — VEHICLE MAINTENANCE INSTRUCTIONS

SECTION	X. New vehicle run-in test	35– 37	81– 87
	XI. Second echelon preventive maintenance	38– 39	87–107
	XII. Organizational tools and equipment	40	107–110
	XIII. Trouble shooting	41– 55	110–129
	XIV. Engine — description, data, maintenance, and adjustment in vehicle	56– 60	129–134
	XV. Engine — removal and installation	61– 65	134–161
	XVI. Fuel system	66– 74	161–187
	XVII. Intake and exhaust systems	75– 80	188–194
	XVIII. Cooling system	81– 82	195

★All pertinent information from TB 700-50, dated 9 January 1943, and TB 700-71, dated 28 June 1943 insofar as it applies to this materiel, is superseded by TM 9-775.

LANDING VEHICLE TRACKED MK. I AND MK. II

CONTENTS – Cont'd

PART TWO—VEHICLE MAINTENANCE INSTRUCTIONS
(Continued)

		Paragraphs	Pages
SECTION	XIX. Lubrication system	83–89	195–206
	XX. Ignition system	90–93	207–217
	XXI. Starting system	94–95	217–218
	XXII. Generating system	96–98	219–221
	XXIII. Battery and lighting system	99–111	222–240
	XXIV. Radio interference suppression system	112–113	240–242
	XXV. Clutch assembly	114	242–247
	XXVI. Propeller shafts and universal joints	115–116	247–254
	XXVII. Final drive	117–118	254–261
	XXVIII. Power train	119–125	261–284
	XXIX. Tracks and suspension	126–137	284–309
	XXX. Panels and instruments	138–141	309–319
	XXXI. Controls and linkage	142–144	320–328
	XXXII. Fire extinguishing system	145–151	329–334
	XXXIII. Hull	152–160	334–348
	XXXIV. Turret	161–171	349–363
	XXXV. Auxiliary generator	172–175	363–368

PART THREE — ARMAMENT

SECTION	XXXVI. Introduction	176	369
	XXXVII. Description of guns and mount	177	369–370
	XXXVIII. Operation of guns	178–181	370–371
	XXXIX. Removal and installation of guns	182–183	371–372
	XL. Stabilizer	184–196	373–386
	XLI. Shipment and temporary storage	197–199	387–391
REFERENCES			392–395
INDEX			396–406

… TM 9-775
1–2

PART ONE — VEHICLE OPERATING INSTRUCTIONS

Section I

INTRODUCTION

	Paragraph
Scope	1
Glossary of nautical terms	2

1. SCOPE*.

   a. This manual is published for the information and guidance of the using arm and services.

   b. The purpose of this manual is to provide technical information required for the identification and care of the Landing Vehicles Tracked, Mk. II (unarmored) LVT (2), Mk. 1 (armored) with turret) LVT (A) (1), and Mk. II (armored) LVT (A) (2). It contains a description of the major units and their functions in relation to the other components of the vehicle, as well as instructions for operation, inspection, minor repair, and unit replacement. The manual is divided into four parts. Part one, "Vehicle Operating Instructions" (secs. I through IX), and part three, "Armament" (secs. XXXIV through XL), contain information chiefly for the guidance of the operating personnel. Part two, "Vehicle Maintenance Instructions" (secs. X through XXXV), contains information intended chiefly for the guidance of the personnel of the using arms doing maintenance work.

   c. In all cases where the nature of the repair, modification, or adjustment is beyond the scope or facilities of the unit, the responsible ordnance service should be informed in order that trained personnel with suitable tools and equipment may be provided, or proper instructions issued.

2. GLOSSARY OF NAUTICAL TERMS.

   a. Glossary. The following nautical terms are defined for the use of the using arm personnel.

   (1) ABAFT: Toward the rear.
   (2) ABEAM: At right angles to the front and rear line of vehicle.
   (3) ABOARD: On or in vehicle.
   (4) AFT: Near rear of vehicle.
   (5) AFTER BODY: Portion of vehicle to rear of center line.

---

*To provide operating instructions with the materiel, this technical manual has been published in advance of complete technical review. Any errors or omissions will be corrected by changes, or, if extensive, by an early revision.

5

LANDING VEHICLE TRACKED MK. I AND MK. II

(6) AFTER RAKE: Overhanging rear section of vehicle.

(7) APPENDAGES: Small portion of vehicle projecting beyond its main outline.

(8) ASTERN: Rear of vehicle.

(9) ATHWART: Across, or from side to side of, the length, direction or course of.

(10) AUXILIARIES: Various pumps, motors, etc., required on vehicle as distinguished from main propulsive machinery.

(11) AWASH: Even with surface of water.

(12) BALLAST: Any weight carried to make vehicle more stable.

(13) BATTENS, CARGO: Strips of wood or steel used to prevent shifting of cargo.

(14) BEAM: Extreme width of vehicle.

(15) BED PLATE: A structure fitted for support of feet of engine.

(16) BELOW: Underneath floor (deck) of vehicle.

(17) BILGE: Rounded portion of vehicle that curves upward from bottom plates, lowest portion of vehicle inside hull.

(18) BOOT TOPPING: Paint applied to outside area of hull.

(19) BOW: Forward end of vehicle.

(20) BREAMING: Cleaning barnacles, paint, etc., from bottom of vehicle with a blowtorch.

(21) BULKHEAD: Any partition used for subdividing the interior of vehicle.

(22) CALK: Operation of jamming material into the contact area of a joint to make it watertight or oiltight.

(23) CANT: Inclination of an object from the perpendicular.

(24) CHECK: Fitting for rope or cable to pass through.

(25) COUNTER: Overhang at rear of vehicle.

(26) DERELICT: A vehicle abandoned and drifting aimlessly at sea.

(27) DRAFT: The depth of vehicle below the waterline measured vertically to the lowest part of vehicle.

(28) DRAG: The amount that rear end of vehicle is below the forward end when vehicle is floating in water with rear end down.

(29) DRAIN WELL: The chamber into which seepage water is collected and pumped by drainage pumps into the sea.

(30) EVEN KEEL: Flotation of vehicle parallel to water line.

(31) FATHOM: Six feet. A nautical unit of length.

(32) FORE: A term used in indicating portions or that part of vehicle at or adjacent to front of vehicle.

(33) FORE AND AFT: In a general direction of the length of the vessel.

INTRODUCTION

(34) FORE RAKE: Forward overhanging portion of vehicle.

(35) FOUL: A term applied to the underwater portion of outside of vehicle when it is more or less covered with sea growth or foreign matter.

(36) FOUND: To fit and bed firmly, also, equipped.

(37) FREEBOARD: The vertical distance from the water line to the top of vehicle at the side.

(38) GANGWAY: The term applied to a place of exit on vehicle.

(39) GEAR: A comprehensive term used in speaking of all implements, apparatus, machinery, etc., which are used in any given operation.

(40) GRATING: Any open iron lattice work used for covering openings, or as platforms.

(41) HARD PATCH: A plate riveted or welded over another plate to cover a hole or brake.

(42) HAWSER: A large rope or cable used in towing or mooring.

(43) HOGGED: A vehicle that is damaged or strained so that the bottom curves upward. Opposite of sagged.

(44) HULK: The body of an old, wrecked, or dismantled vehicle unfit for sea service.

(45) HULL: The framework of a vehicle together with all inside and outside plating but exclusive of equipment.

(46) INBOARD: Toward the center; within vehicle's shell.

(47) KNOT: A nautical mile. Approximately 6080 feet.

(48) LAUNCHING: A term applied to the operation of sliding or lowering a vehicle into the water.

(49) LIMBER HOLE: A hole or slot in a frame or plate for the purpose of preventing water from collecting. Most frequently found on floor plates.

(50) LIST: The deviation of a vehicle from the upright position due to shifting of cargo, or other cause.

(51) LOCKER: Storage compartment on vehicle.

(52) LOG BOOK: A continuous operating record of vehicle.

(53) MAGAZINE: Spaces or compartments for stowage of ammunition.

(54) MIDDLE BODY: That part of vehicle adjacent to center point.

(55) MIDSHIP: At the middle of vehicle's length.

(56) MOORING: A term applied to the operation of securing vehicle to a wharf or dock by means of chains or rope.

(57) ON DECK: On upper or gun deck.

LANDING VEHICLE TRACKED MK. I AND MK. II

(58) OUTBOARD PROFILE: A plan representing the longitudinal exterior of vehicle.

(59) OVERBOARD: Outside, over side of vehicle into the water.

(60) OVERHANG: That portion of front or rear of vehicle which projects beyond a perpendicular at the water line.

(61) PANTING: The pulsation in and out of the front and rear of vehicle when vehicle alternately rises and plunges into the water.

(62) PAYED: (Nautical) Painted, tarred, or greased to resist moisture.

(63) PAYING: Slackening away on a rope or cable.

(64) PITCHING: The alternate rising and falling of front of vehicle in a nearly vertical plane as vehicle meets the crests and troughs of the waves.

(65) PORT: The left-hand side of vehicle as viewed from rear to front. Also an opening.

(66) PROW: That part of front end of vehicle from load water line to top of vehicle.

(67) PUMP DALE: A pipe to convey water from the bilge pump discharge channel through side of vehicle.

(68) QUARTER: A portion of vehicle sides about halfway between rear and middle of vehicle.

(69) ROLL: Motion of vehicle from side to side.

(70) SCUTTLE: Small opening in flooring for admittance of fuel, ammunition, etc.

(71) SCUTTLE BUTT: Designation for a container of the supply of drinking water for use of crew.

(72) SKIN: Outside plating — also applied to inner bottom plating when it is called inner skin.

(73) SOUNDING: Measuring depth of water or other liquid.

(74) STARBOARD: Right side of vehicle looking forward.

(75) STERN: Rear end of vehicle.

(76) TOPSIDE: That portion of the side of the hull which is above the designed water line.

(77) TRIM: The difference in draft at the front of the vehicle from that at the stern.

(78) 'TWEEN DECKS: The space between any continuous flooring.

(79) UNSHIP: To remove anything from its usual place. To take apart.

(80) VEER: To change.

(81) WAKE: The disturbed water left behind by a moving ship.

(82) WATERLOGGED: A vehicle full of water but still afloat.

TM 9-775
3

Section II

DESCRIPTION AND TABULATED DATA

	Paragraph
Description	3
Differences among models	4
Data	5

3. DESCRIPTION.

a. General. The landing vehicles tracked are of all-steel construction, designed so that they can be operated on water as well as on land. Swamps which would be impassable to ordinary vehicles can be traversed with ease. Heavy vegetation is no barrier, and even small trees may be felled.

b. Armament.

(1) On the armored and unarmored cargo carriers, LVT (A) (2) and LVT (2), gun tracks are bolted to the inside of the cargo compartment and across the back of the cab (figs. 3 and 5). The machine guns are secured to skate mounts and may be moved along these gun tracks as desired; therefore, fire is possible in all directions. A firing step is provided for the forward gunners.

(2) On the armored tank LVT (A) (1), a 37-mm gun and caliber .30 machine gun are mounted in the turret (fig. 6). In addition, two manholes in the rear of the turret are equipped with scarf mounts, on each of which a caliber .30 machine gun may be mounted (fig. 8). These scarf mounts permit 360-degree traverse, and elevation and depression of the guns within the limits of normal operation.

c. Hull and Pontoons. Technically, the term "hull" includes the pontoons. Actually, the main or central section extending from front to rear, and consisting of the cab, cargo compartment, and engine room, is called the hull. A pontoon is welded to each side of the hull (fig. 2). Mounted at the rear of each pontoon is the rear idler and bracket (fig. 2). The rear idler is used to adjust the track tension. In front of each pontoon, projecting from the hull, is a track driving sprocket. It is connected through the exterior final drive to the interior final drive on the transmission (fig. 1). Mounted beneath the pontoons are the bogie assemblies, which serve to support the vehicle, and cushion upward and downward movement. Four inset steps are provided in each pontoon to facilitate entering or leaving the vehicle (fig. 2). A towing staple is welded to the front of the hull, and another to the rear of the hull. Grab handles are located where necessary.

d. Cab.

(1) The cabs on the unarmored cargo carrier LVT (2) have two front escape windows (fig. 4). These hinge downward, and may be

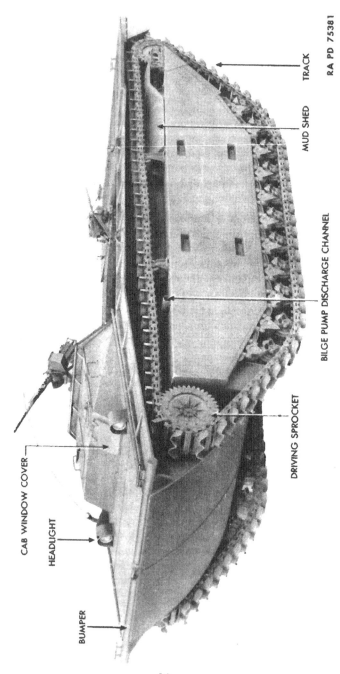

Figure 1 — Landing Vehicle Tracked (Armored) Mk. II (LVT) (A) (2) — ¾ Front View

DESCRIPTION AND TABULATED DATA

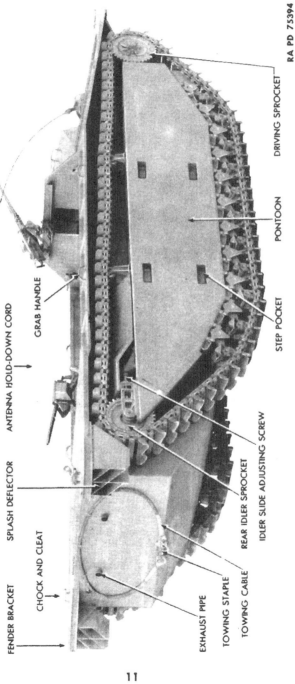

Figure 2 — Landing Vehicle Tracked (Armored) Mk. II (LVT) (A) (2) — ¾ Rear View

TM 9-775
LANDING VEHICLE TRACKED MK. I AND MK. II

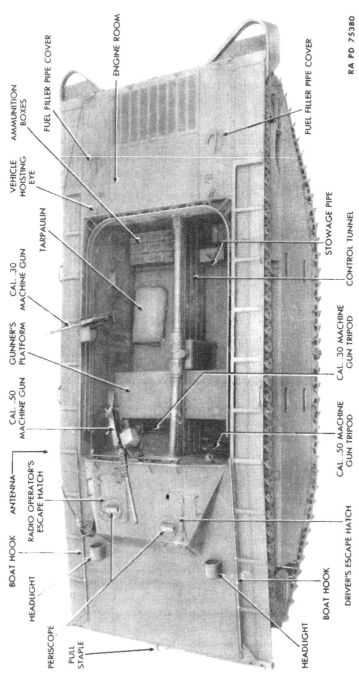

Figure 3 — Landing Vehicle Tracked (Armored) Mk. II (LVT) (A) (2) — Plan View

DESCRIPTION AND TABULATED DATA

opened for ventilation. There is also a small window on each side. All windows are constructed of safety glass.

(2) The cab used on the armored cargo carrier LVT (A) (2) and on the armored tank LVT (A) (1) has two hinged, steel, cab hatch covers in the top. These covers are equipped with fully rotating periscopes (fig. 3). Forward vision is also provided through a direct vision cab window located directly in front of the driver (fig. 1). The steel, hinged cover should be locked in either the open or closed position.

(3) Both the driver's and assistant driver's seats may be adjusted forward and backward, and upward and downward. Safety belts are provided for each seat. Instruments and controls are conveniently located. A drinking water tank is stowed in the cab.

e. Turret. The turret on model LVT (A) (1) may be rotated 360 degrees by means of a hydraulically or hand-operated traversing mechanism. A 37-mm gun and a caliber .30 machine gun are mounted in the turret. Elevation or depression of the guns is controlled by a handwheel located to the lower left of the 37-mm gun. The turret sides, front, and rear are constructed of ½-inch armor plate, the top of ¼-inch armor plate.

f. Engine Room (fig. 3). The vehicles are powered by a 7-cylinder, air-cooled, static-radial, aircraft-type engine located in the engine room at the rear of the hull. Five forward speeds and one reverse speed are provided. Two blowers, one located on each side of the engine room, are used to remove engine room fumes before starting the engine. CAUTION: *These must be operated at least 5 minutes before starting the engine, and must be on in order for the starter and booster switches to operate.* After the engine has started, the blowers may be switched off. Access to the engine is provided by means of a hinged stern cover.

g. Cargo Compartment (fig. 9). The cargo compartment is the open area between the forward bulkhead and the stern bulkhead. Louvers in the stern bulkhead are used to direct flow of air into engine room. Ammunition boxes are stowed on the floor of the cargo compartment. Extending longitudinally through the center of the cargo compartment is the control tunnel. Two propeller shafts, between which is connected a power take-off, transmit power from the engine to the transmission, and are enclosed within the housing at the top of the control tunnel. Instrument controls, connecting with the engine, pass through the lower portion of the tunnel. Removable side plates provide access to the controls within the tunnel. Located in the center of the control tunnel is the power take-off support case, which supports the power take-off. The support case encloses the bilge pump drive shaft, which connects the bilge pump located beneath the center of the floor.

Figure 4 — Landing Vehicle Tracked (Unarmored) Mk. II (LVT) (2) — ¾ Front View

DESCRIPTION AND TABULATED DATA

Figure 5 — Landing Vehicle Tracked (Unarmored) Mk. II (LVT) (2) — Plan View

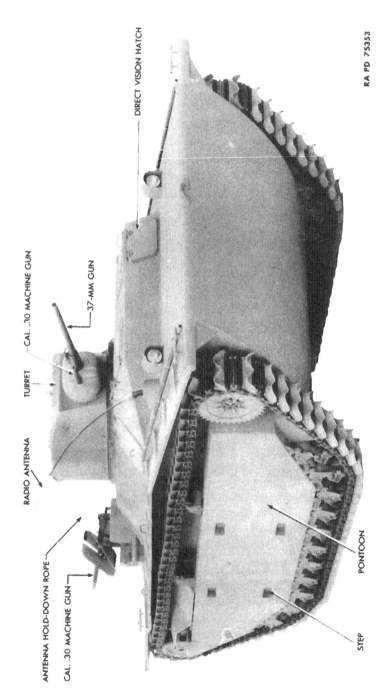

Figure 6 — Landing Vehicle Tracked (Armored) Mk. I (LVT) (A) (1) — ¾ Front View

TM 9-775
3-4

DESCRIPTION AND TABULATED DATA

h. **Tracks.** Tracks used on the vehicle are equipped with grousers which provide traction on land and propel vehicle in water.

4. DIFFERENCES AMONG MODELS.

a. There are four models of the landing vehicles tracked. These differ slightly in external appearance, but are all basically similar. Insofar as operation and control of the vehicle is concerned, all models are identical. Basic hull design and major vehicular components are the same in all models. Certain additional instructions regarding operation and control of the turret and turret armament are necessary for using arm personnel assigned to model LVT (A) (1). These instructions are incorporated in section III (Driving Controls and Operation) and section IV (Turret Controls and Operation). Official designations accorded the vehicles, together with their meanings, follow:

Designation	
LVT (A) (1)	Landing Vehicle Tracked (Armored), with Turret, Mk. I
LVT (A) (2)	Landing Vehicle Tracked (Armored) Mk. II
LVT (2)	Landing Vehicle Tracked (Unarmored) Mk. II

b. **Model LVT (2)** (figs. 4 and 5). There are two LVT (2) models. With the exception of a slight difference in the length of the gun rails, these models are identical. Since the difference is so small, affecting neither operation nor maintenance of the vehicles, the two models will be considered as one model throughout the text of this manual. Characteristics distinguishing the LVT (2) model from other models are that this model is unarmored, the hull construction being of 10-, 12-, and 14-gage sheet steel. In addition, glass windows are incorporated in the front of the cab, and the floor of the cargo compartment has lattice-type wood floor plates to permit drainage of water from the cargo compartment.

c. **Model LVT (A) (2)** (figs. 1, 2, and 3). The external appearance of the LVT (A) (2) model is quite similar to that of the LVT (2) model. However, this model is armored, the hull construction being of armor plate steel. Two cab hatch covers, equipped with rotating periscopes, are located in the top of the cab. There are no glass windows in the front of the cab. Direct, forward vision is provided by means of an opening located in front of the driver. This opening is equipped with a hinged steel cover which is locked in either the open or closed position.

d. **Model LVT (A) (1)** (figs. 6, 7, and 8). The hull and cab design and construction of model LVT (A) (1) is identical to that of model LVT (A) (2). The vehicle differs in that, mounted on top of the cargo compartment, is a turret similar to that used on a

TM 9-775

LANDING VEHICLE TRACKED MK. I AND MK. II

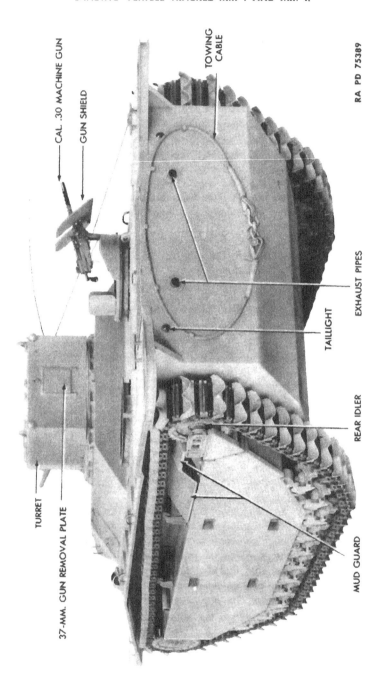

Figure 7 — Landing Vehicle Tracked (Armored) Mk. I (LVT) (A) (1) — ¾ Rear View

TM 9-775
4

DESCRIPTION AND TABULATED DATA

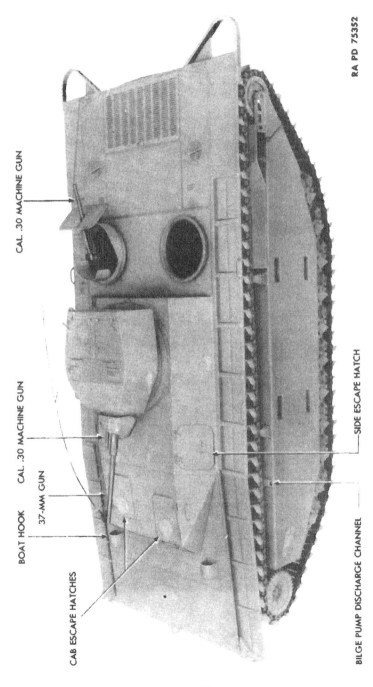

Figure 8 — Landing Vehicle Tracked (Armored) Mk. I (LVT) (A) (1) — Plan View

Figure 9 — Cargo Compartment – Rear Bulkhead Louvers Removed — Rear View

TM 9-775
4–5

DESCRIPTION AND TABULATED DATA

light tank. A 37-mm gun and caliber .30 machine gun are mounted in the turret. Directly behind the turret is a flat plate between the engine compartment and turret, in which are two circular manholes equipped wih scarf mounts for machine guns.

5. **DATA.**

   a. **Vehicle Specifications.**

	LVT (2)	LVT (A) (2)	LVT (A) (1)
Length, over-all	26 ft 1 in.	26 ft 1 in.	26 ft 1 in.
Width, over-all	10 ft 10 in.	10 ft 10 in.	10 ft 10 in.
Height, over-all	8 ft 2 in.	8 ft 3 in.	8 ft 5 in.
Draft	4 ft	4 ft 3 in.	4 ft 2 in.
Track size:			
Length	48 ft 8 in.	48 ft 8 in.	48 ft 8 in.
Width	14¼ in.	14¼ in.	14¼ in.
Crew	3 to 6	4	6
Weight of vehicle:			
Empty	24,400 lb	27,600 lb	31,200 lb
Loaded	30,900 lb	32,800 lb	32,800 lb
Ground pressure			
(lb. per sq in.)	6.8	7.7	8.7
(4-in. penetration):			
Empty			
Loaded	8.6	9.1	9.1
Ground contact			
(4-in. penetration)	3,600 sq in.	3,600 sq in.	3,600 sq in.
Ground clearance:			
Hard ground	18 in.	18 in.	18 in.
Soft ground	15¼ in.	15¼ in.	15¼ in.
Kind and grade of fuel	Aviation	Aviation	Aviation
(Octane rating)	80 octane	80 octane	80 octane
Thickness of armor:			
Hull	None	¼ in.	¼ in.
Cab, front	None	½ in.	½ in.
Cab, sides	None	¼ in.	¼ in.
Turret side	None	None	½ in.
Turret top	None	None	¼ in.
Engine:			
Make	Continental		
Type	Radial—gasoline		
Model	W670-9A		
Brake horsepower	250 bhp		

LANDING VEHICLE TRACKED MK. I AND MK. II

b. Performance.	LVT (2)	LVT (A) (2)	LVT (A) (1)
Approach angle	35 deg	35 deg	35 deg
Departure angle	30 deg	30 deg	30 deg
Towing facilities:			
Front	Eye	Eye	Eye
Rear	Staple	Staple	Staple
Maximum drawbar full (first gear)	18,000 lb	18,000 lb	18,000 lb
Maximum grade ascendability:			
Loaded	63 percent	60 percent	60 percent
Unloaded	75 percent	70 percent	63 percent
Maximum grade descending ability	60 percent	60 percent	60 percent
Maximum allowable engine speed (rpm)	2,400 rpm	2,400 rpm	2,400 rpm
Estimated cruising range:			
Land (at 16 mph)	200 mi	200 mi	200 mi
Water (at 6 mph)	60 mi	60 mi	60 mi
c. Capacities.			
Transmission capacity	24 qt	24 qt	24 qt
Fuel capacity	110 gal	106 gal	106 gal
Final drives (each)	3 qt	3 qt	3 qt
Engine oil tank capacity	5.7 gal	5.7 gal	5.7 gal
d. Radio.			
Type	Navy TCS—sending and receiving		
e. Armament.	1 — Machine Gun, cal. .50 M2, on flex. Mount M35. 1 to 3 Machine Guns M1914A4, on cal. .30, flex. Mount M35.	1 — Machine Gun, cal. .50, M2, on flex. Mount M35. 1 to 3 Machine Guns M1914A4, on cal. .30, flex. Mount M35.	37-mm, in turret. Machine Gun cal. .30, M1919A5, in turret. 2 Machine Guns, cal. .30, M1919A4, on Scarf Mounts Mk. 21.
f. Ammunition.			
Cal. .30	2,000 rounds	2,000 rounds	6,000 rounds
Cal. .50	1,000 rounds	1,000 rounds	—
37-mm	—	—	100 rounds

TM 9-775
6

Section III

DRIVING CONTROLS AND OPERATION

	Paragraph
Instruments and controls	6
Use of instruments and controls in vehicular operation	7
Towing the vehicle	8

6. INSTRUMENTS AND CONTROLS.

a. General. The driver sits on the left side of the cab, to the left of the transmission. The assistant driver sits on the right side of the cab, to the right of the transmission. The instrument panel is directly in front of the driver. The control panel is mounted on the hull to the left of the driver. The radio (Navy TCS, sending and receiving) is mounted on brackets on the right side of the cab (fig. 10), while the radio junction box is located against the lower right side of the forward bulkhead.

b. Steering Levers (fig. 11). Two steering levers are mounted directly in front of the driver. They are for the purpose of steering the vehicle, but are also used to apply the brakes. Each steering lever is provided with a ratchet handle in the top, which controls a latch rod mounted on the lever. By turning the ratchet handle a quarter turn, the latch rod can be lifted or lowered, thus providing a means of locking steering levers at any position.

c. Gearshift Lever (fig. 11). The gearshift lever is mounted on the transmission, convenient to the driver's right hand. It provides a means of shifting transmission gears to obtain the desired ratio between engine speed and vehicle speed. Five speeds forward and one reverse are provided in the transmission (fig. 12). Accidental shifting into first speed or reverse is prevented by a latch mounted on the gearshift lever.

d. Clutch Pedal (fig. 11). The clutch pedal is located in the floor directly in front of the driver, and convenient to the driver's left foot. It provides a means of disengaging and engaging the clutch.

e. Accelerator Pedal (fig. 11). The accelerator pedal is located on the floor, in front of the driver, and convenient to the driver's right foot. It is the means by which the driver controls the speed of the engine.

f. Engine Oil Pressure Gage. The engine oil pressure gage indicates the pressure of the oil circulating throughout the engine lubrication system. It is calibrated in 5-pound stages, from 0 to 100. This gage is located on the upper left-hand side of the instrument panel.

23

TM 9-775
6

LANDING VEHICLE TRACKED MK. I AND MK. II

Figure 10 — Radio Installed

TM 9-775
6

DRIVING CONTROLS AND OPERATION

Figure 11 — Driving Controls

TM 9-775
6

LANDING VEHICLE TRACKED MK. I AND MK. II

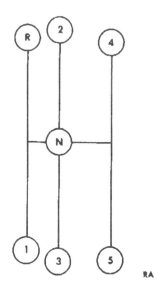

Figure 12 — Gearshift Positions

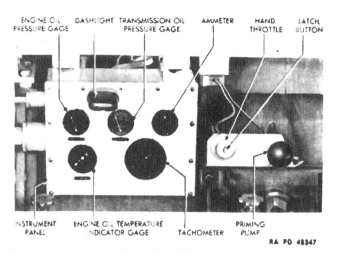

Figure 13 — Instrument Panel, Hand Throttle, and Priming Pump

TM 9-775
6
DRIVING CONTROLS AND OPERATION

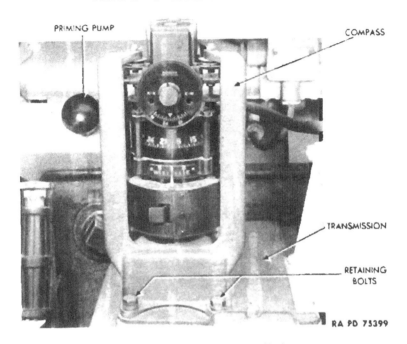

Figure 14 — Compass Installed

g. **Transmission Oil Pressure Gage.** The transmission oil pressure gage indicates the pressure of the oil pumped to the transmission. It registers in 5-pound stages from 0 to 100. This gage is located in the upper center of the instrument panel.

h. **Ammeter.** The ammeter indicates the amount of current in amperes, flowing to the battery or discharging from the battery. It registers from 0 to 60 in 10-ampere stages, in a plus and minus direction. The ammeter is located in the upper right-hand side of the front instrument panel.

i. **Engine Oil Temperature Gage.** The engine oil temperature gage indicates the temperature of the oil circulating throughout the engine. The gage registers temperatures of $40°$ to $250°$ F. This gage is located on the lower left-hand side of the instrument panel.

j. **Tachometer.** The tachometer indicates the revolutions per minute of the engine crankshaft, and registers from 0 to 2,500. It is located on the lower right-hand side of the instrument panel. It is driven by a flexible encased shaft connecting to the tachometer drive in the rear face of the accessory case.

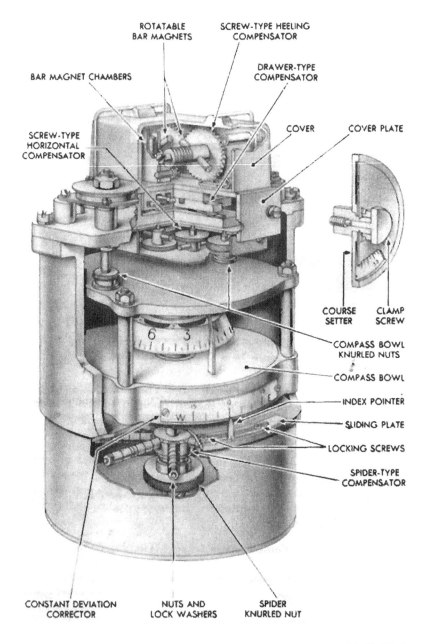

Figure 15 — Compass

DRIVING CONTROLS AND OPERATION

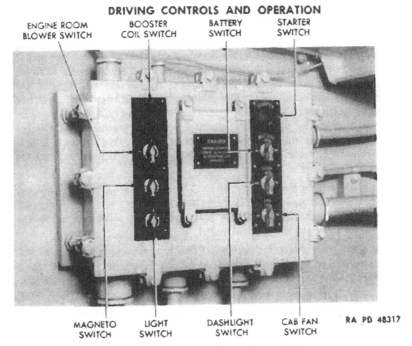

Figure 16 — Control Panel Installed

k. **Priming Pump** (fig. 13). The priming pump provides a means of injecting a spray of fuel into the engine cylinders to facilitate starting. It is located on the right-hand side of a bracket separate from, and to the right of, the front instrument panel.

l. **Hand Throttle** (fig. 13). The hand throttle is mounted to the left and on the same bracket with the priming pump. The hand throttle is used to set the engine at a fixed speed for continuous driving. The throttle is self-locking in any position, and is released by means of a latch button in the center of the control knob.

m. **Compass** (figs. 14 and 15). The compass is mounted on a bracket on top of the transmission to the right of the driver. It is used to determine accurately the direction of travel.

n. **Booster Switch.** The booster switch is a push-type switch, with rubber button for finger contact, and connects the booster coil to the starter circuit to provide additional current during the engine starting period. It is located at the top of the left-hand row of switches on the control panel on the hull to the left of the driver (fig. 16).

o. **Starter Switch** (fig. 16). The starter switch is a push-type switch with rubber button for finger contact, similar to the booster switch, and operates the cranking motor. The starter switch is located

LANDING VEHICLE TRACKED MK. I AND MK. II

at the top of the right-hand row of switches on the control panel.

p. **Blower Switch (Engine Room)** (fig. 16). The blower switch is a two-pole snap-type, used to turn on the two engine room blowers. It is located on the control panel, and is the second switch from the top in the left-hand row. The blow switch has two "ON" and two "OFF" positions. One "ON" position draws air out of the engine compartment; the other "ON" position reverses the motor, and forces outside air into the engine compartment. The blower switch is a safety switch, since it must be on in order for the booster and starter switches to operate.

q. **Dashlight Switch** (fig. 16). The dashlight switch is a two-pole snap type, and operates the dash light mounted on the instrument panel. The dashlight switch is the third switch from the top in the right-hand row on the control panel.

r. **Magneto Switch** (fig. 16). The magneto switch is a two-pole snap type, and operates either one or both magnetos. It is the third switch from the top in the left-hand row on the control panel.

s. **Battery Switch** (fig. 16). The battery switch is a two-pole snap type, and controls the electrical current to all other switches. It must be on before any other switch will operate. It is the second from the top in the right-hand switch row on the control panel.

t. **Fan Switch (Cab)** (fig. 16). The cab fan switch is a two-pole snap type, controlling the ventilating fan in the cab. This switch is located on the control panel, the lowest switch in the right-hand row.

u. **Light Switch** (fig. 16). The light switch is a two-pole snap type, and operates the headlights and the taillights. It is the lowest switch of the left-hand row on the control panel.

v. **Fuel Shut-off Valve Handles** (fig. 9). Two fuel shut-off valve handles are provided, one for each fuel tank, and control the flow of fuel from the tanks to the engine. The valve handles are located in the compartments between the sides of the engine room and the hull. They may be reached by removing the louvers in the stern bulkhead. The fuel shut-off valve handle for the left fuel tank is located behind the outer-left louver; the fuel shut-off valve handle for the right fuel tank, behind the outer right louver.

7. USE OF INSTRUMENTS AND CONTROLS IN VEHICULAR OPERATION.

a. **Before-operation Service.** Perform the services in paragraph 22 before attempting to start engine.

b. **Starting Engine.**

(1) Hand-crank engine 20 crank revolutions before starting. If any obstruction of cranking is felt, caused by fuel or oil in the lower

DRIVING CONTROLS AND OPERATION

cylinders, remove one spark plug from each of the lower cylinders, and drain excess fuel and oil. Install spark plugs and again hand-crank engine. If cranking is still obstructed, notify ordnance personnel.

(2) Open fuel shut-off valves (fig 9).

(3) Switch on main battery switch and engine room blower switch (fig. 16). CAUTION: *Operate blowers for 5 minutes before starting engine.*

(4) See that gearshift lever is in neutral (fig 12).

(5) Unless the engine is already warmed up, prime the engine five strokes of the priming pump (fig. 13). Pull the plunger out slowly, push in quickly. Avoid overpriming, which tends to wash the oil off cylinder walls.

(6) Open the hand throttle slightly, and depress clutch pedal to disengage the clutch (fig. 11). This removes the load of the propeller shaft and transmission from cranking motor.

(7) As soon as the blowers have operated 5 minutes, press in on the booster and starter switches simultaneously (fig 16). Do not switch off the blower switch. This is a safety switch and must be on in order for the booster and starter switches to operate. CAUTION: *Never hold the starter switch closed for more than 30 seconds at a time. Allow starter to cool off before attempting to turn the engine over again.*

(8) Wait 2 seconds, then turn the magneto switch to the "BOTH" position (fig. 16). If the engine fails to start, repeat the process. It may be necessary to continue to prime the engine while starting. After engine starts, press lightly on foot accelerator pedal (fig. 11). CAUTION: *Do not pump accelerator pedal; too much gas when the engine first starts will cause it to stall. If the engine has been overprimed and is flooded, turn it over with the cranking motor, holding the accelerator wide open, but with the magneto switch off.*

(9) As soon as the engine starts, watch the oil pressure gage (fig. 13). If oil pressure does not start building up within 20 seconds at 800 revolutions per minute, shut off the engine, and report the condition to the proper authorities.

(10) Switch off the engine room blower switch (fig. 16).

(11) Warm the engine at 800 to 1,000 revolutions per minute for 15 to 20 minutes, and check operations of instruments while warming up (fig. 13).

(12) When the engine is sufficiently warm, check tachometer readings (fig. 13). Try the magneto switch in the "L" and "R" positions (fig. 16). A 125 to 150 revolution per minute drop in engine speed is permissible when operating on only one magneto. If a greater drop occurs when operated on either magneto, the cause must

LANDING VEHICLE TRACKED MK. I AND MK. II

be investigated. CAUTION: *Do not run engine on one magneto for more than a 30-second interval, as this will cause the inoperative spark plugs to become carbonized.* Never idle engine at less than 450 revolutions per minute. *Never lug engine below 1,500 revolutions per minute at wide-open throttle.* Proper engine operating range is 1,400 to 2,000 revolutions per minute. Maximum economy speed is 1,800 revolutions per minute.

c. Operation of the Vehicle on Land.

(1) PLACING VEHICLE IN MOTION. With the engine at idling speed and all instruments showing normal readings, the driver may now operate the vehicle. Disengage the clutch by pressing the clutch pedal forward as far as it will go and holding it depressed (fig. 11). From neutral, move the gearshift lever as though to shift into third gear. Maintain pressure in this direction long enough to stop the propellor shaft, then, with the clutch still depressed, shift smartly into first speed. Gradually release the clutch pedal, at the same time depressing the foot accelerator pedal (fig. 11). NOTE: *First gear is used only when moving vehicle in building or crossing obstacles.* Except when under fire, do not move the vehicle in or out of close quarters without the aid of personnel outside of the vehicle to serve as guide. When the vehicle has started and is moving with an engine speed of 2,000 revolutions per minute, release the foot accelerator pedal, depress the clutch pedal again, and move the gearshift lever into the third gear position (fig. 12). Release the clutch, and again depress the accelerator to pick up the load of the vehicle. Continue the above procedure until the highest gear is reached which will enable the vehicle to proceed at the desired speed without causing the engine to labor. Do not ride the clutch. The driver's left foot must be completely removed from the clutch pedal while driving, to avoid unnecessary wear on the clutch. NOTE: *If, when shifting to any of the higher speeds, there is a raking of gears, return to neutral, and while still holding the clutch depressed, start the shifting operation, over again.* Do not hurry the shift; a slight delay will even facilitate shifting. Do not attempt to complete a shift that begins with a clashing of gear teeth. There is but one right way to shift into second, third, fourth, or fifth speeds. Maintain a steady pressure toward the desired speed and when a decreased resistance is felt, complete the shift. To place the vehicle in reverse gear, stop vehicle completely, and release the accelerator until the engine slows to its idling speed. Depress the clutch pedal and move the gearshift lever into the reverse position (fig. 12). CAUTION; *Backing the vehicle on land should never be attempted unless an observer is stationed in front to guide the driver.*

(2) STEERING THE VEHICLE. Steering is controlled by the two operating levers located directly in front of the driver (fig. 11). These

DRIVING CONTROLS AND OPERATION

are used for both land and water operation. Steering is accomplished by pulling back on the steering lever on the side toward which it is desired to turn. Pulling back either of the levers slows down or stops the track on that side, while the speed of the other track is increased. When the vehicle is moving in a straight line, steering levers should be all the way forward, and hands should be removed from the levers. Occasionally it may be necessary to use either lever, or both of them, to compensate for uneven terrain or for imperfect trim on water. Use the steering levers firmly but smoothly when making turns. NOTE: *Never pull a steering lever all the way back and hold it there to make a complete turn.* A change of direction of more than 10 degrees must be made by a series of short turns. Failure to observe this rule may result in thrown tracks and damage to the driving mechanism. When a turn is completed, push the steering lever all the way forward with a smooth, continuous movement.

(3) STOPPING THE VEHICLE. To stop the vehicle, release the accelerator, and when the vehicle has slowed down to approximately 2 to 5 miles per hour, depending upon which gear is being used, depress the clutch and pull back on both steering levers at the same time. Move the gearshift lever into neutral position (fig. 12). If the vehicle is to remain stopped, turn the ratchet handles in the tops of the steering levers a quarter turn. This locks the steering levers in the full-rear position (fig. 11). If the halt is of short duration, set the hand throttle for a tachometer reading of 800 revolutions per minute, and permit the engine to run during the halt.

d. **Operation of the Vehicle from Land to Water.** When going from land into water, it is advisable to bring the vehicle to a complete stop near the water's edge, so that the condition of the bank at the water's edge or obstacles in the water may be observed. This is not always practical, particularly where the approach to the water is swampland and continued motion is highly desirable. If the approach is steep, use accelerator cautiously and enter water as slowly as possible. However, keep in second gear and keep moving. If water is not deep enough to permit grousers to clear bottom, cross the body of water in second gear. If the water is deep, shift into third gear. It is important to maintain headway when entering or running through heavy surf. Unless the vehicle is moving fast enough to provide directional control, it is entirely possible for high waves to turn the vehicle broadside and even capsize it. Often, when moving through surf, it will be found that "grousers" are "grounded" only between waves. Because of this the vehicle must be operated in second gear so there will be plenty of reserve power for immediate use to prevent stalling.

e. **Operation on Water.** Operating and steering the vehicle on water is accomplished in the same manner as on land (subpar. c,

LANDING VEHICLE TRACKED MK. I AND MK. II

above) except for some small differences in technic. In deep water always operate the vehicle in third gear. Be on the alert for unseen sand bars, rocks, logs, or pilings. At the moment of contact with such obstacles, release the accelerator, disengage the clutch, and shift into second gear before proceeding. One movement of the steering lever will complete any degree of turn. It is not necessary to change direction with a series of short turns, as on land. To speed up turns, let up on the accelerator before starting a turn, then pull back on the steering lever and accelerate. When coming out of a turn, it is necessary to release the steering lever before coming around completely, since the momentum gained will carry the vehicle around the last half of the turn. Keep engine speed under control at all times by proper operation of the accelerator. Calmness or roughness of the water will determine correct acceleration. Maintain vehicle headway at all times, especially in rough water. As in boat operation, head the vehicles into the seas as much as possible, to minimize the possibility of capsizing.

f. *Operation of the Vehicle from Water to Land.* To beach the vehicle, disengage the engine clutch at the instant the grousers come in contact with the ground, and shift to a lower gear without losing headway. Maintain headway when beaching in heavy surf, to minimize the possibility of capsizing. However, when a wave is coming in under the rear of the vehicle, it is absolutely necessary to temporarily ease up on the accelerator to keep the vehicle from diving. Always attempt to beach the vehicle completely, before changing direction or shifting gears. Where the land contact line is clearly defined, it is advisable to come in as close as possible and shift into second gear without waiting for ground contact. If shore is unusually steep or uneven, and the condition of the water (or surf) permits, shift into first gear and proceed with the landing until safely beached.

g. *Stopping the Engine.* After completing a run, operate the engine at 800 revolutions per minute for 5 minutes to assure a gradual and uniform cooling of the various engine parts. At the end of this cooling period, turn the magneto switch to the "OFF" position and stop the engine (fig. 16). Close fuel shut-off valves (fig. 9). Turn battery switch to the "OFF" position (fig. 16).

8. TOWING THE VEHICLE.

a. *Towing to Start Vehicle.* Never attempt to start the vehicle by towing or coasting unless it is absolutely necessary. This procedure may cause serious damage to the engine and transmission.

b. *Towing Disabled Vehicle.* Disconnect the propeller shaft at the transmission, if tracks are on the disabled vehicle, and wire the propeller shaft securely to prevent it from whipping. NOTE: *If tracks are off the disabled vehicle, it is unnecessary to disconnect the*

TURRET CONTROLS AND OPERATION

propeller shaft. Shift the transmission into fifth gear when towing disabled vehicle on its tracks. This insures adequate circulation of transmission oil. Connect disabled vehicle to towing vehicle by means of towing bar or cable. If the tow bar is used, connect it to the towing clevises. If the towing cable is used and no driver is available for the disabled vehicle, hook the two eyes of the towing cable to the towing vehicle, running the cable through the towing staple of the disabled vehicle. There are certain precautions to be observed in towing the disabled vehicle.

(1) Changes of directions should always be made by a series of slight turns, so that the vehicle being towed is as nearly as possible "tracking" behind the towing vehicle.

(2) Avoid soft, muddy ground, since the tracks of the towed vehicle will slip on such a surface. When it is necessary to cross a muddy area, the driver should be careful to straighten out both vehicles before entering the area, as it is more difficult to pull a vehicle at an angle, than when following in tow.

(3) The maximum speed when towing should not be more than 8 to 10 miles per hour, and then only with an operator for steering the vehicle being towed.

Section IV

TURRET CONTROLS AND OPERATION

	Paragraph
General	9
Description	10
Operation	11

9. GENERAL.

a. A turret is used only on the LVT (A) (1) model of the landing vehicles tracked. This section is applicable to the personnel of the using arms assigned to that model.

10. DESCRIPTION.

a. **Turret Doors.** There are two doors in the turret, each of which is equipped with two snap latches, a latch handle, open locking lever, and locking lever handle. A periscope is mounted in the right turret door to provide rear view indirect vision for the commander (fig. 17).

b. **Periscopes.** Three periscopes are used in the turret; two for the commander, and one for the gunner. One of the commander's periscopes is located in the top of the turret in front of the com-

TM 9-775
10

LANDING VEHICLE TRACKED MK. I AND MK. II

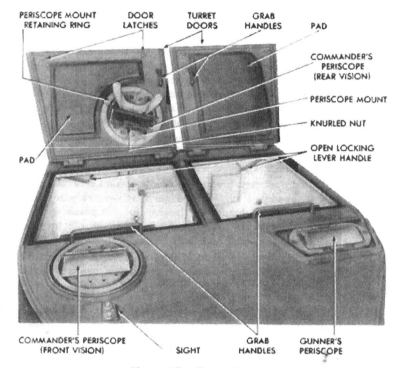

Figure 17 — Turret Doors   RA PD 48279

mander's seat, and the other in the right turret door. Both of these periscopes have a traverse of 360 degrees, an elevation of 26 degrees, and a depression of 27 degrees from the vertical. The gunner's periscope is in a fixed mount, connected by suitable linkage with the gun mount. A telescopic sight is built into the right-hand side of this periscope. A knurled knob at the rear of the periscopes provides for raising or lowering.

c. Turret Traverse Controls. The 360-degree traverse of the turret can be controlled either manually or hydraulically. A clutch lever at the top of the traversing gear box assembly, located between the two seats (fig. 18), provides for selection of the traversing method. The turret can be locked in position with the spring-loaded, positive-cam type turret lock located at the junction of the turret ring and the turret, to the left of the 37-mm gun mount and toward the front of the turret (fig. 227). The manual control crank (fig. 18) is located to the right of the traversing gear box assembly, convenient to the commander's left hand. The hydraulic traversing control handle is mounted on the left side of the turret, beneath the control box

TURRET CONTROLS AND OPERATION

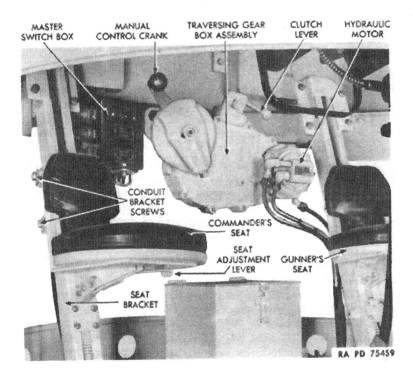

Figure 18 — Turret — Inside Rear View

(fig. 215). An auxiliary control handle on a bracket on the right side of the turret roof is provided for the commander. It is connected to the regular control handle by means of linkage which passes along the turret roof and down the left wall to the handle.

d. **Stabilizer.** Description of the stabilizer, which is used to guide the positioning of the gun so that the gunner can aim and fire accurately while the vehicle is in motion, is covered in part three (par. 184).

e. **Guns.** The guns are described and illustrated in paragraph 177.

f. **Turret Seats.** Two identical seats are provided in the turret (fig. 18). They are adjustable, and can be raised or lowered as desired. The seat adjustment lever for the commander's seat is under the left side of the seat, and the adjustment lever for the gunner's seat on the right side of the seat.

11. OPERATION.

a. Before attempting operation of the turret, be sure that operations listed in paragraph 22 have been performed.

LANDING VEHICLE TRACKED MK. I AND MK. II

b. **Turret Doors.** To open a turret door, the two snap latches must first be released by pulling down on the latch handles, and swinging the latches free. The door may then be swung up and locked in its open position. To close the turret door, the open locking levers must be released by pulling down on the open locking lever handles from inside the turret. While holding the locking lever handles down, swing the turret door free of the open locking levers, and close the door. CAUTION: *When the vehicle is in motion, the turret doors must be locked in either their open or closed position to avoid accidental opening or closing, which might result in injury to personnel.*

c. **Periscopes.** Periscopes can be raised to the using position, or lowered for complete protection, by loosening the knurled knob at the rear of the periscope, and moving the periscope to the desired position. The gunner's periscope is connected by suitable linkage with the gun mount so that, as the gun is elevated or depressed, the periscope is also elevated or depressed, thus remaining sighted on the target. Traverse indirect vision for the gunner is obtained only by rotating the entire turret.

d. **Turret Traverse Controls.**

(1) MANUAL TRAVERSE. To rotate the turret with the manual control crank (fig. 18), first check to see that the turret lock is disengaged; that is, in its back or "UP" position. If it is engaged, it is in its forward or "DOWN" position. To disengage the lock, rotate the handle back or counterclockwise against the spring pressure until the release button engages, holding the lock in the "OFF" or unlocked position. Move the clutch lever toward the turret wall to engage the manual gears. The turret can then be rotated by the manual control crank.

(2) HYDRAULIC TRAVERSE. To rotate the turret with the hydraulic mechanism, first check to see that the turret lock is disengaged, then turn the electric toggle switch for the turret motor to its "ON" position (fig. 18). Move the clutch lever toward the center of the turret to disengage the manual gears. The turret is now ready for hydraulic control. Turn the traverse control handle counterclockwise to rotate the turret to the left. Turn handle clockwise to rotate the turret to the right. The degree of turning the handle determines the speed of rotation.

e. **Stabilizer.** Operation of the stabilizer is covered in paragraph 185.

f. **Guns.** Operation of the guns, including sighting and both manual and electrical firing, is described in paragraphs 178 through 181.

g. **Turret Seats.** To raise the seat, pull out the seat adjustment lever, and take the body weight off the seat. To lower it, pull out the seat adjustment lever, and press down with the body weight.

Section V

AUXILIARY EQUIPMENT CONTROLS AND OPERATION

Paragraph
Auxiliary generator 12
Fixed fire extinguishers 13
Portable fire extinguisher 14

12. AUXILIARY GENERATOR.

a. Description. The auxiliary generator is a self-contained unit for charging the battery when the engines are idle, and to provide current supplementary to that of the battery during periods of heavy current draw. The unit consists of an electrical generator and controls, coupled directly to, and driven by, a single-cylinder gasoline engine. The auxiliary generator is mounted on two steel brackets on the floor of the cargo compartment, to the right of the control tunnel, and slightly forward of the rear bulkhead. A 1-gallon fuel tank is mounted on the hull, above the fixed fire extinguishers. A blower, for air circulation, is mounted in front of the generator. Starting and stopping buttons, blower toggle switch, and ammeter, are located on the unit and are accessible by means of a small door provided in the guard. The crankcase of the auxiliary generator engine has a capacity of 3 pints of oil. The oil filler pipe, which is fitted with a bayonet-type gage in the cap, is located on the front of the unit.

b. Operation. Before attempting to operate the auxiliary equipment, be sure that operations listed in paragraph 22 have been performed.

(1) WHEN TO OPERATE. The purpose of the auxiliary generator is to charge the batteries when the engine generator of the vehicle is unable to supply enough current. Always operate the auxiliary generator when the turret traversing mechanism is being operated, or when the gyro control unit is being used. CAUTION: *Always turn on blower mounted in front of the auxiliary generator before starting the generator.*

(2) NORMAL STARTING. Turn on blower switch (fig. 226). Press the starting button, and hold it down until the engine cranks and gains generating speed; then, release the starting button. The engine is equipped with an automatic choke. There is no ignition switch, and ignition current is supplied by the battery during the cranking period.

(3) STARTING IN SEVERE COLD. In extremely cold weather it may be necessary to start the engine with the cranking rope. Remove blower and blower mounting plate by unscrewing thumb screws that secure blower mounting plate and conduit to mounting box (fig. 226), and lay blower carefully aside. Wrap the cranking pulley on front of generator after anchoring knotted end of the rope in pulley slot. Press

the starting button and pull the cranking rope simultaneously. Repeat if the engine does not start. The starting button must be pressed to establish ignition current, even during the rope-cranking procedure.

(4) STOPPING. To stop the auxiliary generator, press the stop button and hold it down until the engine stops. Turn off blower switch.

13. FIXED FIRE EXTINGUISHERS.

a. **Description** (fig. 198). There are two fixed fire extinguishers mounted to the rear of the right fuel tank, between the upper right rear baffle and the hull. Component parts include two 10-pound carbon dioxide cylinders, four shielded nozzles, dual control cables, and the necessary tubing. This system is used to combat fire in the engine room.

b. **Operation.** In case of fire in the engine compartment, turn on the fire extinguisher by removing or breaking the lock wire, and pulling the pull handle forward. One pull handle is located in the upper right rear corner of the cab, just above the right seat (fig. 146). The other handle is located in the upper right rear corner of the cargo compartment, mounted on the front face of the bulkhead. When either handle is pulled, the two 10-pound charged cylinders, working in unison, will flood the engine compartment with carbon dioxide gas, and will extinguish a fire with the engine running up to 1,200 revolutions per minute. If conditions permit, however, the engines should be stopped. Once the system is set in operation, the cylinders will discharge carbon dioxide until empty.

14. PORTABLE FIRE EXTINGUISHER.

a. **Description.** On the floor of the driver's compartment, just to the left of the driver, there is a portable fire extinguisher. The unit consists of a 15-pound carbon dioxide cylinder, a flexible hose with shielded nozzle, and a control knob on the cylinder head. The entire unit fits into a recess in the floor, and is secured by a hand-operated metal clamp.

b. **Operation.** Snap open the clamp (fig. 23), and lift the portable fire extinguisher up and out of its recess on the floor. Direct the shielded nozzle to the base of the flame; pull out the safety wire ring on the control knob counterclockwise to open the valve. Carbon dioxide will be discharged until the cylinder is empty, or until the control knob is closed. Hold the shielded nozzle as close to the flame as possible. Continue the discharge for several seconds after the flame is out.

c. **Precautions.** Since a person will suffocate in an atmosphere of carbon dioxide, use caution in entering any space filled with this gas. Thoroughly ventilate the space into which the gas has been discharged. If it is necessary to enter the space before it is thoroughly ventilated, do

OPERATION UNDER UNUSUAL CONDITIONS

so by holding the breath and remaining only a very short time. If the man entering the gas-filled space is overcome by the carbon dioxide, he must be rescued from the space within 5 minutes, and taken out into fresh air. Apply artificial respiration.

Section VI
OPERATION UNDER UNUSUAL CONDITIONS

	Paragraph
Preventive maintenance	15
Operation in extreme cold	16
Operation in extreme heat	17
Desert operation	18
Submersion of engine in salt water or fresh water	19
Operation in mud	20

15. PREVENTIVE MAINTENANCE.

a. Operation of vehicle under any unusual conditions requires attention to preventive maintenance (sec. VII). Failure to give this extra service will result in unsatisfactory performance or damage. Refer to FM 17-59 for information on decontamination.

16. OPERATION IN EXTREME COLD.

a. *Temperature Ranges.* Low temperatures have been divided into two ranges; minus 10° F. to minus 30° F., and to below minus 30° F. Engine and lubricants undergo changes in their physical properties below minus 30° F. In many cases accessory equipment for supplying heat to engine, fuel, oil, and intake air is required. Engine and vehicle lubrication at temperatures above minus 10° F. is covered in the lubrication section of this manual (sec. VIII). Instructions in the following subparagraphs supplement this information.

b. *Protection of Vehicle.* The greatest danger in cold weather operation results from engine lubrication failure due to thickened oil, causing slow or no circulation of oil. Whenever possible, keep the vehicle in a heated enclosure when it is not being operated. Always use the protection of a shed or other enclosed space, if available. During a halt or overnight stop, if enclosed space is not available, close the engine room as tightly as possible, and cover with a tarpaulin.

(1) LUBRICANTS. In extremely cold weather, remove oil from vehicle during overnight stops, and keep oil in a warm place until vehicle is to be operated again. If warm storage is not available, heat the oil to a temperature where the bare hand can be inserted without

LANDING VEHICLE TRACKED MK. I AND MK. II

burning, before installing in the vehicle. Attach a tag marked "Engine Oil Drained" to the control panel to warn personnel that crankcase is empty. Chassis lubricants prescribed for use at 10° F. will furnish satisfactory lubrication at temperatures as low as minus 30° F. For sustained temperatures below minus 30° F., greases comparable to special low temperature grease or O. D. No. 00 grease should be used. Greases normally used cannot be applied at temperature below 0° F. except in heated buildings. In an emergency, when heated buildings are not available, use oil, and inspect and lubricate frequently.

(2) GENERATOR AND CRANKING MOTOR. Check the brushes for good contact, commutator for cleanliness, and bearings for free movement. The large surges of current which occur in starting a cold engine require good contact between brushes and commutator.

(3) WIRING AND BATTERY. Clean and tighten all connections, especially battery terminals. Be sure that there are no short circuits. Before starting the engine eliminate any ice on the wiring or other electrical equipment. Efficiency of the battery decreases sharply at low temperatures, and becomes practically nil at minus 40° F. The battery must be fully charged, and have a hydrometer reading between 1.275 and 1.300. A fully charged battery will not freeze at any temperature likely to be encountered, but a fully discharged battery will freeze at 5° F.

c. Starting.

(1) PREPARATION FOR STARTING. Before starting the engine, see that the battery tests above 1.225. NOTE: *For temperatures below minus 30°F. the battery should test approximately 1.280.* Cranking motor connections and battery terminal connections must be clean and tight. Clean and adjust magneto breaker points (par. 91 b (1)). Clean magneto contact points. Clean terminal blocks inside and out, especially of oil film. Magneto ground wires must be well insulated and tight at connections. Make sure that all ignition harness connections are tight. Check the valve clearances to make sure they are correctly adjusted (par. 65 a (2)). Engine oil must conform to lubrication specifications (par. 27).

(2) STARTING PROCEDURE. Set throttle about $\frac{1}{10}$ open. Operate the priming pump until the back pressure can be felt, then pump the plunger three to five more strokes. Follow the regular starting procedure (par. 7), and press the booster coil switch and starter switch simultaneously. As soon as engine starts to fire, release starter and boost coil switches. Use the priming pump a few strokes to keep the engine running. Allow engine to run at 1,000 to 1,100 revolutions per minute (throttle $\frac{1}{10}$ open if correctly adjusted). Pump throttle slightly, if necessary, to keep running. Individual engines may require slightly different throttle setting and priming speeds. After a cold engine

OPERATION UNDER UNUSUAL CONDITIONS

has started, pay particular attention to oil pressure and engine revolutions per minute. If oil pressure does not show signs of raising from zero, even at low engine speed, within 30 seconds after starting, stop the engine and locate the trouble. After the engine has started, follow the warm-up instructions (par. 7 b).

d. **Cold Weather Accessories.** A number of common accessories may be used to aid in starting engines in cold weather. Tarpaulin, tents, or collapsible sheds are useful for covering the vehicle, particularly the engine. Extra batteries and facilities for changing batteries quickly are useful in starting. Steel drums can be used for heating oil. Insulation of the fuel line will help prevent formation of ice inside the line.

17. OPERATION IN EXTREME HEAT.

a. Due to the fact that the engine is air-cooled, high atmospheric temperatures will cause an increase in engine and transmission temperatures. Pay particular attention to the efficiency of the cooling system (par. 82). Be sure no leaves, sticks or dirt obstruct the air intake passages on the outside of the vehicle; see that cylinder baffles are properly installed. Watch the engine oil temperature gage carefully, since engine oil temperatures higher than 200° F. may cause trouble.

18. DESERT OPERATION.

a. When operating in sandy or desert area, trouble caused by high temperature and sand-laden air may be expected. Clean oil filters (par. 84) and fuel filters (par. 70) frequently. Clean oil screens (par. 89) and fuel screen (par. 67 c) frequently. In particularly sandy areas it may be necessary to service the air cleaners (par. 72) every 4 hours. Clean breathers and vents frequently. When filling gasoline and oil tanks, be careful that dust does not fall into tanks. Use cloth over filler openings and air cleaner openings to prevent dirt and dust from entering during servicing operations.

19. SUBMERSION OF ENGINE IN SALT WATER OR FRESH WATER.

a. **General.** Precautions preventing corrosion or rusting of engine parts are comparatively simple, and have been covered in detail in the section on storage and shipment. If, however, an engine has been submerged in either salt or fresh water, the problem of arresting corrosion is difficult, and calls for immediate attention. Corrosive action that has already started must be stopped and further corrosion prevented.

b. **Temporary Protection.** In order to arrest corrosion completely, water should be removed from every part of the unit. Due to

LANDING VEHICLE TRACKED MK. I AND MK. II

the inaccessibility of many parts of an engine, it is impossible to remove all water without complete disassembly of the engine. Since proper equipment for disassembly is not usually available at the scene of submersion, temporary steps must be taken to arrest corrosion until such time as the engine can be disassembled completely and each part cleaned. Coat all parts of engine with oil to prevent air reaching them. It is so important that no air be kept from contacting wet steel parts that it often is advisable to allow the engine to remain under water until oil or some protective compound is obtained, providing, of course, this can be done within a reasonable time.

c. **Permanent Protection.** As soon as possible, the vehicle should be delivered to higher echelons for disassembly and cleaning of all accessories, engine, and other components which may be adversely affected by submersion.

d. **Precautions.** Inspect each salvage part to learn not only the extent of damage caused by corrosion, but to find defects caused by the sudden cooling action of water, in cases where the engine was at operating temperature immediately prior to submersion. If the engine is submerged in salt water for any length of time, aluminum or magnesium parts will probably be destroyed.

20. OPERATION IN MUD.

a. Grousers on the track enable the vehicle to maintain sufficient traction for travel in mud. The most frequent difficulty in mud is stalling. Avoid stalling by shifting down to a lower gear to maintain traction.

Section VII

FIRST ECHELON PREVENTIVE MAINTENANCE SERVICE

	Paragraph
Purpose	21
Before-operation service	22
During-operation service	23
At-halt service	24
After-operation and weekly service	25

21. PURPOSE.

a. To insure mechanical efficiency, it is necessary that the vehicle be systematically inspected at intervals each day it is operated, as well as weekly, so that defects may be discovered and corrected before they result in serious damage or failure. Certain scheduled mainte-

TM 9-775
21–22

FIRST ECHELON PREVENTIVE MAINTENANCE SERVICE

nance services will be performed at these designated intervals. The services set forth in this section are those performed by driver or crew before operation, during operation, at halt, after operation, and weekly.

b. Driver preventive maintenance services are listed on the back of "Driver's Trip Ticket and Preventive Maintenance Service Record," W.D. Form No. 48, covering vehicles of all types and models. Items peculiar to specific vehicles, but not listed on W.D. Form No. 48, are covered in manual procedures under the items to which they are related. Certain items listed on the form that do not pertain to the vehicle involved are eliminated from the procedures as written into the manual. Every organization must thoroughly school each driver in performing the maintenance procedures set forth in manuals, whether they are listed specifically on W.D. Form No. 48, or not.

c. The items listed on W.D. Form No. 48 that apply to this vehicle are expanded in this manual to provide specific procedures for accomplishment of the inspections and services. These services are listed in an order arranged to facilitate inspection and conserve the time of the driver, and are not necessarily in the same numerical order as shown on W.D. Form No. 48. The item numbers, however, are identical with those shown on that form.

d. The general inspection of each item applies also to any supporting member or connection, and generally includes a check to see if the item is in good condition, correctly assembled, secure, or excessively worn.

(1) The inspection for "good condition" is usually an external visual inspection to determine if the unit is damaged beyond safe or serviceable limits. The term "good condition" is further explained by the following: not bent or twisted, chafed or burned, broken or cracked, bare or frayed, dented or collapsed, torn or cut.

(2) The inspection of a unit to see that it is "correctly assembled" is usually an external visual inspection, to see if it is in its normal, assembled position in the vehicle.

(3) The inspection of a unit to determine if it is "secure" is usually an external visual examination, hand-feel, or a pry-bar check for looseness. Such an inspection should include any brackets, lock washers, lock nuts, locking wires, or cotter pins used in assembly.

(4) "Excessively worn" will be understood to mean worn, close to or beyond serviceable limits, and likely to result in a failure if not replaced before the next scheduled inspection.

22. BEFORE-OPERATION SERVICE.

a. This inspection schedule is designed primarily as a check to see that the vehicle has not been tampered with or sabotaged since the After-operation Service was performed. Various combat conditions

LANDING VEHICLE TRACKED MK. I AND MK. II

may have rendered the vehicle unsafe for operation, and it is the duty of the driver to determine whether or not the vehicle is in condition to carry out any mission to which it is assigned. This operation will not be entirely omitted, even in extreme tactical situations.

b. *Procedures.* Before-operation Service consists of inspecting items listed below according to the procedure described, and correcting or reporting any deficiencies. Upon completion of the service, results should be reported promptly to the designated individual in authority.

(1) ITEM 1, TAMPERING AND DAMAGE. Look over vehicle, special equipment, and armament for damage that may have occurred from falling debris, shell fire, sabotage, or collision, since parking vehicle. Open engine compartment covers and inspect for signs of tampering, sabotage, or damage, such as loosened mountings, fuel and oil lines, or control linkage.

(2) ITEM 2, FIRE EXTINGUISHERS. Examine cylinders of fixed system for loose mountings and closed valves. Inspect lines and nozzles for looseness or damage. See that nozzles are not clogged. Inspect portable extinguisher for damage and security. If used, or if valves have been opened, report for recharge or exchange.

(3) ITEM 3, FUEL AND OIL. Check amount of fuel in both tanks, using measuring stick; add, as necessary. Check oil level in oil supply tank, transmission, differential, and final drives. If necessary, add oil to proper level. Turn handle of Cuno fuel filters one complete turn. NOTE: *At this time turn on engine compartment blowers for at least 5 minutes before starting engine.*

(4) ITEM 4, ACCESSORIES AND DRIVES. Inspect for tampering or damage, especially loose generator drive belts.

(5) ITEM 6, LEAKS, GENERAL. Look under vehicle for any indications of fuel or oil leaks. Check engine accessory mountings, oil filters, oil tank, oil coolers, and oil and fuel lines, for indications of leaks. Trace all leaks to their source, and correct or report them.

(6) ITEM 7, ENGINE WARM-UP. To prevent hydrostatic lock, crank engine at least two complete turns by hand. If engine will not turn over readily, drain fluid from lower cylinders. Start engine and note any tendency toward hard starting, low cranking speed, and improper or excessively noisy engagement or disengagement. Set hand throttle so that engine will idle at 800 revolutions per minute during warm-up, and proceed with Before-operation Services.

(7) ITEM 8, PRIMER. When starting engine, test for proper operation of primer, and examine primer for leaks, and loose lines or mounting brackets.

(8) ITEM 9, INSTRUMENTS.

(*a*) *Oil Pressure Gage.* If pressure is not indicated within $\frac{1}{2}$ minute after starting, or if pressure is too low, engine must be

FIRST ECHELON PREVENTIVE MAINTENANCE SERVICE

stopped, and trouble corrected or reported. Correct pressure at 800 revolutions per minute is approximately 40 pounds, at governed speed 90 pounds.

(b) Tachometer. Observe if tachometer is operating and indicating approximate engine revolutions per minute.

(c) Ammeter. The ammeter should show a high charge ($+$) rate for first few minutes after starting engine, until current used in starting has been restored to batteries; thereafter, it should register zero or slight charge ($+$), with lights and accessories turned off.

(d) Engine Oil Temperature Gage. Normal operating engine oil temperature is around 150° F. Temperature should increase gradually during warm-up, but should never exceed 190° F.

(e) Transmission Oil Pressure Gage. Gage should indicate 30 to 40 pounds at 1,900 to 2,200 revolutions per minute.

(9) ITEM 11, GLASS. Clean vision devices, and examine for damage or looseness.

(10) ITEM 12, LAMPS, LIGHTS. Clean lenses of all lights and examine for damage or looseness. If tactical situation permits, test lights to see that switches operate and lights burn.

(11) ITEM 13, WHEEL AND FLANGE NUTS. Inspect to see that rear idler wheel, sprocket flange, bogie suspension, return idler assembly, and mounting nuts or screws are present and secure.

(12) ITEM 14, TRACKS. Examine master links to be sure they are secure, and see that track tension is satisfactory. Look for loose or damaged grousers and cross plates.

(13) ITEM 15, SUSPENSIONS. Remove any objects imbedded in bogie suspension system, and between tracks and bogie wheels, sprocket or idler wheel. Inspect bogie torsional arms and hangers for damage and looseness.

(14) ITEM 16, STEERING BRAKE LINKAGE. Inspect linkage to see that all levers and latches, rods, clevises, and pins are in good condition and secure.

(15) ITEM 18, TOWING CONNECTIONS. See that front and rear towing staples are intact, that tow cable and shackles are in good condition, and if not in use, properly mounted on hull.

(16) ITEM 19, HULL, LOAD, AND TARPAULIN. Inspect hull to see that there are no punctures in hull plates or pontoons, that there are no bright spots in finish or camouflage pattern to cause glare, and that vehicle markings are legible. Look for loose or missing hull drain plugs or inspection plates. Examine load for shifting or damage. See that tarpaulin is in good condition and secure.

(17) ITEM 20, DECONTAMINATOR. Be sure decontaminator is present, fully loaded, and securely mounted.

LANDING VEHICLE TRACKED MK. I AND MK. II

(18) ITEM 21, TOOLS AND EQUIPMENT. Check vehicle stowage lists to be sure all items are present. See that these items are in good condition and secure.

(19) ITEM 22, ENGINE OPERATION. Engine should idle smoothly at 800 revolutions per minute. Accelerate engine several times after reaching operating temperature (150° F.), and note any unusual noise, unsatisfactory operating characteristics, or excessive exhaust smoke.

(20) ITEM 23, DRIVER'S PERMIT AND FORM NO. 26. Driver must have operator's permit on his person. Form No. 26, vehicle manual, Lubrication Guide, and W.D. AGO Form No. 478 must be present, legible, and safely stowed.

(21) ITEM 24. AMPHIBIAN SERVICES.

(a) On LVT (2) model (unarmored), remove section of wood floor in cargo compartment. Remove floor drain plates, and inspect bilges for excessive dirt or water; clean and pump out, as necessary.

(b) On LVT (A) (1) (with turret), and LVT (A) (2) (without turret, but armor-plated), the inspection of bilge can be made only through grille at bilge pump.

(c) Before water operations, be sure that hatches are closed securely against seals; that anchor, if in use, is raised and properly stowed; that boat hooks and hand bilge pump are present and secure; and that no excessive water leaks are apparent.

(22) ITEMS 25, DURING-OPERATION CHECK. Both on land and in water, the During-operation Service should start as soon as the vehicle is put in motion, in the nature of a road test.

23. DURING-OPERATION SERVICE.

a. While vehicle is in motion, listen for any sounds such as rattles, knocks, squeals, or hums, that may indicate trouble. Look for indications of smoke from any part of the vehicle. Be on the alert to detect any odor of overheated components or units such as generator, steering brakes or clutch, fuel vapor from a leak in fuel system, exhaust gas, or other signs of trouble. Any time the steering brakes are used, gears shifted, or vehicle turned, consider this a test, and notice any unsatisfactory or unusual performance. Watch the instruments constantly. Notice promptly any unusual instrument indication that may signify possible trouble in the system to which instrument applies.

b. Procedures. During-operation Service consists of observing items listed below according to the procedures following each item, and investigating any indications of serious trouble. Notice minor deficiencies to be corrected or reported at earliest opportunity, usually at next scheduled halt.

TM 9-775
23

FIRST ECHELON PREVENTIVE MAINTENANCE SERVICE

(1) ITEM 26, STEERING BRAKES. Steering brakes should be in released position when vehicle is moving straight ahead. Note any tendency of vehicle to lead to one side. This usually indicates that brake on "leading" side is tight. When stopping vehicle, levers should have a reasonably equal amount of travel. With vehicle stopped and levers pulled back for parking, lever travel should be approximately three teeth, or notches, on quadrant.

(2) ITEMS 28, CLUTCH. Clutch should not grab, chatter, or squeal during engagement, or slip when fully engaged. Pedal free travel should be $\frac{1}{8}$ to $\frac{1}{4}$ inch.

(3) ITEMS 29, TRANSMISSION. Gears should shift smoothly without clashing, operate without excessive noise, and not creep out of mesh during operation.

(4) ITEM 31, ENGINE AND CONTROLS. Driver should be on alert for deficiencies in engine performance, such as lack of usual power, misfiring, unusual noise, stalling, indications of engine oil overheating, or unusual exhaust smoke. Notice whether engine accelerates satisfactorily, and see that controls operate freely and are properly adjusted. If radio interference is reported, driver will cooperate with radio operator to determine its source (par. 46).

(5) ITEM 32, INSTRUMENTS. Observe frequently the readings of all instruments during operation to see whether they are indicating properly.

(a) *Oil Temperature Gage.* Gage should read in normal range, except when operating in extreme hot or cold temperatures. Excessive heat above 190° F. always indicates trouble, and may cause serious damage to engine unless correction is made.

(b) *Oil Pressure Gage.* In case of any unusual drop, or no oil pressure at all, vehicle should be stopped immediately.

(c) *Ammeter.* Ammeter should register high charge rate for first few minutes after engaging transfer case, until generator restores to battery, current used in starting. After this period, ammeter should register slight charge with lights and accessories turned off and engine operating at 800 revolutions per minute.

(d) *Tachometer.* Tachometer should indicate engine speed and accumulating revolutions correctly.

(6) ITEM 33, STEERING GEAR. With levers in released position, vehicle should move straight ahead, except in water operation, when allowance may be made for drift due to current or tide. Note any excessive looseness or bind in lever linkage, or noise from brake shoes, on application of levers.

(7) ITEM 34, RUNNING GEAR. Listen for any unusual noise from tracks, sprockets, idlers, and bogie suspensions.

LANDING VEHICLE TRACKED MK. I AND MK. II

(8) ITEM 36, GUNS, MOUNTINGS, ELEVATING AND TRAVERSING, STABILIZER AND FIRING CONTROLS.

(a) While vehicle is in operation, but before being used in combat, test all mounted guns to be sure they are securely mounted and respond properly to manual elevating, traversing, and firing controls.

(b) On the LVT (A) (1), test hydraulic and stabilizer mechanism and controls to see that they operate properly, and that guns and turret which they control, respond properly. NOTE: *Recoil tests can be made only while firing.*

(9) ITEM 37, AMPHIBIAN SERVICES.

(a) *Bilge Pump.* Be sure bilge pump is in operation, and note any unusual noises that might indicate a damaged or excessively worn pump, drive, or power take-off.

(b) *Leaks.* Be on the alert for indications of water leaks in hull interior.

(c) *Stability.* While operating in water, note any tendency of the vehicle to ride at uneven keel, which may indicate excess water in bilges, or leaks into pontoons. Cargo must be evenly distributed.

24. AT-HALT SERVICE.

a. At-halt services may be regarded as minimum maintenance procedures and should be performed under all tactical conditions, even though more extensive maintenance services must be slighted or omitted altogether.

b. Procedures. At-halt services consist of investigating any deficiencies noted during operation, inspecting items listed below according to the procedures following the items, and correcting any deficiencies found. Deficiencies not corrected should be reported promptly to the designated individual in authority.

(1) ITEM 38, FUEL AND OIL. Check fuel supply with measuring stick to see that it is adequate to operate vehicle to next refueling point. Also check oil level in engine supply tank, and replenish if necessary.

(2) ITEM 39, TEMPERATURES (HUBS, TRANSMISSION, AND FINAL DRIVES). Place hand on each track wheel hub to see whether it is abnormally hot. If wheel hubs are too hot to touch with the hand, bearings may be inadequately lubricated, damaged, or improperly adjusted. Check the transmission and final drives for overheating or excessive oil leaks.

(3) ITEM 40, AXLE AND TRANSMISSION VENTS. Examine vents of transmission and final drives to see that they are present, and not damaged or clogged.

FIRST ECHELON PREVENTIVE MAINTENANCE SERVICE

(4) ITEM 41, PROPELLER SHAFT. Inspect for damage or looseness of housings, flanges, and universal joints, and for abnormal grease leaks at seals. Inspect for looseness, damage, and oil leaks around center bearing and power take-off.

(5) ITEM 42, SUSPENSIONS. Look for loose or damaged bogie wheels, arms, brackets, idler adjustments, and return idlers.

(6) ITEM 43, STEERING BRAKES. Examine levers and linkage for looseness and damage, and investigate any irregularities noted during operation.

(7) ITEM 44, WHEEL AND FLANGE NUTS. Examine to see that nuts are all present and secure.

(8) ITEM 45, TRACKS. Examine tracks, track wheels, and idlers to see that they are secure, not damaged, and that track tension is satisfactory. Clean out stones and trash from tracks and suspension.

(9) ITEM 46, LEAKS, GENERAL. Look in engine compartment for indications of fuel or oil leaks.

(10) ITEM 47, ACCESSORIES AND BELTS. Inspect for looseness, damage, and incorrect alinement. Generator drive belt tension ¾-inch deflection. If radio interference was caused by operation of vehicle, check for loose junction box and shielding connections.

(11) ITEM 48, AIR CLEANERS. If operating under extremely dusty or sandy conditions, inspect air cleaners and breather caps at each halt to see that they are in condition to deliver clean air properly; service, as required.

(12) ITEM 50, TOWING CONNECTIONS. Inspect staples, front and rear, for damage. See that cable is in good condition and secure.

(13) ITEM 51, HULL, LOAD AND TARPAULIN. Examine hull for damage, and vehicle load for shifting or damage. Be sure tarpaulin is in good condition and secure.

(14) ITEM 52, GLASS. Clean vision devices, and inspect for damage or looseness.

(15) ITEM 53, AMPHIBIAN SERVICES. Investigate any irregularities noted during operation. Inspect hull for leaks, bright spots in finish or camouflage pattern that might cause reflections, and see that all hull attachments are in good condition and secure. Inspect bilges for excessive water and dirt. See that all hatches and grilles are secure, and that all hull and pontoon drain plugs are present and tight.

25. AFTER-OPERATION AND WEEKLY SERVICE

a. After-operation Service is particularly important, because at this time, the driver inspects his vehicle to detect any deficiencies that may have developed, and corrects those he is permitted to handle. He should report promptly, to the designated individual in

LANDING VEHICLE TRACKED MK. I AND MK. II

authority, the results of his inspection. If this schedule is performed thoroughly, the vehicle should be ready to roll again on a moment's notice. The Before-operation Service, with a few exceptions, is then necessary only to ascertain if the vehicle is in the same condition in which it was left upon completion of the After-operation Service. The After-operation Service should never be entirely omitted, even in extreme tactical situations, but may be reduced to the bare fundamental services outlined for the At-halt Service, if necessary.

h. *Procedures.* When performing the After-operation Service, the driver must remember and consider any irregularities noticed during the day in the Before-operation, During-operation, and At-halt Services. The After-operation Service consists of inspecting and servicing the following items. Those items of the After-operation Service that are marked by an asterisk (*) require additional Weekly services, the procedures for which are indicated in subparagraph (h) of each applicable item.

(1) ITEM 55, ENGINE OPERATION. Before stopping engine, accelerate and decelerate; note any unusual noise or irregular performance, and investigate any deficiencies noted during operation. Allow engine to cool off for approximately 5 minutes at idling speed before stopping.

(2) ITEM 56, INSTRUMENTS.

(a) *Oil Pressure Gage.* Before stopping engine, check oil pressure. Normal pressure is 40 pounds at idle speed and 90 pounds at 2,400 revolutions per minute.

(b) *Engine Oil Temperature.* Temperature should be approximately 150° F. under normal operation, and should never exceed 190° F.

(c) *Transmission Oil Pressure.* Normal pressure is about 45 pounds under normal operation.

(d) *Tachometer.* Observe if tachometer is operating and recording the engine revolutions per minute.

(e) *Ammeter.* Reading should be zero or a slight charge (+) at 800 revolutions per minute. With engine stopped and all electrical switches off, reading should be zero.

(3) ITEM 354, FUEL AND OIL. Check fuel supply in both tanks. Check oil level in engine oil supply tank, transmission, differential, and final drives; replenish as necessary. NOTE: *Transmission oil should be allowed to cool for 15 or 20 minutes before checking level.*

(4) ITEM 58, GLASS. Clean all vision device lenses and mirrors and inspect for damage or looseness.

(5) ITEM 59, LAMPS (LIGHTS). Clean all light lenses, and ex-

FIRST ECHELON PREVENTIVE MAINTENANCE SERVICE

amine for looseness and damage. If tactical situation permits, turn on all switches to see that all lights are operating properly.

(6) ITEM 60, FIRE EXTINGUISHERS. Inspect tanks for tight mountings, damage, and leakage of valves or lines. Inspect nozzles for damage and proper aiming.

(7) ITEM 61, DECONTAMINATOR. Inspect decontaminator for damage and security of mounting.

(8) ITEM 62, BATTERY.

(a) Inspect battery for cleanliness, good condition, secure mountings and connections, proper electrolyte level, and leaks. See that vent caps are clean and secure.

(b) Weekly. Clean battery and case; check level of electrolyte. Add distilled water, as necessary, to bring level ½ inch above top of plates. Any clean water is preferable to allowing cells to run dry. NOTE: *Do not add water until battery has been allowed to cool off after operation.*

(9) ITEM 63, ACCESSORIES AND BELTS.

(a) Inspect carburetor, generator and regulator, cranking motor, magnetos, oil and fuel pumps, governor, oil coolers, blowers, etc., to see that they are in good condition and secure.

(b) Weekly. Tighten all accessory mountings and connections. Wipe off excess dirt and grease. Adjust generator drive belts to ¾-inch deflection. Clean out all insects, dirt, and trash from in and around oil cooler core air passages and blower ducts.

(10) ITEM 64, ELECTRICAL WIRING (*Weekly Only*). Inspect all accessible high and low voltage wiring and conduits or shielding to see that it is in good condition, clean, and securely connected and supported.

(11) ITEM 65, AIR CLEANERS, BREATHERS, AND VENTS. Remove air cleaner oil reservoir. Clean and refill to proper level with SAE 50 oil. Extreme dust or sand conditions may require more frequent cleaning. Examine for leakage, general condition, and security of mountings. Examine breather caps and vents for security of mounting; clean, as required.

(12) ITEM 66, FUEL FILTERS.

(a) Turn cleaner handle one complete turn. Examine filters for damage and leaks.

(b) Weekly. Remove drain plugs and drain off all accumulations of dirt and sludge. Replace plugs securely, and recheck for leaks.

(13) ITEM 67, ENGINE CONTROLS. Inspect accelerator and hand throttle controls for loose mountings, damage, excessive wear, and free operation.

TM 9-775

LANDING VEHICLE TRACKED MK. I AND MK. II

(14) ITEM 68, TRACKS.

(a) Remove all stones and other foreign matter from between links and grousers, and from between tracks and suspensions. Inspect for looseness, damage, and excessive wear.

(b) *Weekly.* Tighten any loose links, grousers, or cross bars securely. Adjust track tension so there is no clearance between inner plate and idler wheel sliding bracket, and $\frac{1}{8}$-inch clearance between inner and outer plate on adjusting screw.

(15) ITEM 69, SUSPENSIONS. Inspect bogie wheels and return idlers for excessive wear, damage to tires, and loose or damaged torsion arms or mounting brackets. Inspect sprockets and rear idler wheels for damage, excessive tooth wear, loose mounting or assembly nuts and screws, and excessive lubricant leaks.

(16) ITEM 70, STEERING BRAKES. Inspect brake levers for security, excessive wear or damage, and for freeness of operation of latches and pawls.

(17) ITEM 71, PROPELLER SHAFT AND CENTER BEARING. Inspect these items for damage, excessive wear, and lubricant leaks. Investigate any unusual noise or vibration noted during operation. Examine the clutch release mechanism for excessive wear, binding, flat rollers, loose mountings, and proper free pedal travel of $\frac{3}{8}$ to $\frac{3}{4}$ inch.

(18) ITEM 72, VENTS.

(a) Inspect transmission and final drive vents to see that they are in good condition, secure, and not clogged.

(b) *Weekly.* Remove vents, clean thoroughly, and replace securely.

(19) ITEM 73, LEAKS, GENERAL. Examine fuel and oiling systems, transmission, differential, final drives, and propeller shaft center bearing for leaks.

(20) ITEM 74, GEAR OIL LEVELS. Remove filler plugs and check levels of all gear boxes. Oil levels should be from $\frac{1}{2}$ inch below, to edge of filler hole. Add specified oil, as necessary, to bring to correct level.

(21) ITEM 77, TOWING CONNECTIONS. Examine towing staples, front and rear, and towing cable, to be sure they are secure and not damaged.

(22) ITEM 78, HULL. Inspect hull for damage from shell fire or collision. Check bottom of hull for damage. Inspect escape hatches for alinement and proper operation of locking device, and doors over driving compartment for damage and proper operation.

(23) ITEM 80, VISION DEVICES. Examine protectoscope prisms and windows to see that they are in good condition, clean, secure in

FIRST ECHELON PREVENTIVE MAINTENANCE SERVICE

holders, and that holders are securely mounted. See that lever and locking devices operate freely and are not excessively worn. See that periscope prisms are in good condition and secure, that their traversing, elevating, and locking devices are free, and not excessivly worn. Check spare prisms and windows and their stowage boxes to see that they are in good condition, clean, and secure. CAUTION: *Prisms should be cleaned only with a soft cloth or brush. Never use a coarse or dirty cloth, as this will damage surface of the glass.*

(24) ITEM 81, TURRET AND GUNS (MOUNTINGS, ELEVATING, STABILIZER, AND TRAVERSING AND FIRING CONTROLS). On the LVT (A) (1) model, inspect all mounted guns to see that they are secure, clean, adequately lubricated, and in condition for immediate use. See that gun elevating mechanism and firing controls operate properly, and that mechanism responds to controls. Test stabilizer and turret traversing hydraulic and manual controls, and see that mechanism responds. Check oil level in reservoir, and examine all exposed electrical wiring and oil lines for looseness or damage. Inspect packing glands, oil lines, piston and cylinder assembly, and drain plugs for leaks. Investigate any deficiencies noted during operation.

(25) ITEM 82, TIGHTEN.

(a) Tighten sprocket, idler, bogie suspension, and any other assembly or mounting nuts or screws where inspection indicates the necessity.

(b) Weekly. Tighten all vehicle assembly bolts, such as transmission, differential and final drives, engine mountings, universal joints, control linkage, brackets and track links, accessories, gun mountings, tools and equipment, and attaching brackets.

(26) ITEM 83, LUBRICATE AS NEEDED.

(a) If due, lubricate points indicated on Lubrication Guide as required on daily or 8-hour interval.

(b) Weekly. If due, lubricate points indicated on Lubrication Guide as required on weekly or hourly interval.

(27) ITEM 84, CLEAN ENGINE AND VEHICLE.

(a) Remove all empty shell cases and refuse. Clean all oil from driving compartment and floor of vehicle. Remove excessive mud and dirt from tracks and suspension system. See that grilles in engine compartment are not obstructed.

(b) Weekly. Wash exterior of vehicle and remove all dirt and mud. If washing is impractical, wipe as clean as possible, and watch for bright spots that would cause glare. Open engine compartment, and wipe out all excess dirt; if compressed air is available, blow all dirt out of cylinder cooling fins and oil cooler cores.

TM 9-775
25-26

LANDING VEHICLE TRACKED MK. I AND MK. II

(28) ITEM 85, TOOLS AND EQUIPMENT. Check vehicle stowage lists to see that all tools belonging to vehicle are in place and properly stowed. Check equipment against stowage list, and see that it is in serviceable condiion.

(29) ITEM 86, AMPHIBIAN SERVICES.

(a) Examine hull for damage, indications of punctures, or broken welds. Inspect compartment hatches and grilles for good condition, alinement, and security, and proper operation of hinges and latches.

(b) Weekly. Clean out bilges and test operation of bilge pump. NOTE: *On LVT (2), remove wood cargo compartment floorboards to perform this operation. Examine anchor, boat hooks, and hand bilge pump for good condition and secure mounting. See that bulkheads are tight and not leaking, and that hull and pontoon drain plugs are present and tight.*

Section VIII

LUBRICATION

	Paragraph
Lubrication guide	26
Detailed lubrication instructions	27

26. LUBRICATION GUIDE.

a. War Department Lubrication Guide No. 141 (figs. 19 and 20) prescribes lubrication maintenance for these tracked landing vehicles.

b. A Lubrication Guide is placed on, or is issued with, each vehicle and is to be carried with it at all times. In the event a vehicle is received without a Guide, the using arm should immediately requisition a replacement from the Commanding Officer, Fort Wayne Ordnance Depot, Detroit 32, Michigan.

c. Lubrication instructions on the Guide are binding on all echelons of maintenance and there should be no deviations, except as indicated in subparagraph d below.

d. Service intervals specified on the Guide are for normal operating conditions. Reduce these intervals under extreme conditions, such as prolonged operation in sand or dust, or immersion in water, any one of which may quickly destroy the protective qualities of the lubricant.

e. Lubricants are prescribed in the "Key" in accordance with three temperature ranges: above $+32°F.$, $+32°F.$ to $0°F.$, and below $0°F.$ Determine the time to change grades of lubricants by

56

TM 9-775
26–27

LUBRICATION

maintaining a close check on operation of the vehicle during the approach to change-over periods, especially during initial action. Sluggish starting is an indication of thickened lubricants, and the signal to change to grades prescribed for the next lower temperature range. Ordinarily it will be necessary to change grades of lubricants only when air temperatures are consistently in the next higher or lower range, unless malfunctioning occurs sooner due to lubricants being too thin or too heavy.

27. DETAILED LUBRICATION INSTRUCTIONS.

a. Lubrication Equipment.

(1) Each vehicle is supplied with lubrication equipment adequate to maintain the materiel. This equipment will be cleaned both before and after use.

(2) Operate lubrication guns carefully, and in such manner as to insure a proper distribution of the lubricant.

b. Points of Application.

(1) Lubrication fittings, grease cups, oilers, and oil holes are readily indentifiable on the vehicle. Be sure to wipe clean such lubricators and the surrounding surface before lubricant is applied.

(2) Where relief valves are provided, apply new lubricant until the old lubricant is forced from the vent. Exceptions are specified in notes on the lubrication guide.

c. Cleaning. Use SOLVENT, dry-cleaning, or OIL, fuel, Diesel, to clean or wash all parts. Use of gasoline for this purpose is prohibited. After washing, dry parts thoroughly before applying lubricant.

d. Lubrication Notes on Individual Units and Parts. The following instructions supplement those notes on the Lubrication Guide which pertain to lubrication and service of individual units and parts.

(1) AIR CLEANERS. Daily, check level and refill engine and auxiliary generator air cleaner oil reservoirs to bead level with used crankcase oil or OIL, engine, SAE 30, or Navy Symbol 9250, above +32°F. and SAE 10, or Navy Symbol 9110, from +32°F. to 0°F. From 0°F. to −40°F., use OIL, hydraulic, or Navy Symbol 2075. Below −40°F., remove oil and operate dry. Every 40 hours (daily under extreme dust conditions), remove air cleaners and wash all parts.

(2) VENTS. Keep final drive gear case vents clean at all times. Vents must be cleaned and kept open. Inspect each time oil is checked, and each time vehicle is operated under extremely dirty conditions.

(3) AUXILIARY GENERATOR CRANKCASE. Daily, check level and refill to "FULL" mark on gage with OIL, engine, SAE 30, or Navy

TM 9-775
LANDING VEHICLE TRACKED MK. I AND MK. II

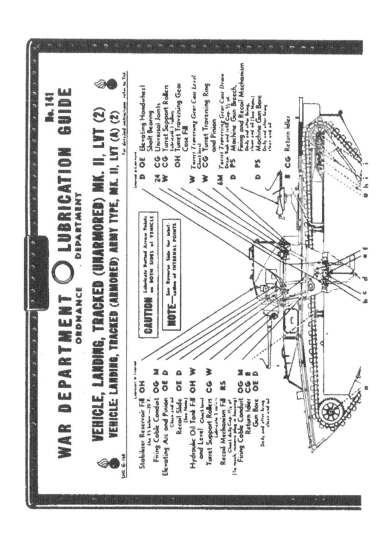

LUBRICATION

TM 9-775
27

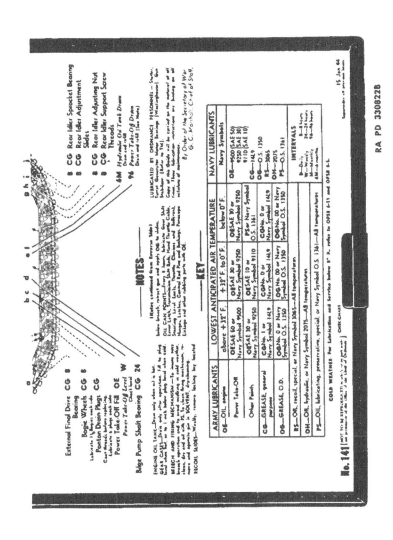

Figure 19 — Lubrication Guide — Internal Points

LANDING VEHICLE TRACKED MK. I AND MK. II

Symbol 9250, above 32°F., or SAE 10 or Navy Symbol 9110 below +32°F. Every 48 hours, remove drain plug, and completely drain case. Drain only when oil is hot. Replace plug and fill to "FULL" mark on gage with correct oil for temperature requirements.

(4) BREECH AND FIRING MECHANISM. Daily, and before and after action, clean and oil all moving parts and unpainted surfaces with OIL, engine, SAE 30, or Navy Symbol 9250, above +32°F., SAE 10 or Navy Symbol 9110 from +32°F. to 0°F. OIL, lubricating, preservative, special, or Navy Symbol O.S. 1361 below 0°F. To insure easy breech operation, and to avoid misfiring in cold weather, clean, dry and oil with OIL, lubricating, preservative, special, or Navy Symbol O.S 1361. To clean firing mechanism, remove and operate pin in SOLVENT, dry-cleaning.

(5) CLUTCH PILOT AND HUB BEARINGS. Whenever clutch is disassembled for any other purpose, remove, clean, and repack.

(6) ENGINE OIL TANK. Daily, check level and refill to "FULL" mark with OIL, engine, SAE 50 or Navy Symbol 9500 above +32°F., or SAE 30 or Navy Symbol 9250 from +32°F. to 0°F. Below 0°F., refer to OFSB 6-11. Every 24 hours, remove drain plug from bottom of tank, and completely drain tank. Drain only when engine is hot. After thoroughly draining, replace drain plug, and refill tank to "FULL" mark on gage with correct lubricant to meet temperature requirements. Run engine a few minutes, and recheck oil level. Be sure pressure gage indicates oil is circulating.

(7) GEAR CASES (TRANSMISSION, DIFFERENTIAL, AND FINAL DRIVE GEAR CASES). Weekly, check level with vehicle on level ground, and if necessary, add lubricant to within 1 inch of plug level when cold, or to plug level when hot. Every 96 hours, drain and refill. Drain only after operation when gear lubricant is warm. Refill with OIL, engine, SAE 50, or Navy Symbol 9500, above +32°F., SAE 30 or Navy Symbol 9250 below +32°F.

(8) OIL FILTER. Daily, check operation of automatic turning mechanism. Weekly, remove, clean, and inspect element.

(9) RECOIL SLIDES. Weekly, remove locking key located below breech, retract gun, and apply GREASE, O.D., No. 0, or Navy Symbol O.S. 1350, above 32°F., No. 00 or Navy Symbol O.S, 1350 below +32°F., to slides.

(10) UNIVERSAL JOINTS AND SLIP JOINTS. Use GREASE, general purpose, No. 1, or Navy Symbol 14L9, above 32°F., No. 0 or Navy Symbol 14L9 below +32°F. Apply grease to universal joint until it overflows at relief valve, and to the slip joint until lubricant is forced from the vent at the universal joint end of the spline.

(11) OILCAN POINTS. Every 8 hours, lubricate gearshift lever latch, transmission shift rails, hand throttle cable, steering lever locks,

LUBRICATION

throttle clevises, and bell crank, hinges, latches, control rod pins and bushings, periscope linkage, and other rubbing parts with OIL, engine, SAE 30, or Navy Symbol 9250, above +32°F., SAE 10, or Navy Symbol 9110, from +32°F. to 0°F.; OIL, lubricating, preservative, special, or Navy Symbol O.S. 1361, below 0°F.

e. **Points to be Lubricated by Ordnance Personnel Only.**

(1) CRANKING MOTOR. Once a year, disassemble cranking motor, clean reduction gears, and coat with GREASE, graphited, light.

(2) TURRET GENERATOR MOTOR BEARINGS (Westinghouse). Once a year, disassemble unit, inspect bearings for wear, and renew if necessary. Clean generator end bearing with SOLVENT, dry-cleaning, and repack with GREASE, special, high temperature. CAUTION: *Do not wash motor end bearing with SOLVENT, dry-cleaning, as this bearing is of the sealed type, and cannot be repacked.*

(3) GUN STABILIZER. When charging the system with oil, it is very important for proper operation of the stabilizer, that all air trapped in the system be removed. The following procedure, therefore, must be adhered to: .

(a) Use OIL, hydraulic, or Navy Symbol 2075.

(b) Heat oil from 150°F. to 200°F., if possible.

(c) Oil may be poured directly into the reservoir, or pumped in with pump-type oilcan by removing the filler plug; or it may be added under a small amount of pressure. To get this pressure, provide a filler can with 3-foot feed line, shut-off valve at reservoir connection end, and ⅜-inch union below shut-off valve. Remove oil supply line from reservoir. Connect filler can feed line to oil reservoir line.

(d) Make certain that the turret switch is in the "OFF" position.

(e) Loosen the oil return line, remove small hexagon plugs, and loosen two bleeder valves on cylinder.

(f) Add oil to system until a flow free of bubbles is obtained from the return line. Tighten this connection permanently.

(g) After a solid flow of oil is obtained from bleeder valves, tighten finger-tight.

(h) Loosen top bleeder valve. Push breech slowly to extreme "UP" position, and tighten after a solid flow of oil is obtained.

(i) Loosen lower bleeder valve. Push breech slowly to extreme "DOWN" position, and tighten after a solid flow of oil is obtained.

(j) Repeat steps (h) and (i).

(k) Remove pressure supply line and connections, if used, and reconnect oil reservoir supply line.

(l) Work gun up and down slowly until no more signs of air appear in oil reservoir.

TM 9-775
LANDING VEHICLE TRACKED MK. I AND MK. II

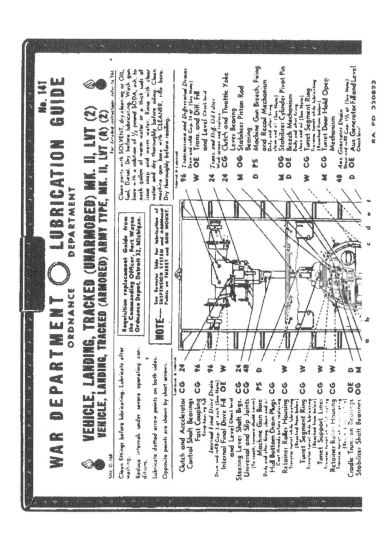

TM 9-775
27

LUBRICATION

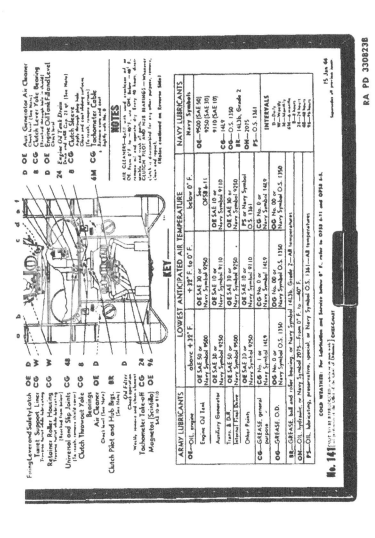

Figure 20 — Lubrication Guide — Suspension System and Additional Points on Turret and Gun Mount

LANDING VEHICLE TRACKED MK. I AND MK. II

(*m*) Run the pump approximately 10 minutes.

(*n*) Loosen both bleeder valves with pump running.

(*o*) After a solid flow of oil is obtained, tighten both valves permanently, and stop motor.

(*p*) Recheck and fill oil reservoir approximately ⅔ full.

(4) TEST FOR AIR IN STABILIZER SYSTEM. To determine if the system is free of air, lock gun in fixed position, and turn turret switch "ON". If oil level drops, air is trapped in the system. To remove trapped air, turn the turret switch to the "OFF" position, disengage hand elevating mechanism, and work gun slowly up and down from 5 to 10 minutes. Then repeat the above test. If air still remains trapped in system, repeat purging procedure (step (3), (*h*) through (*k*) above).

f. Reports and Records. Report unsatisfactory performance of materiel to the Ordnance Officer responsible for maintenance. A record of lubrication may be maintained in the Duty Roster (W.D., A.G.O. Form No. 6).

Section IX

TOOLS AND EQUIPMENT STOWAGE ON THE VEHICLE

	Paragraph
Vehicle tools	28
Vehicle equipment	29
Vehicle spare parts	30
Gun tools	31
Gun equipment	32
Gun spare parts	33
Care of tools and equipment	34

28. VEHICLE TOOLS.

Description	Federal Stock No.	Stowage Position
Adapter, socket wrench, ½ F to ⅜-in. M sq drive	41-A-20-225	In tool box
Bar, jimmy, ⅝-in. x 18 in. (2)	41-B-255	In tool box
Bar, socket extension, ½-in. sq drive, 10-in.	41-B-309	In tool box
Blade, hacksaw, 12-in. (3)	41-B-1165	In tool box
Box, tool, 9½- x 10-in. x 21-in.	—	On brackets in engine room

TM 9-775
28

TOOLS AND EQUIPMENT STOWAGE ON THE VEHICLE

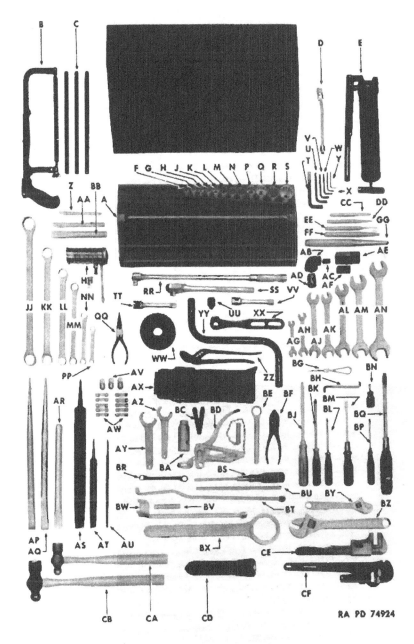

Figure 21 — Tools in Tool Box
(SEE NEXT 2 PAGES FOR LEGEND)

RA PD 74924

65

TM 9-775

LANDING VEHICLE TRACKED MK. I AND MK. II

A—BOX, TOOL, 9½ x 10 x 21-IN.
B—FRAME, HACKSAW, ADJUSTABLE, 8 TO 12 IN.
C—BLADE, HACKSAW, 24 TEETH—3
D—NOZZLE, LUBRICANT GUN (WITH ZERK FITTING)
E—GUN, LUBRICANT, HAND TYPE (STEWART-WARNER 6577E)
F—WRENCH, SOCKET, ½-IN. SQ. DRIVE, 7/16-IN. HEX.
G—WRENCH, SOCKET, ½-IN. SQ. DRIVE, ½-IN. HEX.
H—WRENCH, SOCKET, ½-IN. SQ. DRIVE, 9/16-IN. HEX.
J—WRENCH, SOCKET, ½-IN. SQ. DRIVE, 19/32-IN. HEX.
K—WRENCH, SOCKET, ½-IN. SQ. DRIVE, 5/8-IN HEX.
L—WRENCH, SOCKET, ½-IN. SQ. DRIVE, 11/16-IN. HEX.
M—WRENCH, SOCKET, ½-IN. SQ. DRIVE, ¾-IN. HEX.
N—WRENCH, SOCKET, ½-IN. SQ. DRIVE, ¾-IN. HEX.
P—WRENCH, SOCKET, ½-IN. SQ. DRIVE, 7/8-IN. HEX.
Q—WRENCH, SOCKET, ½-IN. SQ. DRIVE, 1-IN. HEX.
R—WRENCH, SOCKET, ½-IN. SQ. DRIVE, 1 1/16-IN. HEX.
S—WRENCH, SOCKET, ½-IN. SQ. DRIVE, 1¼-IN. HEX.
T—WRENCH, SAFETY SCREW, 3/8-IN. HEX.
U—WRENCH, SAFETY SCREW, ¼-IN. HEX.
V—WRENCH, SAFETY SCREW, 3/16-IN. HEX.
W—WRENCH, SAFETY SCREW, 5/32-IN. HEX.
X—WRENCH, SAFETY SCREW, 1/8-IN. HEX.
Y—WRENCH, SAFETY SCREW, 3/32-IN. HEX.
Z—CHISEL, COLD, 3/8-IN. CUT
AA—CHISEL, COLD, ½-IN. CUT
BB—CHISEL, COLD, ¾-IN. CUT
CC—PUNCH, CENTER, 3/8 x 4½-IN. LONG
DD—PUNCH, PRICK, 3/8 x 5¼-IN. LONG
EE—PUNCH, PIN, 3/8 x 3/16 x 6 1/8-IN. LONG
FF—PUNCH, DRIVE PIN, ½ x ¼ x 6½-IN. LONG
GG—PUNCH, DRIVE PIN, ¾ x 3/8 x 10-IN. LONG
HH—OILER, TRIGGER TYPE, 1 PT.
JJ—WRENCH, BOX, DOUBLE HEAD (ALLOY-S), 1 1/16 x 1¼-IN.
KK—WRENCH, BOX, DOUBLE HEAD (ALLOY-S), 1 3/16 x 1-IN.
LL—WRENCH, BOX, DOUBLE HEAD (ALLOY-S), ¾ x 7/8-IN.
MM—WRENCH, BOX, DOUBLE HEAD (ALLOY-S), 5/8 x 11/16-IN.
NN—WRENCH, BOX, DOUBLE HEAD (ALLOY-S), ½ x 9/16-IN.
PP—WRENCH, BOX, DOUBLE HEAD (ALLOY-S), 3/8 x 7/16-IN.
QQ—PLIERS, NEEDLE NOSE, SIDE CUTTING, 6 IN. LONG
RR—HANDLE, FLEXIBLE, ½-IN. SQ. DRIVE, 18 IN.
SS—RATCHET, WRENCH, REVERSIBLE, ½-IN. SQ. DRIVE, 9 IN.
TT—EXTENSION, ½-IN. SQ. DRIVE, 6-IN. LONG
UU—ADAPTER, ½-IN. SQ. F. TO 3/8-IN. M.
VV—EXTENSION, ½-IN. SQ. DRIVE, 6-IN. LONG
WW—TAPE, FRICTION, ¾-IN. WIDE, 30 FT ROLL
XX—WRENCH, BOX, 5/16-IN. (SPECIAL FOR FAST'S COUPLING)
YY—CRANK, ENGINE
ZZ—PLIERS, WATER PUMP ADJUSTING, 9½-IN. LONG
AB—ELBOW, GALVANIZED, 1 3/8 x ¾-IN.
AC—NIPPLE, GALVANIZED, ¼ x CLOSE

RA PD 74924B

Legend for Figure 21

TOOLS AND EQUIPMENT STOWAGE ON THE VEHICLE

AD—NIPPLE, GALVANIZED, 1 3/8 x 3/4-IN.
AE—PLUG, HULL DRAIN REMOVING
AF—PLUG, BOGIE WHEEL AXLE, 5/8-IN. SQ. x 1 1/2-IN. LONG
AG—WRENCH, ENGINEER'S, DOUBLE HEAD (ALLOY-S), 3/8 x 7/16-IN.
AH—WRENCH, ENGINEER'S, DOUBLE HEAD (ALLOY-S), 1/2 x 9/16-IN.
AJ—WRENCH, ENGINEER'S, DOUBLE HEAD (ALLOY-S), 19/32 x 25/32-IN.
AK—WRENCH, ENGINEER'S, DOUBLE HEAD (ALLOY-S), 5/8 x 3/4-IN.
AL—WRENCH, ENGINEER'S, DOUBLE HEAD (ALLOY-S), 11/16 x 7/8-IN.
AM—WRENCH, ENGINEER'S, DOUBLE HEAD (ALLOY-S), 15/16 x 1-IN.
AN—WRENCH, ENGINEER'S, DOUBLE HEAD (ALLOY-S), 1 1/16 x 1 1/4-IN.
AP—BAR, JIMMY, 5/8 x 18-IN. LONG
AQ—BAR, JIMMY, 5/8 x 18-IN. LONG
AR—CHISEL, COLD, 7/8-IN. CUT, 12-IN. LONG (RIVET CUTTER)
AS—FILE, MILL BASTARD, 14-IN. LONG
AT—FILE, MILL BASTARD, 8-IN. LONG
AU—FILE, ROUND BASTARD, 8-IN. LONG
AV—BULBS, TAILLIGHT
AW—FUSES, 15 AMP
AX—KIT, FABRIC ROLL TYPE
AY—WRENCH, BOX, SEGMENTED, 45 DEG, 1 x 1/8-IN.
AZ—WRENCH, BOX, SEGMENTED, 45 DEG, 1 x 1/4-IN.
BA—WRENCH, SOCKET, EXTRA LONG, 1 x 13/16-IN.
BC—FEELER GAGE, 0.124 x 0.010-IN.
BD—VALVE LIFTER
BE—WRENCH, BOX, 7/8-IN.
BF—PLIERS, 6-IN.
BG—SCREWDRIVER, NON-MAGNETIC, FOR COMPASS
BH—SCREWDRIVER, DOUBLE OFFSET
BJ—SCREWDRIVER (CROSS RECESS TYPE), 8-IN. BLADE
BK—SCREWDRIVER (CROSS RECESS TYPE), 6-IN. BLADE
BL—SCREWDRIVER (CROSS RECESS TYPE), 4-IN. BLADE
BM—SCREWDRIVER, STRADDLE SPLIT
BN—SCREWDRIVER, SPECIAL PURPOSE, 1 1/2-IN. BLADE
BP—SCREWDRIVER, SPECIAL PURPOSE, 3 3/4-IN. BLADE
BQ—SCREWDRIVER, COMMON, 8-IN. BLADE
BR—WRENCH, BOX, 7/16 x 1/2-IN.
BS—SCREWDRIVER, 10 1/2-IN. OVERALL
BT—WRENCH, BOX, 1/2-IN. (EXTRA LONG HANDLE)
BU—CROSSBAR, 3/8-IN. DIAM.
BV—WRENCH, SOCKET, PRONG TYPE, 9/16-IN. (FOR SLOTTED NUT)
BW—WRENCH, HALF-ROUND VERTICAL SPANNER, 2 3/8-IN.
BX—WRENCH, BOX, 2 3/8-IN.
BY—WRENCH, ADJUSTABLE, SINGLE END, 8-IN.
BZ—WRENCH, ADJUSTABLE, SINGLE END, 12-IN.
CA—HAMMER, MACHINIST'S BALL PEEN, 12 OZ.
CB—HAMMER, MACHINIST'S BALL PEEN, 24 OZ.
CD—FLASHLIGHT (PLASTIC)
CE—WRENCH, ADJUSTABLE, SCREW, 11-IN.
CF—WRENCH, PIPE, ADJUSTABLE, 14-IN.

RA PD 74924C

Legend for Figure 21 — Cont'd.

LANDING VEHICLE TRACKED MK. I AND MK. II

Description	Federal Stock No.	Stowage Position
Chisel, cold, $\frac{5}{16}$- x $\frac{3}{8}$- x 5-in. long	—	In tool box
Chisel, cold, $\frac{7}{16}$- x $\frac{1}{2}$- x 6-in. long	41-C-1110	In tool box
Chisel, cold, $\frac{5}{8}$- x $\frac{3}{4}$- x 7-in. long	41-C-1124	In tool box
Cutter, rivet, $\frac{3}{4}$- x $\frac{7}{8}$- x 12-in. long	—	In tool box
Compressor, valve spring	41-C-1405	In tool box
Cross bar	41-C-2559-27	In tool box
Extension, $\frac{1}{2}$-sq drive, 10-in. long (2)	—	In tool box
Extractor, fuse	—	In tool box
File, American std., mill, bastard, 8-in. long	41-F-1155	In tool box
File, American std., bastard cut, 12-in. long	41-F-1158	In tool box
File, American std., bastard cut, 8-in. long	41-F-1304	In tool box
Frame, hacksaw, adjustable, 8-in. to 12-in.	41-F-3394	In tool box
Gage, feeler, 0.124 and 0.010	41-G-407	In tool box
Hammer, machinists, ball peen, 12-oz	41-H-522	In tool box
Hammer, machinists, ball peen, 24-oz.	41-H-524	In tool box
Handle, $\frac{3}{8}$-in. rod	—	In tool box
Handle, ratchet, reversible, $\frac{1}{2}$-sq drive	41-H-1505	In tool box
Handle, hinge, $\frac{1}{2}$-in. sq drive, 18-in	41-H-1502-50	In tool box
Oiler, 1-pt trigger-type	—	In tool box
Pliers, combination, slip joint	41-P-1652	In tool box
Pliers, needle nose	41-P-1991	In tool box
Pliers, water pump	41-P-2100	In tool box
Punch, center, $\frac{3}{8}$- x 4$\frac{7}{8}$-in. long	41-P-3185	In tool box
Punch, drive pin	41-P-3538	In tool box
Punch, starting, $\frac{1}{2}$- x $\frac{1}{4}$- x 6$\frac{1}{2}$-in. long	41-P-3642-400	In tool box
Punch, pin $\frac{3}{8}$- x $\frac{5}{16}$- x 6$\frac{1}{8}$-in. long	41-P-3604-10	In tool box
Punch, prick, $\frac{3}{8}$- x 5$\frac{1}{4}$-in. long	41-P-3782	In tool box
Remover, drain plug, 1-in. square	—	In tool box
Screwdriver, common, 10$\frac{1}{2}$-in. blade	41-S-1010	In tool box
Screwdriver (cross-recess type), 4-in. blade (No. 2)	41-S-1638	In tool box

TOOLS AND EQUIPMENT STOWAGE ON THE VEHICLE

Description	Federal Stock No.	Stowage Position
Screwdriver (cross-recess type), 6-in. blade (No. 3)	41-S-1640	In tool box
Screwdriver (cross-recess type), 8-in. blade (No. 4)	41-S-1642	In tool box
Screwdriver, nonmagnetic, for compass	41-S-1067-700	In tool box
Screwdriver, off-set, 1/4- x 4 1/4- x 3/8-in.	41S-1397	In tool box
Screwdriver, 3/8- x 7/16- x 8-in.	41-S-1106	In tool box
Screwdriver, special purpose, 1-in. blade	41-S-1062	In tool box
Screwdriver, special purpose, 4-in. blade	41-S-1102	In tool box
Screwdriver, straddle, split (Mayhew No. 1520 or equal)	—	In tool box
Sling, cable, lifting	—	In tool box
Socket, 1- x 13/16-in. extra long	—	In tool box
Socket, prong-type, for slotted nuts, 9/16-in.	—	In tool box
Wrench, adjustable, auto-type, 11-in.	41-W-448	In tool box
Wrench, adjustable, Crescent type 8-in.	41-W-486	In tool box
Wrench, adjustable, Crescent type 12-in.	41-W-488	In tool box
Wrench, box, 45° angle, 1 1/8-in.	—	In tool box
Wrench, box, 45° angle, 1 1/4-in.	—	In tool box
Wrench box, 7/16- x 1/2-in.	41-W-599	In tool box
Wrench, box, 2 5/8-in.	41-W-637-590	In tool box
Wrench, box, special, 1/2-in.	—	In tool box
Wrench, box, 15°, 3/8- x 7/16-in.	41-W-620	In tool box
Wrench, box, open-end (alloy-S) 3/8-in. x 7/8-in.	—	In tool box
Wrench, box, open-end (alloy-S) 1/2-in. x 9/16-in.	41-W-662	In tool box
Wrench, box, open-end (alloy-S) 5/8-in. x 11/16-in.	41-W-602	In tool box
Wrench, box, open-end (alloy-S) 3/4-in. x 7/8-in.	41-W-604	In tool box
Wrench, box, open-end (alloy-S) 15/16-in. x 1-in.	41-W-608	In tool box

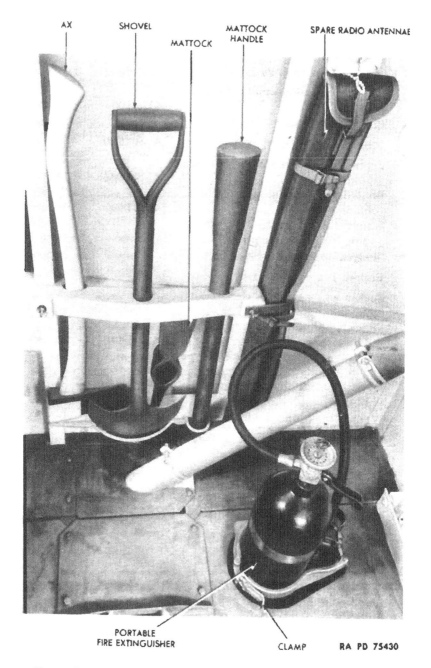

Figure 22 — Equipment Stowed Behind Driver's Seat

TOOLS AND EQUIPMENT STOWAGE ON THE VEHICLE

Description	Federal Stock No.	Stowage Position
Wrench, box, open-end (alloy-S) $1\frac{1}{16}$-in. x $1\frac{1}{4}$-in.	41-W-610-25	In tool box
Wrench, engineers, open-end (alloy-S) $\frac{3}{8}$-in. x $\frac{7}{16}$-in.	41-W-991	In tool box
Wrench, engineers, open-end (alloy-S) $\frac{1}{2}$-in. x $\frac{9}{16}$-in.	41-W-622	In tool box
Wrench, engineers, open-end, $\frac{1}{2}$- x $\frac{19}{32}$-in.	41-W-1003	In tool box
Wrench, engineers, open-end (alloy-S) $\frac{19}{32}$-in. x $\frac{25}{32}$-in.	41-W-1007-20	In tool box
Wrench, engineers, open-end (alloy-S) $\frac{5}{8}$-in. x $\frac{3}{4}$-in.	41-W-1008	In tool box
Wrench, engineers, open-end (alloy-S) $\frac{11}{16}$-in. x $\frac{7}{8}$-in.	41-W-1020	In tool box
Wrench, engineers, open-end (alloy-S) $\frac{9}{16}$-in. x 1-in.	41-W-1021	In tool box
Wrench, engineers, open-end (alloy-S $1\frac{1}{16}$-in. x $1\frac{1}{4}$-in.	41-W-1024-5	In tool box
Wrench, fast coupling	—	In tool box
Wrench, pipe, adjustable, 14-in.	41-W-1663	In tool box
Wrench, safety screw, $\frac{3}{32}$-in.	41-W-2449	In tool box
Wrench, safety screw, $\frac{1}{8}$-in.	41-W-2450	In tool box
Wrench, safety screw, $\frac{5}{32}$-in.	41-W-2451	In tool box
Wrench, safety screw, $\frac{3}{16}$-in.	41-W-2452	In tool box
Wrench, safety screw, $\frac{1}{4}$-in.	41-W-2454	In tool box
Wrench, safety screw, fast coupling	—	In tool box
Wrench, socket, $\frac{1}{2}$-in. sq. drive, $\frac{7}{16}$-in. hexagonal	41-W-3005	In tool box
Wrench, socket, $\frac{1}{2}$-in. sq. drive, $\frac{1}{2}$-in. hexaggonal	41-W-3007	In tool box
Wrench, socket, $\frac{1}{2}$-in. sq. drive, $\frac{9}{16}$-in. hexagonal	41-W-3009	In tool box
Wrench, socket, $\frac{1}{2}$-in. sq. drive, $\frac{19}{32}$-in. hexagonal	41-W-3011	In tool box
Wrench, socket, $\frac{1}{2}$-in. sq. drive, $\frac{5}{8}$-in. hexagonal	41-W-3013	In tool box
Wrench, socket, $\frac{1}{2}$-in. sq. drive, $1\frac{1}{16}$-in. hexagonal	41-W-3015	In tool box
Wrench, socket, $\frac{1}{2}$-in. sq. drive, $\frac{3}{4}$-in. hexagonal (2)	41-W-3017	In tool box

LANDING VEHICLE TRACKED MK. I AND MK. II

Description	Federal Stock No.	Stowage Position
Wrench, socket, ½-in. sq. drive, ⅞-in. hexagonal	41-W-3023	In tool box
Wrench, socket, ½-in. sq. drive, 1 5/16-in.	41-W-3025	In tool box
Wrench, socket, ½-in. sq. drive, 1-in.	41-W-3027	In tool box
Wrench, socket, ½-in. sq. drive, 1 1/16-in.	41-W-3029	In tool box
Wrench, socket, ½-in. sq. drive, deep, 1⅛-in.	41-W-3031	In tool box
Wrench, socket, ½-in. sq. drive, 1¼-in.	41-W-3029-10	In tool box
Wrench, spanner, 2⅜-in. O. D., ½-in. rd	—	In tool box

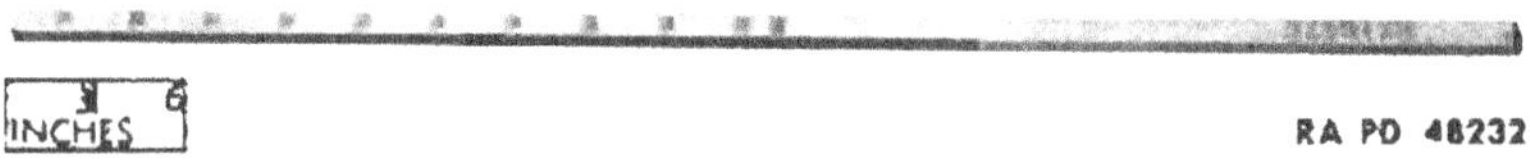

Figure 23 — Fuel Supply Measuring Stick

29. VEHICLE EQUIPMENT.

Description	Stowage Position
Antenna, MP 37 (spare)	Under bracket behind driver's seat
Ax, single-bit, 5-lb.	On forward bulkhead behind driver's seat
Book, O.O. Form 7255	Under bracket on forward bulkhead
Extinguisher, fire, 15-lb CO_2 (portable)	In well on cab floor, left of driver
Flashlight (plastic)	In tool box
Guide, lubrication	Under bracket on forward bulkhead
Gun, lubricant, hand-type	In tool box
Handle, mattock	On forward bulkhead behind driver's seat
Hook, boat, 8-ft. (2)	On forward end of hull, adjacent to catwalks

TM 9-775
29-30

TOOLS AND EQUIPMENT STOWAGE ON THE VEHICLE

Description	Stowage Position
Instruction sheet for compass	Under bracket on forward bulkhead
Lantern, hand, type J-1S, w/mounting bracket (2)	In bracket right side of forward propeller shaft guard
Line, bow painter, 30-ft., 3-in. circumference sisal	Left front floor of cab
Line, Sternfast, 30-ft., 3-in. circumference sisal	Left front floor of cab
Line, tow, 1-in. x 15-ft. galvanized wire rope (shackles) (at each end to fit towing staple)	On stern of hull
Manual, technical, for LVT (A) (2)	Under bracket on forward bulkhead
Mattock, pick, M1 (w/o handle)	On forward bulkhead behind driver's seat
Paulin, w/grommets and rope lashings	On cargo compartment floor
Periscope M6 (except LVT (2))	In periscope mount
Radio set, Model TCS (Navy) (transmitting and receiving with interphone amplifier)	Mounted in right side of cab
Shovel, short handle	On forward bulkhead behind driver's seat
Stick, fuel supply measuring	Left rear corner of cargo compartment
Tank, water, 5-gal (Q.M. Std.)	Right front of cab, ahead of radio
Tape, friction, ¾-in. x 30-ft. (2)	In tool box
Tube, flexible nozzle	
Wire, soft iron, 14 gage, 10-ft.	

30. VEHICLE SPARE PARTS.

Elbow, galvanized, ¾-in. x ¼-in.	In tool box
Head (for periscope M6) (except LVT (2)) (6)	In cargo compartment
Link assembly, track, inside (4)	In cargo compartment
Manual, spare parts, illustrated (for vehicle)	Under bracket on forward bulkhead
Nipple, galvanized, ¼-in. x close	In tool box
Nipple, galvanized, 1⅜-in. x ¾-in.	In tool box

73

LANDING VEHICLE TRACKED MK. I AND MK. II

Description	Stowage Position
Outside link and track channel assembly (Food Machinery No. 51X1097C) (2)	In cargo compartment
Periscope M6 (except LVT (2))	In cargo compartment
Pin, coupling (for track links) (8)	In cargo compartment
Plug, bogie wheel axle, ⅝ sq x 1½ long	In cargo compartment

31. GUN TOOLS.

Roll, tool, M12 (w/o contents)
  Screwdriver, common 3-in. blade
  Wrench, combination, M2 (1 for cal. .50 machine gun)
  Wrench, combinaton, M6 (1 for cal. .30 machine gun)
  Wrench, socket, front barrel bearing plug (1 for cal. .30 machine gun) .... Gun tool chest

32. GUN EQUIPMENT.

a. Ammunition.

Description	Stowage Position
Cal. .30 rounds (in belts C-3951 and boxes M1-D44070) (2,000): 80 percent AP, 20 percent tracer	In cargo compartment
Cal. .50 rounds (in boxes M2-D73913) (1,000): 80 percent AP, 20 percent tracer	In cargo compartment
37-mm rounds:	
HE (56)	14 in turret, 42 in cargo compartment
AP (24)	6 in turret, 18 in cargo compartment
Can (24)	6 in turret, 18 in cargo compartment

b. Cal. .30 Machine Gun.

Description	Stowage Position
Bag, empty cartridge, cal. .30	In cargo compartment
Belt, ammunition, 250 rounds (8)	In cargo compartment
Box, ammunition, cal. .30, M1 (8)	In cargo compartment
Brush, chamber cleaning, M6	In cargo compartment
Brush, cleaning, cal. .30, M2 (3)	In cargo compartment
Can, tubular (w/o contents)	In cargo compartment

TOOLS AND EQUIPMENT STOWAGE ON THE VEHICLE

Description	Stowage Position
Case, cleaning rod, M1	In cargo compartment
Case, cover group	In cargo compartment
Chest, steel, M5 (w/o contents)	In cargo compartment
Cover, machine gun, cal. .30 (gun and mount)	In cargo compartment
Cover, tripod mount, M2	In cargo compartment
Cover, tripod mount, over-all	In cargo compartment
Extractor, ruptured cartridge, Mk. IV	In cargo compartment
Mount, machine gun, cal. .30, or cal. .50 (M35)	In cargo compartment
Mount, tripod, machine gun, M2	In cargo compartment
Oiler, rectangular, 12-oz	In cargo compartment
Reflector, barrel, cal. .30	In cargo compartment
Rod, cleaning, jointed, cal. .30, M1	In cargo compartment

c. **Cal. .50 Machine Gun.**

Description	Stowage Position
Bag, metallic belt link	In cargo compartment
Box, ammunition, cal. .50, M2 (10)	In cargo compartment
Brush, cleaning, cal. .50, M4 (4)	In cargo compartment
Case, cleaning rod, M15	In cargo compartment
Chute, metallic belt link, M1	In cargo compartment
Cover, machine gun, cal. .50 (gun and mount)	In cargo compartment
Cover, tripod mount, M1, for machine gun, cal. .50, M3	In cargo compartment
Extractor, ruptured cartridge	In cargo compartment
Manual, field, for machine gun, cal. .30, M1919A4	Under bracket on front bulkhead
Manual, field, for machine gun, cal. .50, M2	Under bracket on front bulkhead
Mount, machine gun, cal. .30, or cal. .50 (M35)	In cargo compartment
Mount, tripod, cal. .50, M3	In cargo compartment
Oiler, filling, oil buffer	In cargo compartment
Rod, jointed, cleaning, M7	In cargo compartment

33. GUN SPARE PARTS.

a. **General.**

Description	Stowage Position
Case, spare bolt, M2 (w/o contents) (3)	In cargo compartment
Cover, spare barrel	In cargo compartment

LANDING VEHICLE TRACKED MK. I AND MK. II

Description	Stowage Position
Cover, spare barrel, M13, 45-in.	In cargo compartment
Envelope, spare parts, M1 (w/o contents)	In cargo compartment
Envelope, spare parts, M1 (w/o contents) (2)	In cargo compartment
Roll, spare parts, M13 (w/o contents)	In cargo compartment

b. **Cal. .30 Machine Gun.**

Description	Stowage Position
Band, lock, front barrel bearing	In cargo compartment
Band, lock, front barrel bearing plug	In cargo compartment
Barrel	In cargo compartment
Bolt group, consisting of:	
Bolt, assembly, B147299	In cargo compartment
Extractor, assembly, C121076	In cargo compartment
Lever, cocking, B131317	In cargo compartment
Pin, cocking, lever, A20567	In cargo compartment
Pin, firing, assembly, C9186	In cargo compartment
Rod, driving spring, assembly, B147222	In cargo compartment
Sear, C64137	In cargo compartment
Spring, driving, B212654	In cargo compartment
Spring, sear, assembly, A131265	In cargo compartment
Cover group, consisting of:	
Cover, assembly, C9801	In cargo compartment
Lever, feed belt, B17503	In cargo compartment
Pawl, feed belt, C8461	In cargo compartment
Pin, belt feed pawl, assembly, B131255	In cargo compartment
Pivot, belt feed lever, group assembly, B110529	In cargo compartment
Slide, feed bolt, assembly, B131262	In cargo compartment
Spring, feed belt pawl, B147224	In cargo compartment
Spring, cover extractor, B17513	In cargo compartment
Extension, barrel, group, consisting of:	
Extension, barrel, assembly, C64139	In cargo compartment
Lock, breech, B147214	In cargo compartment
Pin, breech lock, assembly, B131253	In cargo compartment
Spring, locking barrel, B147230	In cargo compartment

TOOLS AND EQUIPMENT STOWAGE ON THE VEHICLE

Description	Stowage Position
Frame, lock, group, consisting of:	
Accelerator, C64142	In cargo compartment
Frame, lock, assembly, C9182	In cargo compartment
Pin, accelerator, assembly, B131253	In cargo compartment
Pin, trigger, A20503	In cargo compartment
Plunger, barrel, assembly, B131251	In cargo compartment
Spring, barrel plunger, A135057	In cargo compartment
Spring, trigger pin, B147231	In cargo compartment
Trigger, C8476	In cargo compartment
Lever, cocking	In cargo compartment
Lever, feed belt	In cargo compartment
Pawl, feed belt	In cargo compartment
Pawl, holding belt	In cargo compartment
Pin, accelerator, assembly	In cargo compartment
Pin, belt holding pawl, split	In cargo compartment
Pin, cocking lever	In cargo compartment
Pin, firing, assembly	In cargo compartment
Pin, trigger	In cargo compartment
Plug, front barrel bearing	In cargo compartment
Spring, belt holding pawl	In cargo compartment
Spring, locking barrel	In cargo compartment
Spring, sear, assembly	In cargo compartment
Trigger	In cargo compartment

c. Cal. .50 Machine Gun.

Description	Stowage Position
Barrel, assembly	In cargo compartment
Disk, buffer	In cargo compartment
Extension, firing pin assembly	In cargo compartment
Extractor, assembly	In cargo compartment
Lever, cocking	In cargo compartment
Pin, cotter, belt feed lever pivot stud	In cargo compartment
Pin, cotter, cover pin	In cargo compartment
Pin, cotter, switch pivot (2)	In cargo compartment
Pin, firing	In cargo compartment

TM 9-775
33

LANDING VEHICLE TRACKED MK. I AND MK. II

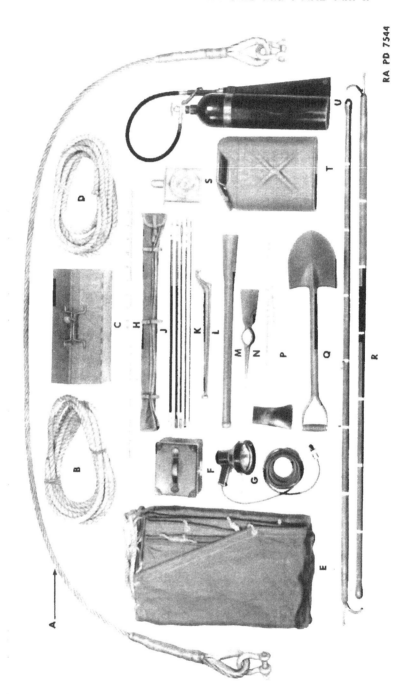

Figure 24 — Lay-out of Vehicle Tools and Equipment

TOOLS AND EQUIPMENT STOWAGE ON THE VEHICLE

A—LINE TOW, 1-IN. x 15 FT. GALVANIZED WIRE ROPE
B—TIE LINE 30 FT. x 3-IN. CIRCUMFERENCE
C—BOX, TOOL, 9½-IN. x 10-IN. x 21-IN.
D—TIE LINE 30 FT. x 3-IN. CIRCUMFERENCE
E—TARPAULIN, WITH GROMMETS AND ROPE LASHINGS
F—BOX, SIGNAL AND SPOTLIGHT
G—LIGHT, COMBINATION SIGNAL AND SPOT
H—STICK, CALIBRATED HARDWOOD FUEL MEASURING
J—BAG ANTENNA, FABRIC ROLL TYPE
K—ANTENNA SET, M49, M50, M51, M52, M53
L—WRENCH, REAR IDLER ADJUSTING
M—HANDLE, MATTOCK PICK
N—MATTOCK PICK
P—AX (CHOPPING, SINGLE BIT, 5 LB.)
Q—SHOVEL, SHORT HANDLE
R—HOOK, BOAT, 8 FT. (2)
S—LANTERN, HAND TYPE J-13
T—TANK, WATER, 5-GALLON
U—EXTINGUISHER, FIRE, 15-LB., CO, (PORTABLE)

Legend for Figure 24

LANDING VEHICLE TRACKED MK. I AND MK. II

Description	Stowage Position
Plunger, belt feed lever	In cargo compartment
Rod, driving spring, w/spring, assembly	In cargo compartment
Slide, belt feed group, consisting of:	
Arm, belt feed pawl, B8914	In cargo compartment
Pawl, feed belt, assembly, B8961	In cargo compartment
Pin, belt feed pawl, assembly, B8962	In cargo compartment
Slide, belt feed, assembly, B261110	In cargo compartment
Spring, belt feed pawl, A9351	In cargo compartment
Slide, sear	In cargo compartment
Spring, belt feed lever plunger	In cargo compartment
Spring, belt holding pawl	In cargo compartment
Spring, cover extractor	In cargo compartment
Spring, locking barrel	In cargo compartment
Spring, sear	In cargo compartment
Stud, bolt	In cargo compartment

34. CARE OF TOOLS AND EQUIPMENT.

a. An accurate record of all tools, accessories, and equipment must be kept, in order that their location and condition may be known at all times. Items lost or unserviceable will be replaced immediately. All tools and equipment must be cleaned and in proper condition for further use before being returned to their location. Care must be used in fastening the tools carried on the outside of the vehicle, and frequent inspections and oiling are necessary to prevent corrosion.

PART TWO — VEHICLE MAINTENANCE INSTRUCTIONS

Section X

NEW VEHICLE RUN-IN TEST

	Paragraph
Purpose	35
Correction of deficiencies	36
Run-in test procedures	37

35. PURPOSE.

a. When a new or reconditioned vehicle is received at the using organization, it is necessary for second echelon personnel to determine whether or not the vehicle will operate satisfactorily when placed in service. For this purpose, inspect all accessories, subassemblies, assemblies, tools, and equipment to see that they are in place and correctly adjusted. In addition, they will perform a run-in test of at least 50 miles as directed in AR 850-15, paragraph 25, table III, according to procedures in paragraph 37, below.

36. CORRECTION OF DEFICIENCIES.

a. Deficiencies disclosed during the course of the run-in test will be treated as follows:

(1) Correct any deficiencies within the scope of the maintenance echelon of the using organization before the vehicle is placed in service.

(2) Refer deficiencies beyond the scope of the maintenance echelon of the using organization to a higher echelon for correction.

(3) Bring deficiencies of serious nature to the attention of the supplying organization.

37. RUN-IN TEST PROCEDURES.

a. Preliminary Service.

(1) FIRE EXTINGUISHERS. Examine cylinders of portable and fixed extinguishers to see if they are in good condition, and see that nozzles are not clogged. If valves are damaged, or appear to have been opened, report for exchange or refill. Inspect lines and remote control mechanism of fixed system for looseness, binding, and damage. See that cylinders, and line and nozzle mountings are secure.

(2) FUEL AND OIL. Fill fuel tanks (include auxiliary generator on LVT (A) (1) model). Check engine oil supply, and add as necessary to fill tank to correct level. If there is a tag attached to oil filler cap

LANDING VEHICLE TRACKED MK. I AND MK. II

concerning engine oil in tank, follow instructions on tag before driving vehicle.

(3) FUEL FILTER. Inspect filter for leaks, damage, and secure mountings and connections. Turn cleaner handle one complete turn, remove sediment bowl drain plug, and run off all dirt and water. If any appreciable amount of dirt and water is present, clean filter bowl and element with dry-cleaning solvent, clean screens in fuel tank filler openings, and drain off accumulated dirt and water from bottom of fuel tanks. Drain only until fuel runs clean.

(4) BATTERY. Make hydrometer and voltage test of battery, and if necessary, add clean water to about ½ inch above plates.

(5) AIR CLEANERS. Examine air cleaners to see if they are in good condition and secure. Remove and wash elements in dry-cleaning solvent. Fill reservoirs to correct level with fresh oil, and reassemble securely. Be sure all gaskets are in good condition, and that duct and air horn connections are tight.

(6) ACCESSORIES AND BELTS. See that accessories such as carburetor, generator, regulator, cranking motor, magnetos, oil and fuel pumps, blowers, fan and shroud, baffles, oil coolers, and auxiliary generator (if so equipped), are in good condition, correctly assembled, and secure. See if generator drive belt is adjusted to have ¾-inch finger-pressure deflection. Clean out all insects, dirt, and trash, from oil cooler core air passages and blower ducts.

(7) ELECTRICAL WIRING. Examine all accessible wiring, conduits, and shielding to see if they are in good condition, securely connected, and properly supported.

(8) TRACKS. Remove stones and other foreign objects from between links and grousers, and between tracks and suspension units. Inspect links, grousers, cross bars, and adjusting screw assemblies, for looseness and damage, and see that track tension is adjusted so there is no clearance between inner plate and idler wheel sliding bracket, and ⅛-inch between inner and outer plate on adjusting screw.

(9) WHEEL AND FLANGE NUTS. Inspect to see that all rear idler wheel, sprocket flange, bogie wheel, and return idler, assembly and mounting nuts and screws are present and secure.

(10) TOWING CONNECTIONS. See that front and rear tow hooks and lifting links are in good condition and secure.

(11) HULL AND TARPAULIN. Inspect hull for damage, particularly punctures in hull plates and pontoons. Look for loose or missing hull drain plugs and inspection plates. Examine bulkheads and floor plates for looseness or damage. Inspect tarpaulin to be sure it is in good condition, and see that it is properly installed or stowed. Inspect escape hatches and doors in driving compartment to see if they are in good condition, alined to openings; if they are open, close, and latch properly.

NEW VEHICLE RUN-IN TEST

(12) VISION DEVICES. Inspect periscope prisms and windows (mounted or spares), to see if they are in good condition and clean. See that mounted units are secure in holders, and that holders are properly mounted. Test each periscope to be sure it will elevate, traverse, and depress through full range. CAUTION: *Clean periscopes only with a soft cloth or brush.*

(13) LUBRICATION. Perform a complete lubrication of the vehicle, covering all intervals according to instructions on Lubrication Guide, paragraph 27, except gear cases, and units lubricated or serviced in steps (1) to (12). Check all gear case oil levels, and add as necessary to bring to correct levels; change only if condition of oil indicates the necessity, or if oil is not of proper grade for existing atmospheric temperature. NOTE: *Perform steps (14) to (17) during lubrication.*

(14) SUSPENSIONS. Inspect torsional arms, spring-end brackets, and bumpers; see if they are in good condition, correctly assembled, and secure.

(15) STEERING LINKAGE. Inspect all shafts, arms, rods, connections, levers, and grips, to see that they are in good condition, correctly and securely assembled and mounted, and operate without excessive looseness or binding. Be sure lever locking devices operate properly.

(16) PROPELLER SHAFT, CENTER BEARING, AND CLUTCH RELEASE MECHANISM. Inspect propeller shafts, center shaft bearing, and universal joints, to see if they are in good condition, correctly and securely assembled, alined, and that joints or bearing do not leak excessively. Examine clutch release yoke, roller, linkage and mountings, for damage; be sure they operate without looseness or binding. There should be $1/8$-inch clearance between roller and collar.

(17) TRANSMISSION AND FINAL DRIVE VENTS. Examine all vents to be sure they are in good condition, secure, and not clogged. If necessary, remove vents and clean elements.

(18) PRIMER. When starting engine in step (19), observe if action of primer system is satisfactory. Notice if there are any fuel leaks at primer pump or connection.

(19) ENGINE WARM-UP. To prevent hydrostatic lock, turn engine over at least two complete revolutions (40 revolutions of crank), to clear lower cylinders of any accumulated liquid. If engine will not turn over readily, drain fluid from lower cylinders. After blowers have been in operation for 5 minutes, start engine, noting if cranking motor action is satisfactory, and any tendency toward hard starting. Set hand throttle to idle engine at 800 revolutions per minute during warm-up.

(20) INSTRUMENTS.

(a) *Oil Pressure Gage.* Immediately after engine starts, observe if oil pressure is satisfactory. NOTE: *Normal pressure at 800 revolu-*

LANDING VEHICLE TRACKED MK. I AND MK. II

tions per minute is 40 pounds; at governed speed, 90 pounds. Stop engine if oil pressure is not indicated in 30 seconds.

(b) *Ammeter.* Ammeter should show a slight positive (+) charge. High charge may be indicated until generator restores to battery, current used in starting, or longer if battery is low or electrical load heavy.

(c) *Engine Oil Temperature Gage.* Engine oil temperature should rise gradually during warm-up, to normal operating range, 160°F. to 175°F.

(d) *Transmission Oil Pressure Gage.* Gage should indicate 30 to 40 pounds at governed speed.

(e) *Tachometer.* Tachometer should indicate the engine revolutions per minute, and record accumulating engine revolutions.

(21) ENGINE CONTROLS. Observe if engine responds properly to controls and if controls operate without excessive looseness or binding.

(22) LAMPS (LIGHTS). Clean lenses and inspect all units for looseness and damage. If tactical situation permits, open and close all light switches to see if lamps respond properly.

(23) LEAKS, GENERAL. Look under vehicle, and within engine and driving compartments, for indications of fuel or oil leaks. Trace any leaks found to source, and correct or report them.

(24) TOOLS AND EQUIPMENT. Check tools and on vehicle stowage lists, paragraphs 28 through 34, to be sure all items are present, and see that they are serviceable, and properly mounted or stowed.

(25) AMPHIBIAN SERVICES.

(a) *Power Take-off and Drive.* Remove deck plate and bilge pump drive cover, and inspect power take-off unit and drive shaft to see if they are in good condition, secure, and not leaking oil. Check oil level in power take-off, and add as necessary to bring to proper level.

(b) *Bilge Pump and Outlet Pipe.* Examine pump for looseness or damage. Clean outlet pipe, and see that it is properly supported and not damaged, and that connections are tight.

(c) *Decks, Hatches, and Ventilators.* Examine decks and hatch covers to see if they are in good condition; see that hatch covers are alined with openings; that seals are in good condition and secure; and that fasteners hold hatch covers tightly against seals. Examine deck ventilators to engine compartment for looseness, damage, and proper operation; and clean out screens. Be sure vent covers seal tightly when closed, and vents are not clogged.

b. **Run-in Test.** Perform the following procedures, steps (1) to (12) inclusive, during the road and water test of the vehicle. On

TM 9-775
37

NEW VEHICLE RUN-IN TEST

vehicles which have been driven 50 miles or more in the course of delivery from the supplying to the using organization, reduce the length of the land and water test runs to the least distance necessary to make observations listed below. Otherwise make a land run of 20 miles, and a water run for 30 minutes. CAUTION: *Avoid continuous operation of the vehicle at engine speeds approaching the maximum indicated on the caution plate, during test.*

(1) DASH INSTRUMENTS AND GAGES. Do not move vehicle until engine oil temperature has reached 135°F. Maximum safe operating temperature is 190°F. Observe readings of ammeter, tachometer, and oil pressure, oil temperature and transmission oil pressure gages, to see if they are indicating the proper function of the unit to which each applies.

(2) STEERING BRAKES. Test brakes, applying levers evenly, to be sure they will stop vehicle effectively without side pull, chatter, or squeal; see that lever locking devices function properly to hold levers in applied position. Test levers separately in both land and water operation to see if vehicle steers satisfactorily, with normal pulling effort applied to levers and that steering brakes start to meet resistance at approximately three notches on quadrant. Vehicle should travel straight on land or water with levers in released position.

(3) CLUTCH. Observe if clutch operates smoothly, without grab, chatter, or squeal on engagement, or slippage when fully engaged under load. See that pedal has ⅛- to ¾-inch free travel before meeting resistance. CAUTION: *Do not ride clutch pedal when not in use, and do not engage or disengage a new clutch severely or unnecessarily until driven and driving disks have become properly worn in.*

(4) TRANSMISSION. Gearshift mechanism should operate smoothly and easily, gears should operate without unusual noise, and not slip out of mesh.

(5) ENGINE. Be on the alert for any unusual engine operating characteristics, unusual noise, such as lack of pulling power or acceleration, backfiring, misfiring, stalling, overheating, or excessive exhaust smoke, or vibration that may indicate loose engine mountings. Observe if engine responds properly to all controls, and if controls appear to be excessively loose or binding. Have radio turned on, and listen for noise from engine interference.

(6) UNUSUAL NOISES. Be alert throughout road or water test for noise from hull or attachments, suspensions, running gear, or tracks that might indicate looseness, damage, excessive wear, or inadequate lubrication.

(7) HALT. Halt vehicle at 10-mile intervals or less, on land operation for services in steps (8) and (9), below.

LANDING VEHICLE TRACKED MK. I AND MK. II

(8) TEMPERATURES. Cautiously hand-feel each bogie wheel, sprocket, idler, and return idler hub, for abnormal temperatures. Examine transmission, differential, and final drives for indications of overheating, or excessive leaks at seals or gaskets.

(9) LEAKS. With engine running, and fuel and oil systems under pressure, look under vehicle and within engine and driver's compartments for indications of fuel or oil leaks.

(10) GUNS; ELEVATING AND TRAVERSING MECHANISM AND STABILIZER.

(a) LVT (2) and (A) (2) Models. Inspect gun mounts and tracks, manual traversing and elevating mechanism, and firing controls, to see that they are in good condition, clean, and correctly and securely assembled and mounted. Test firing controls to be sure they operate, and test manual elevating and traversing controls to be sure mechanism responds properly, without excessive looseness or binding.

(b) LVT (A) (2) Model. Inspect gun mounts, manual and power traversing mechanism, manual elevating and stabilizer mechanism and manual and electric firing controls to see if they are in good condition, clean, correctly and securely assembled and mounted. Test all controls to be sure manual and power mechanism responds properly. Inspect power turret traversing system, including, motor, pump, reservoir, and lines for oil leaks, and be sure oil reservoir is filled to proper level. Traverse turret to be sure it travels the full 360-degrees, without excessive noise or binding, and observe if automatic brake functions when hand control grip is released. Determine whether operation of traversing mechanism causes any noise in radio equipment.

(c) Stabilizer and Recoil Control. Inspect the stabilizer control-unit gear box, connecting oil lines, cylinder and piston, wiring, and control box to see that they are in good condition, secure, correctly assembled, and not leaking oil. Make an operating test of the stabilizer as outlined in paragraph 185. NOTE: *Defects in the stabilizer system should be referred to ordnance maintenance personnel for attention.* Check the recoil cylinders to see that they are in good condition and not leaking oil. Check to see whether the level of the recoil oil is as specified. NOTE: *Recoil operating checks must be made under firing conditions and in accordance with the instructions (refer to TM 9-250.) Determine if operation of stabilizer equipment causes any noise in radio equipment.*

(11) TRACK TENSION. Recheck track tension to see if inside plate on track adjusting screw is against slide bracket, and there is $\frac{1}{8}$-inch clearance between inner and outer plate.

SECOND ECHELON PREVENTIVE MAINTENANCE

(12) AMPHIBIAN SERVICES.

(a) *In Water.*

1. Bilge Pump. Test bilge pump to see if it operates properly and controls bilge level. Observe any unusual noise, and any excessive flow of water from outlet pipe which would indicate hull leaks.

2. Hull. Stop vehicle at 5-minute intervals during water run, and examine inside of hull for indications of water leaks at seam welds, drain plugs, deck seals, or between bulkheads. With vehicle in motion, note any indication of unbalance that may indicate leak into pontoon compartments, or unsatisfactory water control of bilge pumps.

(b) *After Leaving Water.* Remove all weeds and other foreign material from tracks and suspensions.

c. Vehicle Publications and Reports.

(1) PUBLICATIONS. See that vehicle technical manuals, Lubrication Guide, Standard Form No. 26 and WD AGO No. 478 (MWO and Major Unit Assembly Replacement Record), are in the vehicle, legible, and properly stowed. NOTE: *U.S.A. registration number and vehicle nomenclature must be filled in on Form No. 478, for new vehicles.*

(2) REPORTS. Upon completion of the run-in test, correct or report any deficiencies noted. Report general conditions of the vehicle to designated individual in authority.

Section XI

SECOND ECHELON PREVENTIVE MAINTENANCE

	Paragraph
Second echelon preventive maintenance services	38
MWO and major unit assembly replacement record	39

38. SECOND ECHELON PREVENTIVE MAINTENANCE SERVICES.

a. Regular scheduled maintenance inspections and services are a preventive maintenance function of the using arms, and are the responsibility of commanders of operating organizations.

(1) FREQUENCY. The frequencies of the preventive maintenance services outlined herein are considered a minimum requirement for normal operation of vehicles. Under unusual operating conditions, such as extreme temperatures and dusty or sandy terrain, it may be necessary to perform certain maintenance services more frequently.

LANDING VEHICLE TRACKED MK. I AND MK. II

(2) FIRST ECHELON PARTICIPATION. The drivers should accompany their vehicles, and assist the mechanics while periodic second echelon preventive maintenance services are performed. Ordinarily the driver should present the vehicle for a scheduled preventive maintenance service in a reasonably clean condition; that is, it should be dry, and not caked with mud or grease to such an extent that inspection and servicing will be seriously hampered. However, the vehicle should not be washed or wiped thoroughly clean, since certain types of defects, such as cracks, leaks, and loose or shifted parts or assemblies, are more evident if the surfaces are slightly soiled or dusty.

(3) If instructions other than those contained in the general procedures in step (4) or the specific procedures in step (5) which follow, are required for the correct performance of a preventive maintenance service or for correction of a deficiency, other sections of the vehicle operator's manual pertaining to the item involved, or a designated individual in authority, should be consulted.

(4) GENERAL PROCEDURES. These general procedures are basic instructions which are to be followed when performing the services on the items listed in the specific procedures. NOTE: *The second echelon personnel must be thoroughly trained in these procedures, so that they will apply them automatically.*

(a) When new or overhauled subassemblies are installed to correct deficiencies, care should be taken to see that they are clean, correctly installed, and properly lubricated and adjusted.

(b) When installing new lubricant retainer seals, a coating of the lubricant should be wiped over the sealing surface of the lip of the seal. When the new seal is a leather seal, it should be soaked in SAE 10 engine oil (warm, if practicable) for at least 30 minutes. Then, the leather lip should be worked carefully by hand before installing the seal. The lip must not be scratched or marred.

(c) The general inspection of each item applies, also, to any supporting member or connection, and usually includes a check to see whether the item is in good condition, correctly assembled, secure, or excessively worn. The mechanics must be thoroughly trained in the following explanations of these terms.

1. The inspection for "good condition" is usually an external visual inspection to determine whether the unit is damaged beyond safe or serviceable limits. The term "good condition" is explained further by the following: not bent or twisted, not chafed or burned, not broken or cracked, not bare or frayed, not dented or collapsed, not torn or cut.

2. The inspection of a unit to see that it is "correctly assembled" is usually an external visual inspection to see whether it is in its normal assembled position in the vehicle.

TM 9-775
38

SECOND ECHELON PREVENTIVE MAINTENANCE

3. The inspection of a unit to determine if it is "secure" is usually an external visual examination, a hand-feel, or pry-bar check for looseness. Such an inspection should include any brackets, lock washers, lock nuts, locking wires, or cotter pins used in assembly.

4. "Excessively worn" will be understood to mean worn, close to or beyond serviceable limits, and likely to result in a failure, if not replaced before the next scheduled inspection.

(d) Special Services. These services are indicated by repeating the item numbers in the columns which show the interval at which the services are to be performed, and show that the parts or assemblies are to receive certain mandatory services. For example, an item number in one or both columns opposite a procedure means that the actual tightening of the object must be performed. The special services include:

1. Adjust. Make all necessary adjustments in accordance with the pertinent section of the vehicle operator's manual, special bulletins, or other current directives.

2. Clean. Clean units of vehicle with dry-cleaning solvent to remove excess lubricant, dirt, and other foreign material. After the parts are cleaned, rinse them in clean fluid and dry them thoroughly. Take care to keep the parts clean until reassembled, and be certain to keep cleaning fluid away from rubber or other material which it will damage. Clean the protective grease coating from new parts, since this material is not a good lubricant.

3. Special Lubrication. This applies either to lubrication operations that do not appear on the vehicle Lubrication Guide, or to items that do appear on such charts, but should be performed in connection with the maintenance operations, if parts have to be disassembled for inspection or service.

4. Serve. This usually consists of performing special operations, such as replenishing battery water, draining and refilling units with oil, and changing the oil filter cartridge.

5. Tighten. All tightening operations should be performed with sufficient wrench-torque (force on the wrench handle) to tighten the unit according to good mechanical practice. Use torque-indicating wrench where specified. Do not overtighten, as this may strip threads or cause distortion. Tightening will always be understood to include the correct installation of lock washers, lock nuts, and cotter pins provided to secure the tightening.

(e) When conditions make it difficult to perform the complete preventive maintenance procedures at one time, they can sometimes be handled in sections, planning to complete all operations within the week, if possible. All available time at halts and in bivouac areas

TM 9-775
38

LANDING VEHICLE TRACKED MK. I AND MK. II

must be utilized, if necessary, to assure that maintenance operations are completed. When limited by the tactical situation, items with special services in the columns should be given first consideration.

(1) The numbers of the preventive maintenance procedures that follow are identical with those outlined on WD AGO Form No. 462, which is the Preventive Maintenance Service Work Sheet for Full Track and Tank-like Wheeled Vehicles. Certain items on the work sheet that do not apply to this vehicle are not included in the procedures in this manual. In general, the numerical sequence of items on the work sheet is followed in the manual procedures, but in some instances there is deviation for conservation of the mechanic's time and effort.

(5) SPECIFIC PROCEDURES. The procedures for performing each item in the 50-hour (500 miles) and 100-hour (1000 miles) maintenance procedures are described in the following chart. Each page of the chart has two columns at its left edge, corresponding to the 100-hour and the 50-hour maintenance, respectively. Very often it will be found that a particular procedure does not apply to both scheduled maintenances. In order to determine which procedure to follow, look down the column corresponding to the maintenance due, and wherever an item number appears, perform the operations indicated opposite the number.

100-hr 1000-mi Maint.	50-hr 500-mi Maint.	
		ROAD TEST
		NOTE: *When the tactical situation does not permit a full road test, perform those items which require little or no movement of the vehicle, namely, items 2, 5, 6, 9, 12, 13, 14, and 15. When a road test is possible, it should be for 2 miles, but not over 4 miles.*
1	1	*Before-operation Service.* Perform the Before-operation Service outlined in paragraph 22, to determine whether vehicle is in satisfactory condition to make road test safely; see that vehicle has sufficient fuel and oil. CAUTION: *Observe all starting precautions before starting engine (par. 7 b).*
2	2	*Instruments and Gages.*
		Oil Pressure Gage. Watch oil pressure gage to see that it registers sufficient pressure for safe operation, 40 pounds at idle speed, and 75 to 90 pounds at maximum governed speed. Continue to observe oil pressure throughout road test at various speed ranges. CAUTION: *When gage shows zero, or too low pressure, stop engine immediately.*

SECOND ECHELON PREVENTIVE MAINTENANCE

100-hr 1000-mi Maint.	50-hr 500-mi Maint.	
		Ammeter. With battery fully charged, ammeter reading should show high (+) charge for a short time after transfer case is engaged, then slightly above zero with all lights and electrical accessories switched off. If battery is low, charge will be indicated for a longer period of time. *Speedometer and Odometer.* Inspect speedometer for proper mileage reading, excessive fluctuation of hand, or unusual noises, indicating worn or damaged gears in head or worn drive cable. See that odometer registers accumulating mileage correctly. *Engine Oil Temperature Gage.* Inspect gage to see whether it operates properly, and whether engine oil temperature is normal (not over 190° F.) throughout the road test. CAUTION: *If oil temperature becomes excessive (more than 190° F.), stop vehicle and cool engine properly; determine cause of trouble.* *Transmission Oil Pressure Gage.* Gage should register approximately 40 pounds at normal operating speed.
5	5	*Brakes (Levers, Braking Effect, and Steering Action).* With vehicle stopped, pull back on steering brake levers. If brakes are properly adjusted, lever should meet heavy resistance at approximately a vertical position. If lever travels beyond vertical position, brake should be adjusted. Parking brake lock should be free, and should hold vehicle on reasonable incline, with ¼ to ⅓ ratchet travel in reserve.
6	6	*Clutch (Free Travel, Drag, Noise, Grab, Chatter, and Slip).* Test clutch for grabbing, dragging, chatter, or noise that may indicate faulty adjustment, defective parts, or dry release bearing. Pedal should have ⅜- to ¾-inch free travel before meeting resistance. While running at low speed in high gear, accelerate momentarily, and note any tendency of clutch to slip.
7	7	*Transmission (Lever Action, Vibration, and Noise).* Shift transmission through entire gear range. Observe whether shift lever operates freely, if gears mesh without unusual noise, and whether gears show any tendency to creep out of mesh. Note any unusual vibration or noise that may indicate excessive wear, inadequate lubrication, or loose mountings.

LANDING VEHICLE TRACKED MK. I AND MK. II

100-hr 1000-mi Maint.	50-hr 500-mi Maint.	
9	9	*Engine (Idle, Acceleration, Power, Noise, Smoke, and Governed Speed).* *Idle.* With the vehicle stopped, observe if engine runs smooth at normal idling speed (800 rpm). Throughout road test, observe whether there is any tendency of engine to stall when accelerator is released and hand throttle closed. *Acceleration, Power, Vibration, and Noise.* Test engine for normal acceleration and pulling power in each speed range. While testing in high range, accelerate momentarily from low speed with wide-open throttle, and listen for unusual engine noise, ping, or vibration that might indicate loose, damaged, excessively worn, or inadequately lubricated engine parts or accessories. During the road test, look for excessive smoke from exhaust or engine compartment. An abnormal blue smoke at engine exhaust usually indicates excessive oil consumption. At completion of road test, organization records of vehicle should be checked to see if engine has been consuming an excessive amount of oil. *Governed Speed.* With vehicle in low gear, depress accelerator to toeboard and see that engine speed does not exceed 2,400 revolutions per minute.
10	10	*Unusual Noise (Propeller Shaft and Universal Joints, Differential and Final Drives, Sprockets, Idlers, Wheels and Return Idlers).* During road test, listen for any unusual noise indicating damaged, defective, or loose parts or inadequate lubrication.
12	12	*Gun Elevating and Traversing Mechanism.* On LVT (A) (1), place vehicle in a position where it is tilted laterally (sidewise) about 10 degrees. Traverse turret through its full 360-degree range by both hand and power controls; check indication of binding. With gun pointed forward, elevate it through entire range with hand controls; check for binding, excessive lash, or erratic action. On LVT (2) and (A) (2), see that guns mounted on gun carriage tracks are secure, that they respond to controls properly, and that they are adequately lubricated and clean.
13	13	*Leaks (Engine Oil and Fuel).* Look in engine compartment and driver's compartment for indications of fuel, engine oil, or gear oil leaks.

SECOND ECHELON PREVENTIVE MAINTENANCE

100-hr 1000-mi Maint.	50-hr 500-mi Maint.	
14	14	*Noise and Vibrations (Engine, Mountings, Accessories and Drives, Clutch, and Exhaust).* With engine running and compartment doors open, look for indications of looseness and vibrations in these units, and for exhaust leaks or compression blow-by.
15	15	*Track Tension (Final Road Test).* Observe if track tension is properly adjusted. Inside plate on track adjusting screw should be tight against slide bracket, and have $\frac{1}{8}$-inch clearance between inner and outer plates.

MAINTENANCE OPERATIONS
(Stop Engine—Open Master Battery Switches)

18	18	*Armor (Paint, Markings, and Towing Staples).* Examine exterior of hull for damage and security of assembly and attachment mounting nuts and screws. Look for rust spots, broken welds, and bright spots in finish or camouflage pattern that may cause reflections. See if vehicle markings are all legible. Inspect towing staples for damage and security.
19	19	*Hull (Drain Plugs, Pontoons, and Bottom Plates).* Inspect drain plugs to see if they are in good condition and secure. Examine pontoons for indications of leaks or broken welds. Inspect hull bottom for damage.
21	21	*Track (Links, Lock Pins, Cross Plates, and Grousers).* Inspect to see that these items are in good condition, correctly assembled, and secure. Tighten assembly nuts securely.
22	22	*Idler (Wheels, Adjustment, Take-up Screws and Lock Nuts, and Idler Slide).* Inspect for correct assembly, security of mounting, and excessive wear. Note whether idler bearing seals are leaking excessively Inspect the hub bearings for looseness or end play. Tighten all assembly and mounting bolts securely.
23	23	*Bogie (Torsional Arms, Brackets, and Bumpers).* Inspect to see that these items are in good condition, secure, and not excessively worn.
23		*Tighten.* Tighten all assembly and mounting bolts securely.
24	24	*Wheel (Return Idlers and Tires).* Inspect unit for good condition, correct assembly, and secure mounting. Pay particular attention to see that tire rubber has not separated from rim, and that tires are not cut, torn, or ex-

LANDING VEHICLE TRACKED MK. I AND MK. II

100-hr 1000-mi Maint.	50-hr 500-mi Maint.	
		cessively worn. Inspect for excessive lubricant leaks from wheel bearings. NOTE: *Lubrication will force some lubricant through seal. This does not indicate defective seal.*
24		*Inspect and Tighten.* Jack up ground wheels and feel bearings for looseness and end play. Spin wheels and listen for any unusual noise that might indicate damaged, inadequately lubricated, or excessively worn bearings. Raise track free of support rollers with jack or pry bar, insert a prop to hold the track off roller, and test compensator wheel bearings for end play and bearing wear. Tighten assembly and mounting bolts securely.
25	25	*Sprockets (Hubs, Teeth, and Nuts).* Inspect these items for good condition, correct assembly, and security of mounting bolts or cap screws. Inspect sprocket teeth for excessive wear, and see whether shaft flange gaskets or oil seals are leaking lubricant excessively. If sprocket teeth are excessively worn, sprocket should be replaced or reversed. Tighten assembly and mounting bolts securely.
26	26	*Track Tension.* Check tension by observing clearance between inner and outer plates on idler adjusting screw. Inner plate should be tight against skid bracket, and should have ⅛-inch clearance between it and outer plate. Adjust, if necessary, and tighten lock nuts securely (par. 127).
27	27	*Top Armor (Turret, LVT (A) (1) Only, Paint and Markings, Grilles, Doors, Hatches, Covers, and Latches, and Antenna Mast).* Inspect to see that these items are in good condition and secure, and that hatch hinges and latches operate properly, are not excessively worn, and are adequately lubricated. Examine paint for rust spots or polished surfaces that may cause reflections, and all vehicle markings to see that they are legible.
28	28	*Caps and Gaskets (Fuel and Oil).* Inspect to see that these items are in good condition, secure, and not leaking. Be sure gaskets are in place and serviceable.
30		*Engine Removal.* (when required) Remove engine only if inspections made in items 6, 9, 13, and 14, and a check of records on oil consumption indicate definite need. Clean exterior of engine and dry thoroughly, taking care to keep dry-cleaning solvent away from elec-

TM 9-775
38

SECOND ECHELON PREVENTIVE MAINTENANCE

100 hr 1000 mi Maint.	50 hr 500 mi Maint.	
		trical wiring, terminal boxes, and equipment. NOTE: *The above cleaning, and the services in items 31 to 60, will be performed in best manner on engines that do not require removal.*
31		*Valve Mechanism (Clearances, Lubrication, Rocker Boxes, and Push Rod Housings).* NOTE: *Perform item 31 only when engines are removed from the vehicle.* Adjust valve clearances to 0.010 inch, cold. Also inspect valve tappets, rocker arms and shafts, and valve springs to see that they are in good condition, correctly assembled, and secure. See that oil is going to rocker arms and shafts properly. Inspect rocker arms and shafts for excessive wear, and rocker arm rollers for flat spots. Inspect rocker box covers for condition, for serviceable gaskets, and for condition of push rod housings.
32		*Spark Plugs.* Replace spark plugs with new or properly reconditioned plugs.
34	34	*Generator and Cranking Motor.* Inspect for good condition and security of mounting and wiring connections.
34		*Clean and Tighten.* Remove the commutator inspection cover and examine commutator for good condition; see that brushes are free in brush holders, clean, and not excessively worn; that the brush connections are secure, and that wires are not broken or chafing. Clean commutator end of generator and cranking motor by blowing out with compressed air. Tighten the cranking motor and generator mounting bolts securely.
37	37	*Magnetos (Points).* Inspect magnetos for good condition, security of mounting, and for evidence of oil leaks at mounting pad gaskets. Remove breaker point inspection covers to see that points are in good condition and clean, that breaker points are well alined, and that mating surfaces engage squarely.
37		*Adjust.* Adjust the magneto breaker point gaps to 0.012 inch.
38	38	*Ignition Wiring and Conduits.* Inspect these items for good condition, clean lines, correct assembly, tight connections, security of mountings, and for chafing against other engine parts. Clean all exposed ignition wiring with a dry cloth. NOTE: *Do not disturb connections unless they are actually loose.*

TM 9-775
38

LANDING VEHICLE TRACKED MK. I AND MK. II

100-hr 1000-mi Maint.	50-hr 500-mi Maint.	
39	39	*Booster Coil.* Examine booster coil and conduits to see that they are in good condition, clean, and securely mounted.
40	40	*Engine (Oil Pump, Sump, Oil Screens and Lines, Accessory Case, Crankcase, Fuel Screens and Lines, and Control Linkage).* Inspect to see that these items are in good condition and secure, and that oil is not leaking from oil pumps, sump, lines, accessory case, or crankcase. Inspect fuel lines to engine for leaks. Remove oil pump screens, clean them thoroughly in dry-cleaning solvent, dry, and reinstall.
43	43	*Air Cleaners.* Inspect air cleaner parts to see if they are in good condition. Clean the reservoirs and elements in dry-cleaning solvent. Fill reservoirs to correct level with clean engine oil. Reassemble cleaners, making certain all gaskets are in good condition and in place, giving special attention when mounting to see that cleaners are pressed firmly in place against air horn seals and securely fastened.
44	44	*Carburetor (Throttle, Linkage, Governor, and Primer).* Inspect these items for good condition, correct assembly, and security of mounting. See that carburetor does not leak, that throttle control linkage is not excessively worn and operates freely, and that governor is properly sealed. Also see that lines of priming system are in good condition, secure, and not leaking. Remove screen from carburetor fuel inlet, clean in dry-cleaning solvent, dry, and reinstall. Check bonds on throttle control rod for tightness and good condition.
45	45	*Manifolds (Intake and Exhaust).* Inspect to see that manifolds and gaskets are in good condition, correctly assembled, and secure. Tighten intake pipes at their flanges, clamps, and gland packing nuts. Check for indications of leaks by looking for excessive carbon streaks.
46	46	*Cylinders.* Inspect cylinders to see that they are in good condition and secure; see whether there are indications of oil leakage or blow-by around studs or gaskets. Inspect cylinders to see whether the cooling fins are clogged. CAUTION: *Cylinder hold-down nuts should not be tightened unless there is a definite indication of looseness or leaks. If tightening is necessary, use a torque wrench, and tighten to 27 to 31 foot-pounds tension.*

SECOND ECHELON PREVENTIVE MAINTENANCE

100-hr 1000-mi Maint.	50-hr 500-mi Maint.	
46		*Clean.* Clean excess deposits of dirt or grease from in and around cylinder cooling fins.
47	47	*Engine (Cowling, Air Deflectors, Flywheel Fan, Steady Bar, and Support Beam).* Inspect these items to see that they are in good condition, correctly assembled, and securely mounted. CAUTION: *Be sure inspection covers are all in place.*
47		*Tighten.* Tighten all accessible mounting and assembly bolts or screws securely.
48	48	*Clutch Assembly.* Inspect clutch release mechanism for good condition, security, proper lubrication, and excessive wear. Pedal free travel is $3/8$ to $3/4$ inch; roller clearance is $1/8$ inch. NOTE: *When engines are removed from vehicle, disassemble clutch, clean thoroughly, and inspect for damage and excessive wear.*
50	50	*Accessory Drives.* Inspect generator drive belts to see that they are in good condition and properly alined; are not worn, oil-soaked, or bottoming in pulleys. Adjust belts to $3/4$-inch deflection.
51	51	*Engine Compartment (Bulkhead and Control Linkage).* Inspect to see that engine compartment and bulkhead are in good condition and clean. See that control linkage in the engine compartment is in good condition, and securely connected and mounted. Check bonds on clutch control rod for tightness and good condition.
51		*Clean.* Clean engine compartment as thoroughly as possible. When engines are removed for repairs or replacement, thoroughly clean entire compartment.
52	52	*Engine Oil (Tank, Cooler, and Lines and Fittings).* Inspect these items to see that they are in good condition, correctly assembled, securely mounted, and not leaking. Check oil level with bayonet gage, and inspect sample of oil on bayonet gage for grit, water, or dilution. See that filler cap and gasket are in good condition and sealed properly.
52		*Serve.* Drain engine oil tank, reinstall drain plug, and fill oil tank approximately $1/4$ full (3 gal) with light engine oil. Agitate oil with a clean stick or rod to loosen sediment. Continuing to stir oil, remove drain plug, and drain oil and sediment from tank. When thoroughly drained, reinstall plug and fill tank to proper level with specified engine oil.

LANDING VEHICLE TRACKED MK. I AND MK. II

100-hr 1000-mi Maint.	50-hr 500-mi Maint.	
53	53	*Fuel (Tanks, Lines, and Pumps).* Inspect to see that these items are in good condition, correctly assembled, securely mounted, and not leaking.
	53	*Tighten.* Tighten all fuel tank mountings and brackets securely.
	53	*Serve.* With tanks nearly empty, drain water and sediment from each fuel tank by removing drain plugs and allowing fuel to drain until it runs clean. Tighten plugs securely, and be sure they do not leak.
54	54	*Engine Oil Filters.* Inspect engine oil filter to see if it is in good condition, secure, and not leaking.
54		*Clean and Serve.* Remove the filter element and clean thoroughly in dry-cleaning solvent. Clean all sediment from body and reassemble securely. Remove drain plug from Cuno oil filter, drain off sediment, and replace plug securely.
55	55	*Fuel Filters and Screens.* Inspect to see that these items are in good condition, and that lines to filters are secure and not leaking. Turn handle on filter one complete turn.
55		*Clean and Serve.* Remove drain plug and drain off sediment until fuel runs clean. Remove strainers from fuel tanks and clean screens; inspect screens for damage.
56	56	*Oil Cooler Transmission (Core and Lines).* Examine to see that these items are in good condition, secure, and not leaking. Clean insects and trash from core air passages.
57	57	*Exhaust Pipes and Mufflers.* Inspect to see that these items are in good condition, and securely assembled and mounted.
57		*Tighten.* Tighten all mounting bolts and connections securely.
58	58	*Engine Mountings.* Examine to see that mountings are in good condition and secure.
58		*Tighten.* Tighten all mounting bolts securely. When engine is removed, tighten support ring to crankcase bolts and to engine support assembly bolts while engine is out of vehicle.

TM 9-775
38

SECOND ECHELON PREVENTIVE MAINTENANCE

100-hr 1000-mi Maint.	50-hr 500-mi Maint.	
59	59	*Clutch Release (Yoke, Rollers, Linkage, and Mountings).* Inspect to see that these parts of the clutch-release mechanism are in good condition, correctly assembled, secure, and not excessively worn. Also see that the yoke rollers are adequately lubricated, that they do not have flat spots on their outside diameters, and that they do have $\frac{1}{8}$-inch clearance between rollers and collars.
59		*Tighten.* Tighten all accessible assembly and mounting bolts and screws. When engine is removed, this service should be performed before the engine is reinstalled.
60	60	*Fire Extinguisher System (Tanks, Valves, Lines and Nozzles, and Mountings).* Remove tanks and weigh to determine charge, which should be same as that marked on tag attached. Inspect valves for general condition, and blow out lines and nozzles with compressed air. Reinstall tanks and tighten securely. Inspect control cables and handles to see that they are in good condition, and not corroded or frozen. Inspect lines and nozzles for condition and security of mounting. Disconnect control cables and lubricate, as needed, to assure free operation.
60		*Tighten.* Tighten all assembly and mounting bolts and screws.
61	61	*Engine Installation (Mountings, Lines and Fittings, Wiring, Control Linkage, and Oil Supply).* If engine was removed for repair or replacements, reinstall at this time according to instructions in paragraph 62. Tighten mountings securely, and properly connect all fuel and oil lines, wiring, and control linkage. Be sure oil supply is adequate. Tighten bond strap bolts at engine mountings.
63	63	*Batteries (Cables, Hold-downs, Carrier, Gravity and Voltage, Switch, and Fuel Valves).* Remove batteries and clean with soda wash, if available, and dry thoroughly. Test and record gravity and voltage readings. Clean battery carrier and paint, if corroded. Clean battery terminals, scrape until bright, and make high-rate discharge test. See if all cells are in satisfactory condition (a true test cannot be made if gravity reading is

LANDING VEHICLE TRACKED MK. I AND MK. II

100-hr 1000-mi Maint.	50-hr 500-mi Maint.	
		below 1.225). If cells vary more than 30 percent, report condition. Raise electrolyte level at least ½ inch above plates by adding distilled or clean water. Close master battery switch, and open fuel shut-off valves.
64	64	*Accelerator Linkage.* Examine accelerator and connecting linkage to see that it is in good condition, opens throttle fully, is securely connected, and operates freely.
65	65	*Cranking Motor, Primer, and Instruments.* Observe all starting precautions as outlined in item 1. Start engine and observe primer operation. Note whether action of cranking motor is satisfactory; particularly, that cranking motor drive engages and operates properly without unusual noise, that it has adequate cranking speed, and that engine starts readily. As the engine starts, see that all instruments operate properly, and that oil pressure and ammeter indications are satisfactory.
66	66	*Leaks (Engine Oil and Fuel).* Inspect within engine compartment to see whether oil is leaking from engine, oil filter, or lines; and whether there are any leaks from the fuel system.
67		*Ignition Timing.* Test and set according to instructions (par. 91 e).
68	68	*Regulator Unit (Connections, Voltage, Current, and Cutout).* Inspect unit for good condition, and see that all connections and mountings are secure.
68		*Serve.* With unit at normal operating temperature, connect the low-voltage circuit tester to the regulator, and see whether voltage regulator, current regulator, and cutout properly control generator output. Follow instructions which accompany the test instrument.
69	69	*Engine Idle.* Observe engine for smooth idle at normal idle speed. Adjust engine idle speed to 450 revolutions per minute by means of throttle stop screw (hand throttle all the way in). Move mixture lever in direction which "leans" mixture, until engine idle becomes rough due to misfiring; then, slowly move lever in opposite direction to enrich the mixture until "roughness" disappears, and the engine idles smoothly. Do not turn further than necessary to smooth out idle. If this adjustment increases or decreases the engine idle speed, reset idle speed to 450 revolutions per minute.

SECOND ECHELON PREVENTIVE MAINTENANCE

100-hr 1000-mi Maint.	50-hr 500-mi Maint.	
71	71	*Fighting Compartment LVT (A) (1), Paint, Seats, Safety Straps, Crash Pads, Stowage Boxes, Ammunition Boxes, Clips, and Racks.* Inspect to see that these items are in good condition, securely assembled and mounted; that fighting compartment is clean; that paint is in satisfactory condition; and that adjusting mechanism of seats operates properly, and is adequately lubricated. Make sure that dividers and shell pads are all present and properly installed in ammunition boxes and racks, and that clips have sufficient tension to hold shells securely (on the LVT (2) and (A) (2), check the applicable items in the driver's compartment in the same manner).
72	72	*Turret, Basket, and Locks LVT (A) (1).* Examine to see that these items are in good condition, secure, correctly assembled and adjusted, adequately lubricated, and not excessively worn.
73	73	*Periscopes.* Examine periscope prisms and windows to see that they are in good condition, clean, secure in holders, and that holders are securely mounted. See that lever and locking devices operate freely and are not excessively worn, and that their traversing, elevating, and locking devices are free and not excessively worn. Examine spare periscopes and their stowage boxes to see that they are in good condition, clean, and secure. CAUTION: *Prisms should be cleaned only with a soft cloth or brush.*
74	74	*Clutch Pedal (Free Travel, Linkage, and Return Spring).* Inspect to see that these items are in good condition, secure, correctly assembled and adjusted, adequately lubricated, and are not excessively worn. See that pedal return spring tension is sufficient to return pedal to stop. Adjust pedal free play to ⅛ to ¾ inch.
75	75	*Brakes (Steering Levers, Latches, Linkage, and Shafts).* Inspect steering brake levers, parking locks, linkage, and shafts to see that they are in good condition, securely connected and mounted, and are not excessively worn. Apply steering brake levers, and observe whether they both begin to meet resistance just before reaching a perpendicular position.
75		*Tighten.* Tighten all assembly and mounting bolts securely.

LANDING VEHICLE TRACKED MK. I AND MK. II

100-hr 1000-mi Maint.	50-hr 500-mi Maint.	
77	77	*Differential and Breathers.* Inspect the accessible part of the differential case in the driver's compartment to see that it is in good condition, that all mounting and assembly bolts or cap screws are secure, and that there are no indications of oil leaks. Inspect the breathers to see that they are in good condition, secure, and not clogged.
77		*Serve.* Remove the breathers, clean in dry-cleaning solvent, dry, and reinstall. Tighten all external assembly and mounting nuts and screws.
78	78	*Transmission (Breathers and Seals).* Inspect transmission to see that it is in good condition and securely assembled and mounted; make sure that there are no oil leaks. Remove and clean breather.
78		*Tighten.* Tighten securely all external assembly and mounting bolts and screws.
80	80	*Transmission Controls.* Inspect to see that operating levers on transmission and shift mechanism operate properly, are in good condition, correctly assembled, securely connected and mounted, and not excessively worn.
81	81	*Propeller Shafts (Center Bearing, Universal Joints, Seals, and Flanges).* Inspect propeller shaft to see that it is in good condition and correctly and securely assembled and mounted. See that universal joints are properly alined, and not leaking or excessively worn. Inspect center bearing for good condition and leaks.
81		*Tighten.* Tighten securely universal joint assembly and companion flange bolts.
84	84	*Compass.* Examine compass to see that it is in good condition and secure; look for low fluid level, or indications of bubbles in the fluid bowl. Fill fluid bowl with Ethyl alcohol, if needed.
85	85	*Lamps (Lights) and Switches (Head, Tail, and Internal).* Test to see that switches and lights operate properly. Inspect to see that all lights are in good condition and secure; check for broken lenses and discolored reflectors.
85		*Adjust..* Adjust headlight beams.
86	86	*Wiring (Junction and Terminal Blocks and Boxes).* Inspect to see that all exposed electrical wiring and conduits, terminal blocks, and boxes are in good condition, well supported, and securely connected.

SECOND ECHELON PREVENTIVE MAINTENANCE

100-hr 1000-mi Maint.	50-hr 500-mi Maint.	
87		*Collector Ring (LVT (A) (2), Brushes, Heads, Cylinder, and Cover).* With master battery switch open, remove collector ring cover, and examine to see that these items are all in good condition and clean, that brushes contact cylinder properly under normal spring tension, and that leads are securely connected and not chafing. Reinstall cover securely.
88	88	*Radio Bonding (Suppressors, Filters, Condensors, and Shielding).* Examine and tighten all accessible bonds and connections not previously serviced. If radio noise continues, see paragraph 46.

AUXILIARY GENERATOR

89	89	*Engine (Crankcase, Fan and Housing, Cylinder Shield, Mountings, and Exhaust Pipe).* Inspect to see that these items are in good condition and secure.
90	90	*Spark Plug.* Replace the spark plug with a new reconditioned plug, using a new gasket, and make sure gap is set to 0.025 inch.
91	91	*Magneto (Points, Wiring, and Shield).* Inspect to see that these items are in good condition, correctly assembled, and securely mounted; that interior of magneto and the rotor arm are in good condition and clean; and that breaker points are clean and not uneven or pitted. Adjust the breaker point gap with the points fully open to 0.020 inch.
92	92	*Carburetor and Air Cleaner.* Inspect to see that these items are in good condition, securely mounted, and not leaking. Close fuel supply valve, remove air cleaner element and strainer in carburetor fuel inlet connection, clean in dry-cleaning solvent, and dry thoroughly. Dip upper end of air cleaner element in engine oil, drain, and reassemble. Reinstall carburetor inlet screen.
93	93	*Fuel (Filter, Line, Tank, and Cap).* Examine to see that these items are in good condition, secure, and not leaking. Clean fuel filter sediment bowl and screen; reinstall, using new gasket, if necessary. Open fuel supply valve.
94	94	*Generator (Commutator, Brushes, Control Box, and Wiring).* Remove the brush head cover plate, and examine commutator to see that it is in good condition, clean, and

LANDING VEHICLE TRACKED MK. I AND MK. II

100-hr 1000-mi Maint.	50-hr 500-mi Maint.	
		not excessively worn. See that the brushes are clean, free in their holders, properly spring-loaded, and not excessively worn. Inspect control box and buttons, ammeter, and wiring to see that they are in good condition, correctly assembled and connected, and secure.
94		*Clean.* At each third 100-hour service, clean commutator by placing a strip of very fine flint paper 2/0 over a wood block of correct size, and with engine running slowly, press the flint paper against the commutator until it is clean. Blow out dust with compressed air.
95	95	*Operation (Engine, Generator, Ammeter, and Leaks).* Start engine, observing whether it starts easily and runs at normal speed; listen for any unusual noise. See whether generator output is satisfactory. Check for fuel or oil leaks with engine running.

ARMAMENT

125	125	*Guns, cal. .30 and cal. .50 (LVT (2) and (A) (2), Mounts, Manual Traversing and Elevating Mechanism, and Firing Controls).* Inspect to see that these items are in good condition, clean, adequately lubricated, correctly and securely assembled and mounted, and that mounts and mechanism are not excessively worn. Test firing controls and mechanism for good condition and correct operation, and see that firing mechanism responds. Operate manual elevating controls, and traverse guns over entire track to be sure there is no bind or excessive looseness.
126	126	*Guns, 37-mm (LVT (A) (1), Mounts, Traversing and Elevating Mechanism, and Firing Controls).* Inspect to see that these items are in good condition, well lubricated, and correctly assembled. Check hydraulic pump, electric motor, traversing and elevating mechanisms, controls, lines, wiring, and reservoir for good condition and proper operation. Operate traversing and elevating mechanisms manually through entire range to see that they function properly. Check oil level in reservoir and tighten all assembly mounting bolts and screws securely. Tighten pump packing gland cautiously, as overtightening may score shaft and cause leaks.
127	127	*Stabilizer and Recoil Control LVT (A) (1).* Inspect the stabilizer control-unit gear box, connecting oil lines, cylinder and piston, wiring, and control box to see that these

TM 9-775
38

SECOND ECHELON PREVENTIVE MAINTENANCE

100-hr 1000-mi Maint.	50-hr 500-mi Maint.	
		items are in good condition, secure, correctly assembled, and are not leaking oil. Make an operating test of the stabilizer as outlined in the vehicle maintenance manual, or current bulletins and directives. NOTE: *Defects in the stabilizer system should be referred to ordnance maintenance personnel for attention.* Check recoil cylinders to see whether the level of the oil is as specified, and cylinder not leaking. NOTE: *Recoil operating checks must be made under firing conditions and in accordance with the instructions in the maintenance manual.*
129	129	*Spare Gun Barrels and Parts.* See that these items are present, in good condition, and properly stowed.

ITEMS SPECIAL TO AMPHIBIANS

NOTE: *The following items (a) and (b) (the required minimum maintenance for amphibian vehicles) are substituted, since no item numbers are assigned to Form No. 462.*

a	a	*Hull (Plugs, Decks, Hatches, Ventilators, Compartments, Bulkheads, and Ammunition and Cargo Stowage).* Inspect hull, hull plugs, seals, deck and hatch covers for punctures or open seams that might cause leaks. Check hull for bare spots that might cause rust or reflection. Examine deck to engine compartment ventilators for good condition and proper operation. Check screens to see that they are not clogged, and are properly sealed when covered. Inspect engine compartment and bilges for good condition and cleanliness. Check to see that ammunition, ammunition containers, dividers, and shell pads are properly installed, in good condition, and securely mounted.
b	b	*Bilge Pump (Power Take-off, Drive Shafts, and Outlet Pipes).* Remove deck plate and pump drive cover to inspect pump and pump drive shafts for proper lubrication, good condition, and proper operation. Check oil level and clean out pump outlet pipes.
b		*Tighten.* Tighten mounting and assembly nuts and screws securely.

TOOLS AND EQUIPMENT

130	130	*Tools (Vehicle Kit and Pioneer).* Check to see that all standard vehicle tools, Pioneer tools and their brackets and straps are present, in good condition, and properly stowed or mounted. If tools with cutting edges are not sharp, sharpen, if necessary.

TM 9-775

LANDING VEHICLE TRACKED MK. I AND MK. II

100-hr 1000-mi Maint.	50-hr 500-mi Maint.	
131	131	*Equipment.* Check equipment against vehicle stowage list to see if all items are present, in serviceable condition, and properly stowed or mounted. Be sure safety devices such as hand bilge pump, anchor, water canteens, rations, life belts, etc., are present and properly stowed.
133	133	*Spare Oil Supply.* On all models, LVT (2), (A) (1), and (A) (2), be sure the supply of spare engine and transmission oil is adequate and properly stowed. On LVT (A) (1) only, also check the supply of spare recoil and hydraulic oil.
134	134	*Decontaminator.* Examine to see that decontaminator is in good condition, secure, and fully charged. Make latter check by removing filler plug. NOTE: *The solution must be renewed every 3 months, as it deteriorates.*
135	135	*Fire Extinguisher (Portable).* Inspect to see that extinguisher is in good condition and secure, and that red cap on outlet safety valve is intact. Shake extinguisher to determine if it is fully charged. If red safety blow-off seal on valve head is not intact, it indicates that cylinder has been discharged due to high temperature.
136	136	*Publications and Form No. 26.* Check to see whether vehicle manuals, parts lists, Lubrication Guide, and Accident Report Form No. 26 are present, legible, and properly stowed.
137	137	*Vehicle Lubrication.* If due, lubricate in accordance with Lubrication Guide (sec. VIII), and current lubrication directives, using only clean lubricant and omitting items that have had special lubrication during this service. Replace damaged or missing fittings, vents, flexible lines, or plugs.
138	138	*Modifications (MWO's).* Inspect the vehicle to determine whether all Modification Work Orders have been completed and entered on WD, AGO Form No. 478. Enter any replacement of major unit assembly made at time of service.
139	139	*Final Road Test.* Make a final road test, rechecking items 2, 5, 6, 7, 9, 10, 11, 12, 13, 14, and 15. Recheck all gear cases and engine oil supply tank to see that lubricant is at correct level and not leaking. Confine this road test to the minimum distance necessary to make satisfactory observations. NOTE: *Make needed corrections or report any deficiencies noted during final road test to designated authority.*

ORGANIZATIONAL TOOLS AND EQUIPMENT

39. MWO AND MAJOR UNIT ASSEMBLY REPLACEMENT RECORD.

a. **Description.** Every vehicle is supplied with a copy of AGO Form No. 478 which provides a means of keeping a record of each MWO completed or major unit assembly replaced. This form includes spaces for the vehicle name and U. S. A. registration number, instructions for use, and information pertinent to the work accomplished. It is very important that the form be used as directed and that it remain with the vehicle until the vehicle is removed from service.

b. **Instructions for Use.** Personnel performing modifications or major unit assembly replacements must record clearly on the form a description of the work completed and must initial the form in the columns provided. When each modification is completed, record the date, hours and/or mileage, and MWO number. When major unit assemblies, such as engines, transmissions, transfer cases, are replaced, record the date, hours and/or mileage, and nomenclature of the unit assembly. Minor repairs and minor parts and accessory replacements need not be recorded.

c. **Early Modifications.** Upon receipt by a third or fourth echelon repair facility of a vehicle for modification or repair, maintenance personnel will record the MWO numbers of modifications applied prior to the date of AGO Form No. 478.

Section XII

ORGANIZATIONAL TOOLS AND EQUIPMENT

	Paragraph
Tools for maintenance	40

40. TOOLS FOR MAINTENANCE.

a. **Standard Tools and Equipment.** Common hand tools and equipment available to second echelon are listed in SNL N-19

b. **Organizational Special Tools and Equipment.**

Engine

Description	Mfg. No.	Fed. Stock No.	Ord. Drawing No.
Disk, timing, with pointer, engine	A3487	41-D-1266	B-144896
Indicator, piston top dead center	A3247	41-I-73-100	B-144307
Sling, engine lifting		41-S-3832	C-76360
Stand, engine transport	15041-A	41-S-4942-22	C-102887
Stand, engine inspection	15037	41-S-4942-11	C-76233
Wrench, engineer's double, 9/16-in. opening, 45- and 90-degree angle		41-W-1059	B-169934
Wrench, box socket, double, hexagonal, cylinder base nuts, 1/2-in. offset		41-W-562	A-163172

TM 9-775

LANDING VEHICLE TRACKED MK. I AND MK. II

Intake and Exhaust Valves

Description	Mfg. No.	Fed. Stock No.	Ord. Drawing No.
Gage, thickness, 0.010- and 0.124-in.....		41-G-404	A-139829
Wrench, valve adjusting............		41-W-3812-440	A-163166
Wrench, push-rod housing, gland retaining nut........................		41-W-1986-100	A-163170
Wrench, intake pipe, gland nut..........		41-W-1536-500	B-108078
Wrench, push-rod housing, gland nut....		41-W-3812-530	A-163168

Clutch

Description	Mfg. No.	Fed. Stock No.	Ord. Drawing No.
Puller, slide hammer, clutch spindle......		41-P-2957-40	C-66645B
Adapter, clutch companion flange puller		41-A-14-100	C-66645A

Generator Tools

Description	Mfg. No.	Fed. Stock No.	Ord. Drawing No.
Wrench, insulated, generator control box	EC-T-5100	41-W-1536-400	

Cranking Motor

Description	Mfg. No.	Fed. Stock No.	Ord. Drawing No.
Wrench, crowfoot, starter attaching, 9/16-in............................	MTM-3-505	41-W-871-45	B-248213

Magneto

Description	Mfg. No.	Fed. Stock No.	Ord. Drawing No.
Wrench, magneto, with gage...........	S11-400	41-W-1555	A-171152

Power Train

Description	Mfg. No.	Fed. Stock No.	Ord. Drawing No.
Bar, alining, final drive shaft...........	MTM-L-10	41-B-19-540	B-296182
Pins, short pair, 11/16-in. tapered, alining final drive unit to hanger brackets	MTM-L-17		
Pins, short pair, 1-in. tapered, alining final drive unit to hanger brackets.....	MTM-L-18		
Replacer, final drive oil seal............	MTM-L-11	41-R-2392-550	B-296184
Replacer, oil seal, final drive outer.....	MTM-L-12	41-R-2397-555	B-296193
Replacer, oil seal, final drive inner......	MTM-L-13	41-R-2397-550	B-296194

Brake

Description	Mfg. No.	Fed. Stock No.	Ord. Drawing No.
Wrench, brake adjusting, 1-1/16-in......		41-W-642-200	A-158000

Tracks

Description	Mfg. No.	Fed. Stock No.	Ord. Drawing No.
Fixture, track connecting...............	MTM-L-26		
Handle, replacer, seals and bearings, idler	MTM-L-94	41-H-1395-990	A-380560
Pin, tapered, transmission mounting yoke alining	MTM-L-15	41-P-560	
Pin, tapered, transmission, mounting yoke alining	MTM-L-16	41-P-560-10	
Puller, bearing and grease seal, idler wheel	MTM-L-76	41-P-2900-18	B-296188

ORGANIZATIONAL TOOLS AND EQUIPMENT

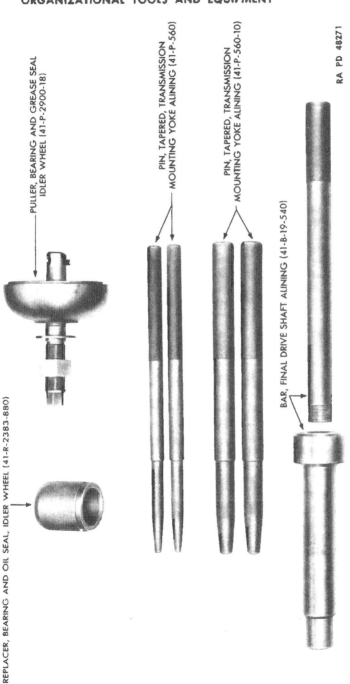

Figure 25 — Special Tools

LANDING VEHICLE TRACKED MK. I AND MK. II

Description	Mfg. No.	Fed. Stock No.	Ord. Drawing No.
Replacer, grease seal, idler sprocket shaft bearing cap	MTM-L-51	41-R-2397-730	A-380567
Replacer, bearing and oil seal, idler wheel	MTM-L-75	41-R-2383-880	A-380552
Replacer, bearing and spacer, idler sprocket	MTM-L-52	41-R-2384-40	A-380551
Replacer, bogie grease seal	MTM-L-3	41-R-2397-440	A-380566
Spreader, track chain link	MTM-L-25		
Wrench, bogie wheel axle plug	MTM-L-5		
Wrench, bogie wheel grease seal, installing	MTM-L-4		
Wrench, rear idler sprocket slide screw adjusting hook spanner	MTM-L-50		

Miscellaneous

Driver, bilge pump shaft seals	MTM-L-95		
Wrench, fuel tank filler pipe upper and lower gland nuts	MTM-L-21		

Section XIII

TROUBLE SHOOTING

	Paragraph
General	41
Engine	42
Starting system	43
Ignition system	44
Batteries and generating system	45
Radio interference suppression	46
Fuel system	47
Engine oiling system	48
Lighting system	49
Clutch	50
Propeller shaft	51
Transmission and final drive assembly	52
Tracks and suspension	53
Hydraulic traversing mechanism	54
Hull and pontoons	55

41. GENERAL.

a. This section contains trouble shooting information and tests which can be made to help determine the causes of some of the troubles that may develop in vehicles used under average climatic conditions

TM 9-775
41

TROUBLE SHOOTING

(above 32° F.). Each symptom of trouble given under the individual unit or system is followed by a list of possible causes of the trouble. The tests necessary to determine which one of the possible causes is responsible for the trouble are explained after each possible cause.

42. ENGINE.

a. **Engine Will not Turn.**

(1) HYDROSTATIC LOCK OR SEIZURE. Remove the rear spark plugs from the two lower cylinders, and attempt to turn the engine with the hand crank to check for excess fuel or oil in combustion chambers. If engine turns, the lock will be relieved. If the engine does not turn, seizure due to internal damage is indicated; notify higher authority.

(2) CRANKING MOTOR (par. 43 a).

(3) INCORRECT OIL VISCOSITY. Drain and refill with proper grade of oil (par. 27).

b. **Engine Turns But Will not Start.**

(1) INOPERATIVE FUEL SYSTEM. Open fuel valves, remove outlet line at the fuel pump, and, with the ignition switch off, turn the engine with the cranking motor. If free flow of fuel is not evident, fuel is not reaching carburetor (par. 47 a).

(2) INOPERATIVE IGNITION SYSTEM. Turn on magneto switch and booster switch. Remove a cable from a spark plug, hold cable terminal 1/4-inch from the cylinder casting, and crank the engine. If a spark does not jump the 1/4-inch gap, the ignition is inadequate (par. 44).

(3) SLOW CRANKING SPEED (par. 43 c).

c. **Engine Does not Develop Full Power.**

(1) IMPROPER IGNITION (par. 44 b).

(2) OIL TEMPERATURE TOO HIGH (par. 48 c).

(3) IMPROPER VALVE ADJUSTMENT. Check clearance and adjust if necessary (par. 65 a (2)).

(4) USE OF IMPROPER TYPE OF FUEL. See paragraph 5 a for fuel specifications.

(5) PREIGNITION. If the proper octane fuel is being used, and the ignition system is functioning satisfactorily, spark plugs of improper heat range may be a cause of trouble; otherwise, internal engine troubles would be indicated. Notify higher authority.

(6) CLOGGED FUEL STRAINER IN CARBURETOR. Clean strainer (par. 70).

(7) AIR LEAKS AT CARBURETOR OR MANIFOLD FLANGES. With engine running at 800 revolutions per minute, apply a small amount of oil at carburetor gaskets and manifold flanges. If oil is sucked in, there is evidence of a leak. Notify higher authority.

LANDING VEHICLE TRACKED MK. I AND MK. II

(8) LOW ENGINE COMPRESSION OR IMPROPER VALVE TIMING. If the engine does not develop full power with fuel reaching the combustion chambers, adequate ignition, and sufficient oil in the engine lubrication system, low compression, or improper valve timing would be indicated. Notify higher authority.

(9) INCORRECT GOVERNOR SETTING. Disconnect governor linkage at the carburetor, and check for sprung linkage or stuck throttle. If the throttle and linkage operate freely, start the engine and accelerate; if speed of 2,400 revolutions per minute is reached, the governor is faulty.

d. Engine Misfires.

(1) FAULTY IGNITION SYSTEM (par. 44 h).
(2) LOW ENGINE COMPRESSION (subpar. c (8), above).
(3) INCORRECT CARBURETOR ADJUSTMENT. Adjust carburetor (par. 67).
(4) CLOGGED FUEL TANK CAP VENTS. Open vents or replace cap.
(5) RESTRICTED FUEL FLOW (par. 47 a (4)).
(6) WATER IN FUEL. Remove the two drain plugs at the bottom of the carburetor, and inspect for water.

e. Excessive Oil Consumption.

(1) OIL VISCOSITY TOO LOW. Drain and refill with proper grade of oil (par. 27).
(2) EXTERNAL OIL LEAKS. Inspect for external oil leakage at oil line connections.
(3) SUPERCHARGER OIL SEAL RINGS OR PISTON RINGS DAMAGED. Notify higher authority.

f. Engine Will not Stop.

(1) FAULTY CARBURETOR IDLE CUT-OFF. Replace (par. 67).
(2) DEFECTIVE MAGNETO SWITCH; SWITCH GROUND WIRE BROKEN; GROUND WIRE FROM MAGNETOS TO SWITCH BROKEN (par. 44 h).
(3) THROTTLE NOT FULLY CLOSED. Close throttle.
(4) OVERHEATED COMBUSTION CHAMBERS. Check oil temperature gage for evidence of high oil temperature. Determine if all cooling surfaces of the engine are free of dirt and oil. Remove obstructions from the air intake grille and engine oil cooler. Test for clogging in the oil cooler (par. 48 c (4)).

g. Tachometer Inoperative.

(1) BROKEN OR KINKED CABLE. Replace cable.
(2) FAULTY TACHOMETER HEAD. Replace head.
(3) TACHOMETER ADAPTER DRIVE GEAR BROKEN. Replace adapter.

43. STARTING SYSTEM.

a. Cranking Motor Will not Operate.

(1) MASTER SWITCH NOT ON. Turn switch on.

TM 9-775
43

TROUBLE SHOOTING

(2) BATTERIES RUN DOWN. Close main battery switch. Insert the test prods of voltmeter lead into master switch box receptacle outlet and observe reading. Depress starter switch. If the voltage reading is below 12 volts, test the batteries with a hydrometer and a battery high-rate discharge tester (par. 100). If this test indicates that the batteries will not retain a charge, or are too low to recharge quickly with auxiliary generator, replace the batteries (par. 100). If batteries test satisfactorily, clean and tighten all battery connections.

(3) BATTERY MASTER SWITCH INOPERATIVE. Turn on all lights. Depress starter switch. Under this test, the lights will normally dim somewhat, but if they dim excessively, connect the voltmeter positive lead to battery cable terminal on battery master switch, and negative lead to the other terminal. Depress starter switch. If the voltmeter reads more than 0.5 volts, indicating high resistance in the switch, and the batteries test satisfactorily (step (2), above) notify the higher authority.

(4) STARTER SWITCH INOPERATIVE OR FAULTY WIRING. If the panel lights burn, indicating that current is reaching the instrument panel, disconnect the connector on starter solenoid feed wire in the panel. Connect test voltmeter to the green wire leading from the switch, and depress the starter switch lever. If no reading is obtained on test meter, replace the switch assembly. If reading is obtained, test solenoid and wiring (step (5), below).

(5) STARTER SOLENOID INOPERATIVE OR FAULTY WIRING IN BATTERY BOX. Tighten terminal connections on solenoid. Connect test voltmeter positive lead to the green wire terminal on starter solenoid, and voltmeter negative lead to ground. Depress starter switch. If the test voltmeter reads 24 volts, the circuit is complete to solenoid. If no reading is obtained, the circuit is open between the starter switch and the solenoid. If the terminals are secure, and there is no obvious break in the wiring, notify higher authority. NOTE: *In an emergency, current can be supplied to the solenoid from the bus bar by making connections with a short jumper wire. After engine is started, remove jumper.* If the circuit is complete to the solenoid, make sure that the solenoid ground wire is making good contact. If solenoid still does not function, replace the solenoid (par. 109). If solenoid functions but cranking motor does not operate, connect voltmeter positive lead to 24-volt feed cable on solenoid switch terminal, and voltmeter negative lead to ground. If reading of 24 volts is obtained, the circuit is complete to the solenoid switch. If no reading is obtained, test cable connection on solenoid, clean, and tighten. Connect positive voltmeter lead to the cranking motor cable terminal on solenoid switch, and negative lead to ground. Depress starter switch, or use jumper wire to operate solenoid. If no reading is obtained, replace solenoid (par. 109). If voltmeter reading is obtained and cranking motor does

TM 9-775
43—44

LANDING VEHICLE TRACKED MK. I AND MK. II

not operate, test cable connection in rear terminal box. Clean and tighten connections.

(6) CRANKING MOTOR DEFECTIVE. Connect positive lead of voltmeter to cranking motor terminal, and voltmeter negative lead to ground. Depress starter switch and observe voltmeter reading. If reading is obtained, and cranking motor will not function, replace the cranking motor (par. 95).

b. **Cranking Motor Operates and Engages Flywheel But Will not Turn Engine.**

(1) TRANSMISSION IN GEAR. Shift to neutral.
(2) HYDROSTATIC LOCK OR SEIZURE (par. 42 a (1)).
(3) BATTERIES RUN DOWN (subpar. a (2), above).
(4) BENDIX GEAR DAMAGED OR STUCK. Replace cranking motor (par. 95).
(5) INCORRECT ENGINE OIL VISCOSITY. Drain and refill with proper grade of oil.

c. **Slow Cranking Speed.**

(1) HIGH ELECTRICAL RESISTANCE (LOOSE OR CORRODED TERMINALS, WRONG SIZE CABLES OR WIRE, FAULTY CRANKING MOTOR OR SWITCHES). See subpar. a (1) through (6), above.
(2) ENGINE OIL TOO HEAVY. Use proper grade of oil (par. 27).
(3) CRANKING MOTOR WORN OUT (EXCESSIVELY NOISY). Replace cranking motor (par. 95).

44. IGNITION SYSTEM.

a. **Booster Does not Function (Scintilla Magneto).**

(1) FAULTY BOOSTER COIL. Turn on battery master switch and instrument panel light switch. Push accessory circuit breaker reset button to make sure the circuit is complete. If the panel lights do not burn, test batteries and connections (par. 45 a (5)), and make necessary corrections. If the panel lights burn, indicating that there is voltage to the instrument panel, depress the booster switch and listen at rear doors for a buzzing sound produced by the booster coil. If no sound is heard, disconnect the low tension lead from the coil, and test with a voltmeter to ground (hull). If a reading is obtained, the trouble is in the booster coil. If the buzzing sound is heard, and the engine will not start, disconnect the high tension lead from the magneto (fig. 31), and hold the terminal close to the engine with booster switch depressed. If a good spark is produced, depress starter switch. If spark becomes very weak while engine is being cranked, the batteries are low, or the cranking motor circuit has high resistance (loose connections or corroded terminals). (Test. par. 43 a (3), (4), and (5).) If a good spark is obtained during the test and engine will not start, disconnect the spark plug cable from No. 1 cylinder rear plug, hold

TM 9-775
44

TROUBLE SHOOTING

the cable terminal approximately ¼ inch from ground on engine, crank the engine, and observe the spark. If no spark is obtained, replace magneto (par. 91).

(2) SHORTED OR GROUNDED BOOSTER CIRCUIT (INSTRUMENT PANEL TO BOOSTER COIL). If no reading is obtained on voltmeter when testing the primary circuit at the coil (step (1), above), repeat test from booster circuit terminal in main junction box. If a reading is obtained at that point, the trouble is in the wire or connections from junction box to booster coil. Tighten all connections, or replace the coil wire. Fasten a new wire to the old one; pull new wire into conduit as old one is drawn out. If no reading is obtained, separate the connection on the small flexible conduit behind the instrument panel, and ground the terminal marked C in the connector on the conduit. Disconnect white wire in the top section of the main junction box, and connect voltmeter between the white wire and terminal No. 5 in the lower section of the junction box (fig. 115). This will apply voltage from No. 5 terminal to the booster circuit wire in the conduit to the instrument panel. If reading is obtained, the circuit is satisfactory. If no reading is obtained, the wire or connections in the conduit or connector are faulty.

(3) FAULTY BOOSTER SWITCH AND CIRCUITS IN INSTRUMENT PANEL. Separate the connector on the small conduit behind the instrument panel leading to rear terminal box. Connect a test voltmeter between the terminal marked C in the female part of the connector plug and ground (hull). If the panel lights burn, indicating that battery current is reaching the instrument panel, the test voltmeter should show a reading when the booster switch is depressed. If no reading is obtained, the trouble is in the booster switch.

b. **Improper Ignition (Bosch or Scintilla Magneto).**

(1) MAGNETO SWITCH OR WIRING FAULTY. If engine runs unevenly, there may be an intermittent ground in the magneto switch or wiring. Run the engine on first one magneto and then the other to determine which circuit is at fault; then disconnect the faulty circuit in the main junction box (fig. 115). If trouble is eliminated, connect a test ammeter or light between wire and No. 5 terminal in the other section of the junction box to supply voltage to the circuit. Move the magneto switch handle slightly while it is set on the faulty magneto circuit, and observe the light or meter. If an intermittent reading is obtained, indicating that the switch is at fault, replace the magneto switch. If engine will not stop when the switch is in an "OFF" position, separate the connector on the small conduit behind the instrument panel, and ground the terminals marked A and B to the shell of the connector. If the engine stops, the trouble is in the magneto switch. If the engine does not stop with these connections grounded, the trouble is in the wiring. Ground the magneto wire

LANDING VEHICLE TRACKED MK. I AND MK. II

terminals in the rear terminal box to stop the engine and notify higher authority. If the engine does not stop when the terminals in the main junction box are grounded, indicating that the wiring between the main junction box and the magnetos is at fault (subpar. a (2), above), replace the faulty wire.

(2) FAULTY MAGNETO. If no change in engine operation is noticed after disconnecting the magneto ground wires in main junction box (step (1), above), connect the wires, and run engine on one magneto at a time. If the speed of the engine drops off 300 revolutions per minute or more, and the spark plugs are operating properly, the trouble is in the magneto. Check magneto timing, or replace magneto (par. 91).

(3) SPARK PLUGS FAULTY. Uneven operation at idle speed, misfiring at high speed, or loss of power, may be caused by faulty spark plugs. If no appreciable difference in the speed of the engine is noticed when running on first one magneto and then on the other (step (1), above), remove and inspect spark plugs. Replace with new plugs if faulty (par. 92). If a definite drop-off in speed of engine is noticeable when switching from one magneto to the other (step (1), above), remove the spark plug cable from the most accessible spark plug, hold cable ¼ inch from cylinder, and observe spark. If a good spark is obtained, inspect for faulty spark plugs that are operating on the magneto circuit being tested (the rear plugs operate on the right magneto, and the front plugs operate on the left magneto).

45. BATTERIES AND GENERATING SYSTEM.

a. **Batteries Run Down.**

(1) GENERATOR CIRCUIT BREAKER KICKED OUT. Reset generator circuit breaker on master switch box.

(2) FAULTY BATTERIES. Test batteries (par. 100).

(3) EXCESSIVE USE OF ELECTRICAL ACCESSORIES WHEN GENERATOR IS NOT OPERATING. Use auxiliary generator (par. 12).

(4) SWITCHES LEFT ON WHEN NOT IN USE. Turn switches off when not in use.

(5) GROUNDED OR SHORTED CIRCUITS. Test as follows:

(a) Connect Battery for Test. Remove discharged batteries (par. 100). Place one fully charged battery in the battery box against the propeller shaft housing with the terminals toward the rear of the tank, and connect the ground cable to the negative battery post. This position of the battery will allow sufficient room to work in the battery box wiring panel. Make tests (steps (b) and (c), below) with the one 12-volt battery.

(b) Flash Test. With all switches turned off, touch the terminals of both 12-volt and 24-volt cables, one at a time, to the positive post of the battery and watch for flash. If flash is seen on either test, indi-

TM 9-775
45

TROUBLE SHOOTING

cating that cables are bare or terminals grounded, inspect cables and connections. Service if practical, or replace cables (par. 100). If no flash is seen during these tests, close the battery master and radio switches, one at a time, and repeat flash tests. If a flash is seen in either instance, test the faulty circuit for dead shorts. If no flash is seen on these tests, make ammeter test (step (c), below).

(c) *Test With Ammeter.* Connect one lead of test ammeter to the positive post of battery, and the other lead to the 12-volt radio feed cable with radio switch turned on. If no reading is obtained on the test meter, the circuit is clear. Connect ammeter lead to 24-volt cable and observe meter. NOTE: *A reading will be seen, as some of the sending units on the engines are connected directly to the main circuit.* If a reading of more than 1 ampere is obtained, indicating a shorted or grounded circuit, disconnect the leads from the negative (right) end of the shunt, one at a time, starting with the largest cable (regulator lead), until the faulty circuit is located. If a heavy current draw is indicated on the ammmeter before disconnecting the lead, and is corrected when the cable is removed from the terminal, test regulator for faulty circuit breaker. Disconnect battery cable marked "B" in regulator terminal box, and repeat ammeter test. If no reading is obtained, replace the regulator (par. 98). If no change is indicated in ammeter reading, continue to disconnect circuits from shunt and bus bar until the faulty circuit is located; then test the faulty circuit for shorts or grounds.

(6) CIRCUIT BREAKER FAULTY. If the ammeter on the instrument panel shows a heavy discharge when the generator is not running and all switches are off except the battery master switch, disconnect the battery lead marked "B" in the regulator terminal box. If the condition is corrected, the regulator circuit breaker contact points are stuck. If the ammeter on the instrument panel does not show charge until generator is running at high speed, the regulator circuit breaker is adjusted to operate at too high a voltage. In either of these cases, replace the regulator (par. 98).

(7) REGULATOR INOPERATIVE. Start the engine and observe ammeter on the instrument panel. If no charging rate is indicated, connect the battery and armature terminals marked "B" and "A" together in the regulator terminal box, using a short piece of insulated wire. Hold jumper wire across the two terminals and watch the ammeter. If reading is obtained, the regulator is not connecting the generator to the battery. If this test does not reveal the trouble, connect the battery and field terminals together with the jumper wire. If a reading is obained, the regulator is not allowing current to reach the generator field coils, preventing charge. If excessive charge is experienced, and the batteries and circuits test properly, the trouble is caused by improper regulator adjustment. In either case, the regulator is inoperative.

LANDING VEHICLE TRACKED MK. I AND MK. II

(8) GENERATOR INOPERATIVE. If regulator tests have been made and no charge is obtained, connect a test voltmeter between armature terminal marked "A" in regulator terminal box and ground (hull). This test will show if generator is charging. If no voltage reading is shown, leave the voltmeter connected, and connect the battery and field terminals marked "B" and "F" together with the jumper wire. A flash will be seen, and the test voltmeter will show a reading when the jumper wire is connected, if the circuit is complete. Check the ammeter on the instrument panel. If a charge is shown, the trouble has been corrected by flashing the fields, which has increased the magnetism or properly polarized the field coil shoes. If no reading is obtained on the voltmeter, inspect the terminals at the generator for loose or broken connections. If no trouble is observed in the connections or leads, the generator is inoperative.

b. **Ammeter Does not Show Charge.**

(1) GENERATOR CIRCUIT BREAKER OPEN. Reset generator circuit breaker.

(2) AMMETER INOPERATIVE. If the ammeter fails to register a charge, turn on all lights, and see if a discharge is shown. If no discharge is observed, connect a new ammeter temporarily to the leads in the instrument panel. If a reading is obtained, the ammeter is faulty. If no reading is obtained, test wiring from ammeter to shunt for open circuit.

(3) REGULATOR INOPERATIVE. (Subpar. a (7), above).

(4) GENERATOR INOPERATIVE. (Subpar. a (8), above).

(5) LOOSE OR CORRODED CONNECTIONS. Clean and tighten connections.

(6) GENERATOR GROUND STRAP LOOSE OR BROKEN. Inspect ground strap. Tighten or replace.

c. **Ammeter Shows Excessive Charge.**

(1) CURRENT REGULATOR IMPROPERLY ADJUSTED (subpar. a (7), above).

(2) BATTERIES RUN DOWN. Test batteries (par. 100). Recharge with auxiliary generator (par. 12) or replace.

(3) BATTERIES SHORTED INTERNALLY. Test batteries and replace if faulty (par. 100).

d. **Ammeter Shows Discharge With Engine Running.**

(1) GENERATOR NOT OPERATING. (Subpar. a (8), above).

(2) REGULATOR CIRCUIT BREAKER CUT-IN VOLTAGE TOO HIGH. (Subpar. a (6), above).

(3) SHORTED CIRCUITS. (Subpar. a (5), above).

(4) GENERATOR DRIVE BELTS LOOSE OR BROKEN. Tighten or replace belts (par. 97).

TROUBLE SHOOTING

e. **Ammeter Shows Heavy Discharge With Engine Stopped.**

(1) SHORTED CIRCUITS. (Subpar. a (5), above).

(2) REGULATOR CIRCUIT BREAKER POINTS STUCK. (Subpar. a (6), above).

(3) AMMETER HAND STICKING OR AMMETER BURNED OUT. (Subpar. b (2), above).

f. **Ammeter Hand Fluctuates Rapidly.**

(1) GENERATOR DRIVE BELTS LOOSE. Tighten or replace belts (par. 97).

(2) GENERATOR GROUND STRAP LOOSE OR BROKEN. Tighten or replace ground strap.

(3) REGULATOR CIRCUIT BREAKER CUT-IN VOLTAGE TOO LOW OR CONTACTS BURNED. See step (5), below.

(4) REGULATOR LOOSE, NOT PROPERLY GROUNDED, OR VIBRATING AGAINST OTHER EQUIPMENT. Tighten regulator on mountings, inspect ground straps, and relieve interference.

(5) GENERATOR OR REGULATOR FAULTY. If ammeter needle fluctuates rapidly, while generator is running, test all regulator and generator mountings to see if they are tight, and inspect for broken ground straps. If ground straps and mountings are satisfactory, the condition is caused by incorrect setting of regular circuit breaker, worn generator brushes, faulty generator drive belts, or regulator bumping against other equipment. If inspection reveals that the generator drive belts are properly adjusted (par. 97), and there is no interference with the regulator, connect a jumper wire between battery terminal marked "B" and armature terminal marked "A" in the regulator terminal box. If the fluctuation stops with the jumper wire connected, indicating that the regulator circuit breaker points have been vibrating, replace the regulator (par. 98). If fluctuation continues, indicating that the generator is at fault, replace the generator (par. 97).

g. **Auxiliary Generator Will not Crank Electrically.**

(1) BATTERY MASTER SWITCH OPEN. Close battery master switch.

(2) BATTERIES RUN DOWN. Test batteries (par. 100). Start auxiliary generator manually, and recharge (par. 12), or replace batteries (par. 100).

(3) LOOSE CONNECTIONS ON BUS BAR IN BATTERY BOX OR IN PANEL ON AUXILIARY GENERATOR. Tighten connections.

(4) AUXILIARY GENERATOR STARTER SWITCH INOPERATIVE. Replace switch.

h. **Auxiliary Generator Turns But Will not Start.**

(1) LACK OF FUEL. Replenish fuel supply. Clean fuel strainer (par. 70).

(2) LACK OF IGNITION. Service auxiliary generator (par. 170).

LANDING VEHICLE TRACKED MK. I AND MK. II

i. **Auxiliary Generator Runs Unevenly or Stalls Under Load.**

(1) CARBURETOR NOT PROPERLY ADJUSTED. Service auxiliary generator (par. 173).

(2) SHUT-OFF COCK ON FUEL FILTER CLOSED OR CLOGGED. Open shut-off cock or clean out fuel filter.

(3) FUEL LINE CLOGGED. Disconnect line and clean out.

(4) WATER OR DIRT IN FUEL. Drain system and clean.

j. **Auxiliary Generator Runs but Shows no Charge.**

(1) GENERATOR NOT FUNCTIONING. Make tests as explained for main generator (subpar. a (8), above). Replace auxiliary generator if faulty (par. 173).

(2) REGULATOR NOT FUNCTIONING. Make tests as explained for main generator regulator (subpar. a (7), above). Replace regulator if faulty (par. 98).

46. RADIO INTERFERENCE SUPPRESSION.

a. Locate circuit causing noise, as follows (the cooperation of the radio operator will be required):

(1) Operate engine while listening to radio. Crackling noises that cease the instant the ignition is shut off are caused by the ignition circuits.

(2) An irregular clicking that continues a few seconds after the ignition is shut off is caused by the regulator.

(3) A whining noise that varies with engine speed and continues a few seconds, after the ignition is shut off, is caused by the generator.

(4) On LVT (A) (1) models, operation of traversing motor, auxiliary engine, and other accessories will determine which of these devices is causing interference.

b. **Ignition Circuit Noise.**

(1) Make certain ignition system is functioning properly (pars. 90 thru 93). Improper plug gaps, poor adjustment, and worn parts will affect the suppression system.

(2) Inspect shielding on high tension leads. Tighten all connections.

(3) Tighten bonds on engine mountings.

(4) Check condenser in radio junction box, and replace if faulty.

c. **Regulator Noise.**

(1) Check bonds on instrument box and bonds on lines at instrument box.

(2) Check filter in instrument box, and replace if faulty.

(3) Check condenser and connections in radio junction box.

TROUBLE SHOOTING

d. Generator Noise.

(1) Check connections and bond straps. Tighten all shielding on generator-regulator circuits.

(2) Check condenser in regulator-relay box, and replace if faulty.

e. Fan Motor Noise. Tighten all shielding and mounting bolts.

f. Auxiliary Engine Generator Noise (on LVT (A) (1) Only).

(1) Check bond on engine generator.

(2) Check filter, and replace if faulty.

g. Turret Traversing Motor Noise (on LVT (A) (1) Only). Check collector ring. Clean and tighten all connections.

h. Stabilizer Noise (on LVT (A) (1) Only).

(1) Check collector ring. Clean and tighten all connections.

(2) If stabilizer unit is faulty, replace.

47. FUEL SYSTEM.

a. Fuel Does not Reach Carburetor.

(1) LACK OF FUEL. Check gage on instrument panel and replenish fuel.

(2) FUEL VALVES NOT TURNED ON. Turn on fuel valves.

(3) CLOGGED GAS TANK VENTS. Open gas tank vents.

(4) INOPERATIVE FUEL PUMP, OR CLOGGED FUEL FILTER OR LINES. Remove drain plug from fuel filter and check fuel flow from the tanks. If the fuel does not flow freely at filter, clean lines back to fuel tanks. Service fuel filter (par. 70). If fuel flows freely through filter but does not reach carburetor, the fuel pump is inoperative.

b. Fuel Does not Reach Cylinders.

(1) INOPERATIVE PRIMER PUMP. Remove one of the primer lines from one of the cylinders, and operate the primer pump to see if fuel enters the cylinder.

(2) PRIMER PUMP STRAINER CLOGGED. Clean or replace (par. 73).

(3) CARBURETOR STRAINER CLOGGED. Clean or replace (par. 67).

(4) THROTTLE NOT OPENING. Adjust throttle (par. 68).

(5) CARBURETOR JETS CLOGGED. Replace carburetor (par. 67).

(6) LOW FUEL PUMP PRESSURE. Install a fuel pump pressure gage (41-C-500) in the outlet side of the fuel pump. Pressure should read $3\frac{1}{2}$ pounds.

48. ENGINE OILING SYSTEM.

a. Low or No Oil Pressure.

(1) LACK OF OIL. Replenish oil supply (par. 86).

(2) CLOGGED OIL LINES OR INOPERATIVE OIL PUMP. Loosen the inlet connection at the oil pump, and apply compressed air to the

open end of the hose. When free passage to the oil supply tank is established, fill the line with oil and connect the inlet hose to the oil pump. Remove oil pressure relief valve, crank engine with starter, and observe if oil is discharged from relief valve opening. If no oil is discharged, replace the oil pump (par. 88).

(3) LEAKING OIL LINES OR FITTINGS. Tighten or replace.

(4) PRESSURE GAGE INOPERATIVE. If the gage registers no pressure or insufficient pressure, stop the engine immediately and locate the cause.

(5) PRESSURE RELIEF VALVE STUCK OR SET TOO LOW. Watch the oil pressure gage; turn the oil pressure adjusting screw, and check for variation of pressure. If no change is evident, service the relief valve assembly (par. 85).

b. Oil Temperature Too Low.

(1) OIL TEMPERATURE GAGE OR SENDING UNIT INOPERATIVE. Stop the engine, and allow oil to cool. If gage reading is not reduced as oil cools off, disconnect the wire from the sending unit and if reading then drops, replace the sending unit (par. 141). If the gage indicates high oil temperature with wire disconnected from the sending unit, replace the gage.

(2) BYPASS VALVE INOPERATIVE. Remove the cap on top of the bypass valve, take out the valve, and check it for freedom of operation.

c. Oil Temperature Too High.

(1) LOW OIL SUPPLY. Fill with oil (par. 86).

(2) AIR INTAKE GRILLE OR SCREEN CLOGGED WITH FOREIGN MATTER. Remove all obstructions.

(3) ACCUMULATION OF OIL OR DIRT ON EXTERIOR OF ENGINE. Clean engine.

(4) ENGINE OIL COOLER OBSTRUCTED OR COVERED. Loosen the coupling at the outlet from the engine oil cooler to the oil supply tank. If normal flow of oil is not evident, loosen the inlet line at the bottom of the cooler, and check for flow of oil to the cooler. If normal flow of oil is found at this point, the oil cooler is clogged. Remove and clean (par. 85).

(5) LATE IGNITION TIMING. Reset timing (par. 91 e).

(6) INOPERATIVE BYPASS VALVE (subpar. b (2), above).

49. LIGHTING SYSTEM.

a. All Lights Will not Burn.

(1) BATTERY MASTER SWITCH TURNED OFF. Turn switch on.

(2) FAULTY OR DISCHARGED BATTERIES (par. 45 a (5)).

(3) LOOSE OR CORRODED TERMINALS. Clean and tighten connections.

TM 9-775
49

TROUBLE SHOOTING

(4) BATTERY MASTER SWITCH INOPERATIVE (par. 43 a (3)).

b. **All External Lights Will not Burn.**

(1) BATTERY MASTER SWITCH TURNED OFF. Turn switch on.

(2) FAULTY OR DISCHARGED BATTERIES (par. 45 a (5)).

(3) LOOSE OR CORRODED TERMINALS. Clean and tighten connections.

(4) FAULTY LIGHTING SWITCH OR SHORT CIRCUITS. Turn on battery switch and instrument panel light switch. If batteries, connections, and wiring are satisfactory (par. 45 a (5)), the panel lights will burn, indicating that current is reaching the instrument panel. If all external lights fail to operate, turn off main lighting switch, and push light circuit breaker reset button to make sure that it is not kicked out. CAUTION: *Do not hold circuit breaker reset button in, as a short circuit would damage the circuit breaker, or set fire in the wiring.* Turn light switch from one contact to another, slowly. If the switch and wiring connections inside the panel, and the conduit connections on the back of the panel, are satisfactory, some of the lights will burn. When the faulty circuit is contacted with the switch, the circuit breaker will kick out, indicating a short. Test that circuit for trouble and correct (subpar. f (4), below). If no short exists, and all external lights do not function, loose wiring connections or a faulty main lighting switch are indicated. Inspect for loose connections or replace switch.

(5) BATTERY MASTER SWITCH INOPERATIVE (par. 43 a (3)).

c. **All Internal Lights Will not Burn.**

(1) BATTERY MASTER SWITCH TURNED OFF. Turn switch on.

(2) BATTERIES DISCHARGED (par. 45 a (5)).

(3) LOOSE OR CORRODED TERMINALS. Clean and tighten connections.

(4) SHORT CIRCUITS (ACCESSORIES CIRCUIT BREAKER KICKED OUT). The instrument panel lights, dome lights in the fighting compartment, compass light, outlet receptacle on the instrument panel, and the flame detector circuits are all fed through the accessories circuit breaker. If a short exists, causing the circuit breaker to kick out with the instrument panel light switch turned off, the trouble would be in either the dome light or the flame detector circuits. Disconnect the multiple connector, feeding the front wiring harness from the back of the instrument panel. Reset the accessories circuit breaker again and if it remains in contact, inspect all dome light outlets for damaged wiring or short circuits. If the trouble is not found in these units or connections, replace the front wiring harness. If the accessories circuit breaker kicks out after disconnecting the front wiring harness from the instrument panel, the trouble is in the flame detector circuit. Dis-

LANDING VEHICLE TRACKED MK. I AND MK. II

connect the two white wires from No. 1 terminal post in the contro
panel (fig. 187). If the short remains, the trouble is in the instru
ment panel to control panel conduit. Replace the conduit. If the con
dition is corrected, disconnect the brown wire from No. 7 terminal, an
connect the white wires again. If the trouble has not been cleared, th
flame detector circuit in the engine compartment is at fault. Notif
higher authority. If the short circuit or faulty operation exists onl
when the panel light switch is turned on, disconnect the compass ligh
and test again. If the trouble still exists, the wiring in the panel or th
panel light switch is faulty. Inspect wiring in the panel and test fc
complete circuit through the panel light switch with a voltmeter.]
panel light switch is at fault, replace the switch (par. 141).

(5) BATTERY MASTER SWITCH INOPERATIVE (par. 43 a (3)).

d. **All Lights Burn Dim.**

(1) BATTERIES DISCHARGED OR LOOSE OR CORRODED TERMINALS (pa 45 a (5)).

(2) BATTERY MASTER SWITCH CONTACTS BURNED. Replace switcl

(3) LOOSE CONNECTIONS IN INSTRUMENT PANEL FEED WIRES CAUS ING HIGH RESISTANCE. Tighten connections.

e. **One or More Lamp-units Burn Out Continually.**

(1) GROUND STRAPS OR CONNECTIONS LOOSE OR BROKEN. Clea and tighten all connections; replace broken ground straps.

(2) BATTERY GROUND CABLE LOOSE OR BROKEN. Tighten or re place cable.

(3) GENERATOR REGULATOR IMPROPERLY ADJUSTED (par. 98).

f. **Individual Lights or Circuits Inoperative.**

(1) LAMP BURNED OUT. Replace lamp (pars. 101 and 102).

(2) LOOSE CONNECTIONS AT LIGHT. Tighten connections.

(3) BROKEN WIRING OR DAMAGED CONDUIT TO UNIT. Replace cor duit and wiring or report to higher authority.

(4) SHORT CIRCUITS OR GROUNDS (CIRCUIT BREAKERS KICK OUT) If short circuits are encountered, disconnect the conduit connectors c wires from the inoperative unit. If the trouble is eliminated, and th circuit breaker remains in contact when reset, the trouble is in tha unit. Inspect, make corrections if practical, or replace faulty unit (pars. 106 through 108). If the condition prevails after disconnectin the conduit connector or wire at the unit, disconnect the conduit cor taining the inoperative circuit from the back of the instrument pane If the condition is relieved, the trouble is in the conduit. Replac faulty conduit or notify higher authority. If the trouble still exist after disconnecting the conduit from the instrument panel, the troubl is in the panel wiring, light switches, or circuit breakers. Disconnec the faulty circuit from the switch to determine if wiring is at faul If the trouble still exists, replace the faulty switch.

TROUBLE SHOOTING

50. CLUTCH.

a. Clutch Drag. Idle the engine at 800 revolutions per minute. Push the clutch pedal to the fully released position, and allow time for the clutch to stop. Shift the transmission into first or reverse gear. If the shift cannot be made without a severe clash of the gears or if, after engagement of the gears, there is a jumping or creeping movement of the tank with the clutch still fully released, the clutch is at fault.

(1) EXCESSIVE PEDAL CLEARANCE. Adjust clutch linkage (par. 114).

(2) INCORRECT PLATE SEPARATOR STUD CLEARANCE. Adjust (par. 114).

(3) DAMAGED OR MISSING SEPARATOR PINS OR SEPARATOR PIN SPRING. Replace (par. 114).

(4) WARPED OR CRACKED DRIVING OR DRIVEN PLATES. Replace damaged parts (par. 114).

(5) DRIVING PLATE SLOTS BINDING ON DRIVING PINS. Free up (par. 114).

(6) EXCESSIVE DIRT IN CLUTCH ASSEMBLY. Disassemble clutch and clean out (par. 114).

(7) DAMAGED OR WORN DRIVE SPLINE OR BEARING. Disassemble and replace damaged or worn parts (par. 114).

(8) METAL TRANSFER OR BONDING OF CLUTCH FACINGS. Replace damaged plates (par. 114).

b. Clutch Slips.

(1) IMPROPER ADJUSTMENT OF CLUTCH RELEASE LEVER YOKE SHOES. Adjust clutch linkage (par. 114).

(2) LOSS OF SPRING LOAD CAUSED BY EXCESSIVE HEAT OR BROKEN SPRINGS. Notify higher authority.

(3) DIRT IN CLUTCH CAUSING BINDING OF DRIVEN PLATE. Disassemble and clean out (par. 114).

(4) CLUTCH DRIVEN PLATE FACINGS WORN. Replace driven plates par. 114).

(5) BROKEN BOOSTER SPRING. Replace booster spring (par. 114).

c. Complete Failure of Clutch to Engage or Release.

(1) DISCONNECTED CLUTCH LINKAGE OR BINDING OF CLUTCH LINKAGE. Inspect linkage. Replace or connect parts (par. 114).

(2) BROKEN OR DAMAGED CLUTCH PLATES. Replace damaged plates (par. 114).

(3) DAMAGED CLUTCH SPINDLE OR BEARING. Replace damaged parts (par. 114).

(4) EXCESSIVE PEDAL FREE PLAY. Adjust pedal free play (par. 114).

LANDING VEHICLE TRACKED MK. I AND MK. II

51. PROPELLER SHAFT.

a. Backlash.

(1) WORN OR DAMAGED UNIVERSAL JOINT CROSS BEARING. Replace (par. 115).

(2) LOOSE BOLTS AT UNIVERSAL JOINT COMPANION FLANGES. Tighten bolts.

(3) UNIVERSAL JOINT COMPANION FLANGES LOOSE ON TRANSMISSION INPUT SHAFT OR ON MASTER CLUTCH SPINDLE. Tighten flange bolts or spindle nut.

b. Vibration in Propeller Shaft.

(1) WORN OR DAMAGED UNIVERSAL JOINTS. Replace (par. 115).

(2) LOOSE BOLTS AT UNIVERSAL JOINT COMPANION FLANGES. Tighten flange bolts.

52. TRANSMISSION AND FINAL DRIVE ASSEMBLY.

a. Lubricant Leakage.

(1) DAMAGED GASKET AT FILLER PLUG. Replace gasket.

(2) LOOSE DRAIN PLUGS. Tighten.

(3) DAMAGED FLEXIBLE HOSE OR COOLER TUBES. Notify higher authority.

(4) DAMAGED FINAL DRIVE HOUSING COVER GASKET OR LOOSE CAP SCREWS. Tighten cap screws or replace gaskets (par. 117).

(5) DAMAGED BRAKE INSPECTION COVER GASKET. . Replace gasket.

(6) DAMAGED GASKET BETWEEN TRANSMISSION AND FINAL DRIVE HOUSING. Notify higher authority.

(7) WORN OR DAMAGED INPUT OR OUTPUT SHAFT OIL SEALS, OR GASKET. Notify higher authority.

(8) DAMAGED GASKET AT TRANSMISSION INSPECTION PLATE. Replace gasket.

b. Track Will not Move on One Side (Engine Running and Transmission in Gear).

(1) BROKEN FINAL DRIVE SHAFT OR COMPENSATING SHAFT. Replace final drive unit (par. 117).

(2) TOOTH STRIPPED ON FINAL DRIVE SHAFT GEAR OR COMPENSATING GEAR. Replace final drive unit (par. 117).

(3) BROKEN FINAL DRIVE ASSEMBLY PARTS. Notify higher authority.

c. Hard Shifting (Severe Gear Clash).

(1) INCORRECT CLUTCH LINKAGE ADJUSTMENT. Adjust (par. 114).

(2) CLUTCH DRAGGING (par. 50 a).

(3) BINDING OF TRANSMISSION GEARSHIFT LEVER. Free up.

(4) DAMAGED TRANSMISSION PARTS. Notify higher authority.

TROUBLE SHOOTING

d. **Backlash.**

(1) Worn or Damaged Final Drive Assembly. Notify higher authority.

(2) Worn or Damaged Transmission Parts. Notify higher authority.

e. **Poor Steering.**

(1) Steering Brakes Not Properly Adjusted. Adjust (par. 123).

(2) Steering Brake Shoe Lining Worn or Damaged. Notify higher authority.

53. TRACKS AND SUSPENSION.

a. **Bogie Wheel Tire Wear.**

(1) Master Link Pin Bent, Broken, or Missing. Replace link pin (par. 128).

(2) Damaged Track. Replace track (pars. 128 and 129).

(3) Mud Collecting in Track Between End Connectors. Remove mud from connectors.

b. **Thrown Tracks.**

(1) Improper Track Tension. Adjust track tension (par. 127).

(2) Rock Between Track and Idler. Clean out.

(3) Misalinement of Idler Wheel. Tighten bracket bolts.

(4) Idler Shaft Loose in Bracket. Lock idler adjustment (par. 136).

c. **Inoperative Track Return Idlers.**

(1) Mud Between Idlers and Pontoon. Remove mud.

(2) Bearings Seized. Replace idler assembly (par. 137).

(3) Insufficient Lubrication. Lubricate roller periodically.

d. **Inoperative Idler Wheel.**

(1) Bearings Seized. Replace bearings (par. 133).

(2) Insufficient Lubrication. Lubricate bearings periodically (par. 27).

54. HYDRAULIC TRAVERSING MECHANISM.

CAUTION: *Before investigating for troubles in the hydraulic turret traversing mechanism, place shifter lever on gear box in "DOWN" position, and traverse turret by hand to make sure turret will rotate smoothly in both directions (par. 11).*

a. **Traverse Pump Fails to Operate.**

(1) Electric Motor Coupling Sheared. Replace coupling (par. 163).

b. **Oil Leaking from System.**

(1) Shaft Oil Seals Worn Out. Replace unit (par. 163).

LANDING VEHICLE TRACKED MK. I AND MK. II

(2) DEFECTIVE GASKETS. Tighten or replace (par. 163).
(3) LOOSE TUBING AND FITTING CONNECTIONS. Tighten or replace.

c. Unsteady Turret Operation.
(1) LOW OIL LEVEL. Refill reservoir (par. 27).
(2) LOOSE GEAR PUMP SUCTION TUBE CONNECTIONS. Tighten connections.

d. Traverse Pump Runs, but Turret Cannot Be Turned in Either Direction.
(1) TURRET LOCK ENGAGED. Release.
(2) GEAR BOX SHIFTER LEVER IN INTERMEDIATE POSITION. Reset to hydraulic traverse position.
(3) LOW OIL LEVEL. Refill reservoir (par. 27).

e. Turret Will not Stop Rotating When Control Handle Returns to Neutral.
(1) PILOT VALVE PLUNGER STICKING IN OPEN POSITION. Clean.

f. Turret Traverse Speed Low in Either Direction.
(1) OIL LEVEL LOW ALLOWING AIR TO BE SUCKED INTO SYSTEM. Refill reservoir (par. 27).
(2) CHECK VALVES LEAKING. Clean.

g. Turret Traverse Speed Is Low in One Direction.
(1) ONE CHECK VALVE LEAKING. Clean.

h. Turret Creeps in One Direction While Tank is in a Horizontal Position.
(1) PUMP CONTROL SHAFT NOT ADJUSTED TO NEUTRAL POSITION. Adjust.

i. Turret Drifts Excessively When Tank is not in Horizontal Position.
(1) LOW OIL LEVEL IN RESERVOIR. Refill reservoir (par. 27).
(2) LOOSE GEAR PUMP SUCTION TUBE CONNECTIONS. Tighten connections.
(3) CRACKED GEAR PUMP HOUSING. Replace pump.
(4) LEAKING HIGH-PRESSURE RELIEF VALVE. Replace.

55. HULL AND PONTOONS.

a. Vehicle Lists.
(1) UNBALANCED LOAD. Distribute load evenly.
(2) PONTOON DRAIN PLUGS LOOSE, PERMITTING WATER TO ENTER PONTOON. Remove plug, drain pontoon. Install drain plug tightly.

b. Vehicle Rides Too Low in Water.
(1) BILGE PUMP BROKEN. Replace bilge pump (par. 156).
(2) BILGE PUMP DRIVE SHAFT BROKEN. Replace bilge pump drive shaft (par. 156).

TM 9-775
55–56

ENGINE – DESCRIPTION, DATA, MAINTENANCE, AND ADJUSTMENT IN VEHICLE

(3) FLOOR DRAIN PLATE PLUGGED. Clean dirt and debris from floor drain plate.

(4) VEHICLE OVERLOADED. Lighten load. NOTE: *The cargo-carrying capacity of the landing vehicle tracked, exclusive of crew, is as follows:*

Model LVT (2) .... 5,950 lb
Model LVT (A) (1) .... 950 lb
Model LVT (A) (2) .... 4,550 lb

If these weight limits are exceeded, vehicle will ride too low in water.

(5) HOLES IN HULL. Notify ordnance maintenance personnel.

(6) DRAIN PLUGS ON BOTH PONTOONS LOOSE OR BROKEN. Tighten drain plugs. Install new drain plugs.

Section XIV

ENGINE – DESCRIPTION, DATA, MAINTENANCE, AND ADJUSTMENT IN VEHICLE

	Paragraph
Description and data	56
Engine support beam	57
Baffles and cowling	58
Valves and valve push rods	59
Maintenance and adjustment of the vehicle	60

56. DESCRIPTION AND DATA.

a. **General.** The engine is a W670-9A (CO) air-cooled, seven-cylinder, static-radial of the aviation type. It is mounted on an engine support beam which is supported on brackets on each side of the engine room (fig. 26). Two clamps on the front section of the crankcase hold a steady bar, the ends of which are clamped in supporting brackets on the sides of the engine room (fig. 28). Two magnetos, an electric starter, generator drive pulley, oil pumps, and fuel pump are mounted on the accessory case (figs. 26 and 27). The generator, which is mounted on a movable bracket attached to the engine support beam, is located below and to the left of the accessory case. The carburetor is centrally mounted on the lower side of the rear crankcase section. A dual purpose gear-type oil pump supplies oil under pressure to the various engine parts, and incorporates a scavenger pump which draws oil from the engine sump and returns it to the supply tank. A second scavenger pump transfers oil from the two lower rocker boxes to the main engine sump. The flywheel end of the engine is referred to as the "front" of the engine, the acces-

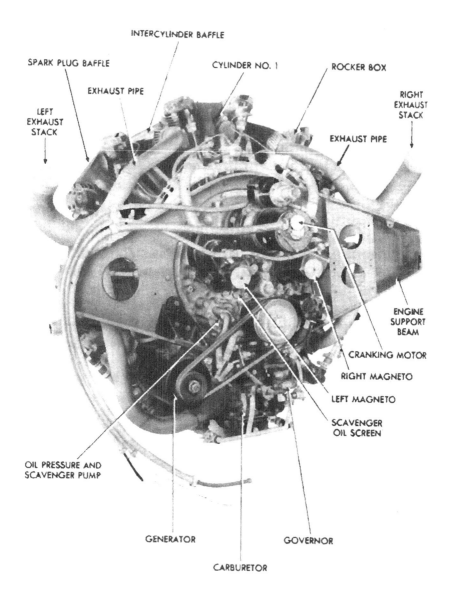

Figure 26 — Left Side of Engine — ¾ Rear View

TM 9-775

ENGINE — DESCRIPTION, DATA, MAINTENANCE, AND ADJUSTMENT
IN VEHICLE

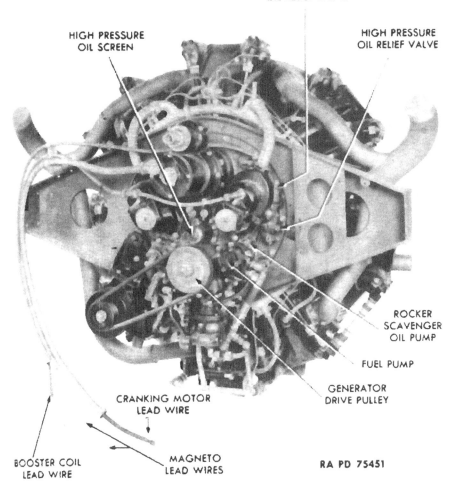

Figure 27 — Right Side of Engine — ¾ Rear View

TM 9-775
56

LANDING VEHICLE TRACKED MK. I AND MK. II

Figure 28 — Right Side of Engine — ¾ Front View

sory end as the "rear." The terms "right" and "left" are used with reference to the engine as viewed from the rear. Horizontal and vertical positions of the engine are referred to with respect to the crankshaft. When the crankshaft is in the horizontal position, the engine is in the vertical position. Direction of rotation is determined by viewing the engine from the rear.

b. Data.

Make	Continental
Type	Single-row, static-radial, air-cooled
Model and series	W670 series 9A
Over-all diameter	42⅜ in.
Weight, with accessories (approx.)	1,100 lb

132

ENGINE — DESCRIPTION, DATA, MAINTENANCE, AND ADJUSTMENT IN VEHICLE

Horsepower	250 at 2,400 rpm
Maximum allowable rpm	2,400
Number of cylinders	7
Bore	5⅛ in.
Stroke	4⅝ in.
Piston displacement	667.86 cu in.
Compression ratio	6.1 to 1

Direction of rotation (viewed from rear of engine):

Crankshaft	Clockwise
Cam	Counterclockwise
Tachometer	Counterclockwise
Fuel pump	Clockwise
Cranking motor	Counterclockwise
Generator	Clockwise
Magnetos	Counterclockwise

Accessory speeds:

Tachometer	1½ crankshaft speed
Generator	39/20 crankshaft speed
Magnetos	⅞ crankshaft speed

Magnetos:
Make and model	Scintilla, model VMN7-DFA or Bosch, model MJT7A 302
Breaker point gap	0.012 in.
Spark plug gap	0.015 in.
Valve clearance (between valve stem and roller, engine cold)	0.010 in.
Carburetor make and model	Bendix Stromberg, model NA-R6B
Firing order (clockwise viewed at accessory case)	1-3-5-7-2-4-6
No. 1 cylinder location	Top

57. ENGINE SUPPORT BEAM.

a. The engine support beam is of all-steel construction, and is bolted to the crankcase. The accessory housing and accessories protrude through a cut-out hole in the center of the engine support beam (fig. 26).

58. BAFFLES AND COWLING.

a. Baffles and cowling on the engine and around the engine room control the flow of air through the engine. On the engine itself are spark plug baffles and right front and rear baffles. Air enters through the five front louvers, passes through the engine, and escapes through the hinged rear cover. If the left or right (front or rear) baffles are removed, air will bypass the engine, causing it to overheat (fig. 33).

133

TM 9-775
59—61

LANDING VEHICLE TRACKED MK. I AND MK. II

59. VALVES AND VALVE PUSH RODS.

a. The valves are operated by a double-track cam ring mounted on the crankshaft, forward of the accessory section, in the rear of the crankcase. The valves are driven by the intermediate cam drive gear. The push rods and valve tappets are returned to position by valve spring action (fig. 51).

60. MAINTENANCE AND ADJUSTMENT OF THE VEHICLE.

a. Maintenance and adjustment operations are performed with the engine removed from the vehicle. Removal and installation procedure is covered in section XV.

Section XV

ENGINE — REMOVAL AND INSTALLATION

	Paragraph
Engine removal	61
Engine installation	62
Engine support beam—removal and installation	63
Baffles and cowling—removal and installation	64
Valves and valve push rods	65

61. ENGINE REMOVAL.

a. **Open Battery and Radio Switches** (fig. 16). Turn battery switch on control panel to "OFF." Open radio switch.

b. **Remove Stern Covers** (figs. 29 and 30). Remove cap screw, bolts, nuts, and toothed lock washers which hold bolted stern cover in position, and remove cover. Open hinged stern cover. In order to provide more room for working in vehicle it may be desirable to remove cap screws (two each side) and cover brace across recessed screws to remove hinged stern cover completely from vehicle.

c. **Remove Angle Support and Upper Front Baffle.** Remove nut, bolt, and toothed lock washer (one each end) which secures angle support (fig. 30). Remove one nut, bolt, and toothed lock washer which holds each side of upper baffle to lower front baffle (fig. 30).

d. **Remove Muffler and Flexible Tubing** (fig. 31). Work forward ends (each side) of flexible tubing off exhaust manifold. Remove four bolts and toothed lock washers which secure each muffler to hull. Lift out muffler and assembled flexible tubing. NOTE: *In some vehicles the flexible tubing may be spot-welded either to the muffler or to the exhaust manifold. In such cases, tap off spot-weld with a chisel and hammer.*

TM 9-775
61

ENGINE — REMOVAL AND INSTALLATION

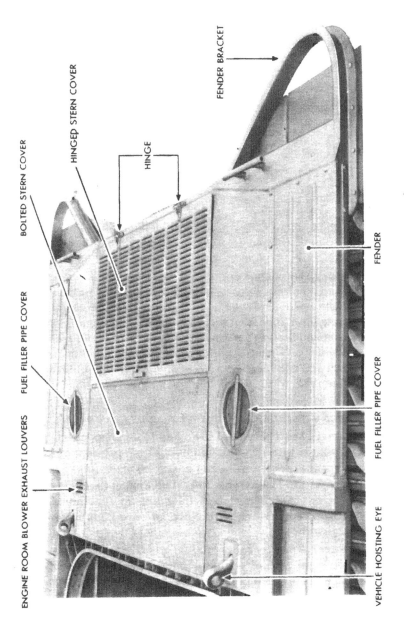

Figure 29 — Rear Deck

TM 9-775

LANDING VEHICLE TRACKED MK. I AND MK. II

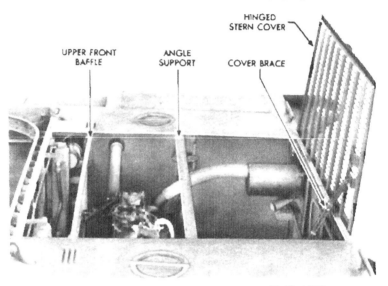

Figure 30 — Rear Deck bolted Stern Cover Removed, Hinged Stern Cover Raised

e. **Disconnect Universal Joints.** Remove eight cap-screws, with flat washers and toothed lock washers, from rear end of each control tunnel side plate. Remove four bolts, nuts, and toothed lock washers securing aft propeller shaft bell housing to stern bulkhead. To facilitate removal, lift out center bulkhead louver. Tap down locking lugs which secure universal joint retaining cap screws (fig. 32). Remove four cap screws and two locking lug plates which secure universal joint of aft propeller shaft to universal joint on engine.

f. **Remove Magneto Guard.** Remove screws which hold magneto guard to engine support beam. Lift off magneto guard.

g. **Disconnect Oil Pressure Line** (fig. 31). Disconnect oil pressure line at fitting beneath right magneto. Plug open end of line with cloth.

h. **Disconnect Fuel Lines** (fig. 31). Disconnect carburetor bypass line at carburetor. Disconnect fuel inlet line to fuel pump at fuel pump. Plug open end of line with cloth, or wire lines in an upright position to engine mount.

i. **Disconnect Primer Line** (fig. 31). Disconnect primer line, located on top right side of engine support beam, at union between flexible tubing and copper tubing. Remove nut, bolt, and toothed lock washer from clamp which holds primer line flexible tubing to engine support beam.

TM 9-775

ENGINE — REMOVAL AND INSTALLATION

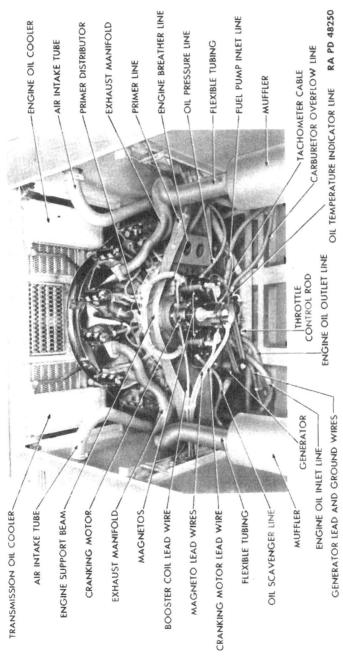

Figure 31 — Engine Installed

137

Figure 32 — Propeller Shaft Bell Housing in Position to Disconnect Universal Joint

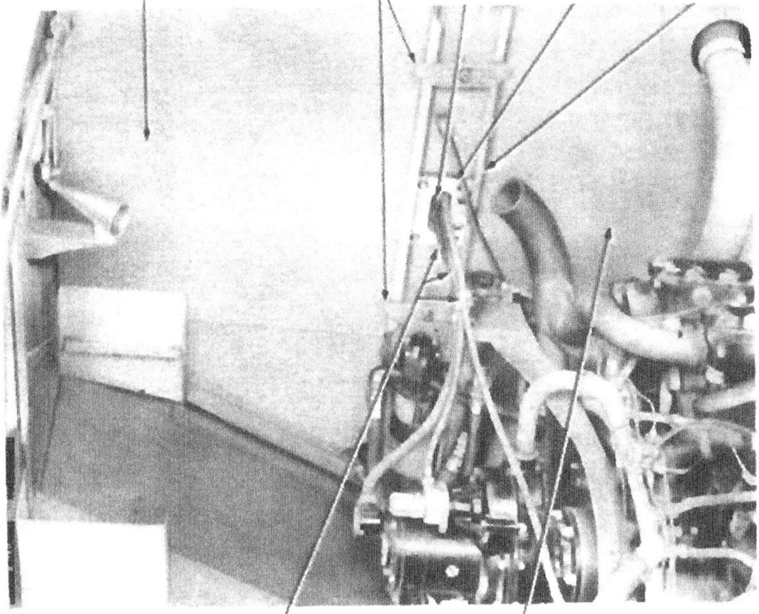

ENGINE – REMOVAL AND INSTALLATION

Figure 33 — Electrical Connections Through Engine Room Baffles

j. **Disconnect Oil Temperature Indicator Line** (fig. 31). Disconnect oil temperature indicator line from its position below tachometer cable adapter. Pull out line with attached bulb. Loosen nut and palnut on stud on which line clamp is mounted. Work line out of clamp and away from engine.

k. **Disconnect Tachometer Cable** (fig. 31). Disconnect tachometer cable knurled nut, slide nut down on shaft, and pull tachometer cable out of tachometer cable adapter. Remove nut which tightens clamp securing tachometer cable to lower cylinder. Leave clamp on cable.

l. **Disconnect Engine Breather Line** (fig. 31). Disconnect engine breather line at accessory housing above right-hand magneto.

m. **Disconnect Oil Lines** (fig. 31). Disconnect oil inlet and oil outlet lines at oil pump. Disconnect oil scavenger line at accessory housing. Plug all open ends of line with cloth, or wire lines in an upright position to engine mount.

n. **Disconnect Generator Lead and Ground Wire** (fig. 31). Remove two screws which secure shielded cover plate to side of generator. Remove two washers and stud nuts which secure generator

TM 9-775

LANDING VEHICLE TRACKED MK. I AND MK. II

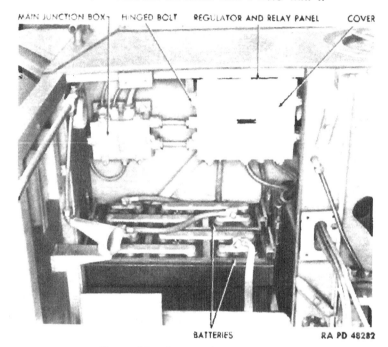

Figure 34 — Battery Compartment

lead wire and ground wire to studs within shielded connection. Loosen conduit knurled nut, and pull conduit and wires out of generator.

o. **Remove Electrical Lines Felt Gasket** (fig. 33). Remove four cap screws, flat washers, and toothed lock washers which secure gasket cover to channel support between upper left front and rear baffle. Pull gasket cover out on electrical lines, and lift off two halves of felt gasket.

p. **Remove Upper Left Rear Baffle** (fig. 33). Loosen stud nuts on which two clamps, upper and lower, are mounted, which hold forward ends of upper left rear baffle in position. Rotate clamps to a vertical position. Lift baffle forward, upward, and out of vehicle.

q. **Disconnect Engine Electrical Connections in Regulator and Relay Panel** (figs. 34, 35, and 36). Loosen 10 hinged bolts securing regulator and relay panel cover. Swing bolts out of clips on cover, and remove cover. Disconnect cranking motor lead wire. Disconnect two magneto lead wires. Disconnect booster coil lead wire. Loosen knurled nuts which secure flexible conduits in which wires are encased, and pull wires out of regulator and relay panel. Remove screws to remove clamps which secure conduits to angle support within battery compartment. Pull conduits out through baffle channel support, and lay them on engine.

TM 9-775
61

ENGINE – REMOVAL AND INSTALLATION

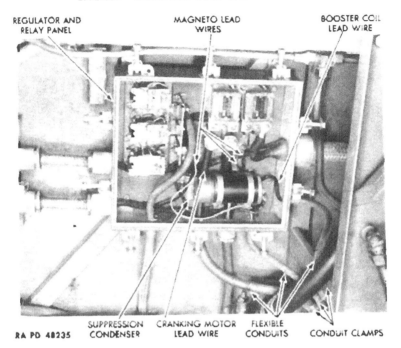

Figure 35 – Regulator and Relay Panel – Cover Removed

r. **Disconnect Throttle Control Rod** (fig. 31). Remove cotter pin and clevis pin which secure rear end of throttle control rod to throttle arm mounted on governor shaft.

s. **Remove Air Intake Tubes** (fig. 31). Loosen hose clamps at each end of air intake tubes (right and left), and remove tubes.

t. **Remove Engine Mounting Bolts and Cap Screws.** Remove engine mounting bolts, nuts, and cotter pins (four each side) securing engine support beam to engine mount (fig. 37). The front bottom nut securing left side of engine support beam also secures engine ground strap. Remove lock wire and two cap screws securing steady bar caps (one each side), and remove caps (fig. 38).

u. **Install Engine Sling** (fig. 39). Place engine lifting sling (41-S-3832) on top of engine. Hook straight arm of hoisting sling to lifting eye located on a stud at right of No. 1 cylinder. Hook crooked arm to lifting eye located on stud at left of No. 1 cylinder. Loop chain around clutch hub, and attach chain hook to chain.

v. **Remove Engine** (fig. 40). Hook a hoist to engine sling, and carefully lift engine from vehicle. As engine is being lifted watch closely to see that all disconnections have been made. Make sure

141

Figure 36 — Engine Electrical Leads Disconnected

TM 9-775
61—62

ENGINE — REMOVAL AND INSTALLATION

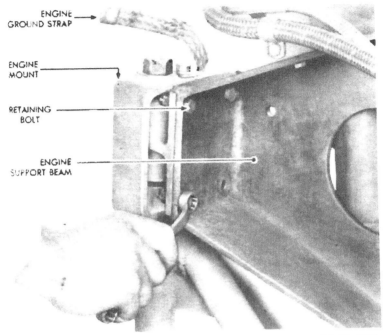

Figure 37 — Removing Engine Support Beam Retaining Bolts

engine works free of bulkhead. Take care no projection of engine catches on any part of engine room while engine is being removed. Lower engine to a previously prepared engine stand (41-S-4942-11). If engine is to be transported, use engine transport stand (41-S-4942-22).

62. ENGINE INSTALLATION.

a. **Install Engine Sling** (fig. 39). Place engine sling (41-S-3832) on top of engine. Hook straight arm of sling to lifting eye located on a stud at right of No. 1 cylinder. Hook crooked arm to lifting eye located on stud at left of No. 1 cylinder. Loop chain around clutch hub, and attach chain hook to chain.

b. **Hoist Engine into Position** (figs. 40 and 41). Hook a chain hoist to engine sling. Lift engine up, then lower it into engine compartment. This is best accomplished by one man controlling hoist, and one or two men guiding engine. Be sure felt lining on lower front baffle is in back of engine fan shroud. As engine is lowered into position, forward end must be tilted slightly downward so universal joint on engine meets universal joint on propeller shaft.

LANDING VEHICLE TRACKED MK. I AND MK. II

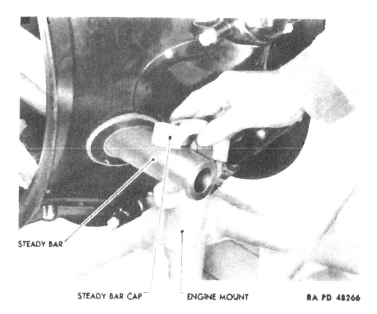

Figure 38 — Engine Mount Steady Bar Cap Removed

c. **Install Engine Mounting Bolts.** Line up holes and install engine mounting bolts, nuts, and cotter pins (four each side) securing engine support beam to engine mount (fig. 37). The front bottom nut securing left side of support beam also secures engine ground strap. Install cap screws securing steady bar caps (one each side) and lace with safety wire (fig. 38).

d. **Install Air Intake Tubes** (fig. 31). The left and right air intake tubes are installed in same manner. Apply a small amount of grease to ends of tube. Work lower end of tube into air horn on carburetor, and upper end into flexible hose extending through baffle from air cleaner. Tighten hose clamps.

e. **Connect Throttle Control Rod** (fig. 31). Install rod end pin and cotter pin which secure rear end of throttle control rod to throttle arm mounted on governor shaft.

f. **Connect Engine Electrical Connections in Regulator and Relay Panel Box** (figs. 35 and 36). Insert three flexible conduits through hole in channel support, being sure conduits pass through metal gasket cover. Connect cranking motor lead wire in regulator and relay panel. Connect magneto lead and ground wires. Connect booster coil lead wire. Tighten knurled nuts which secure three conduits to regulator and relay panel. Install nuts, bolts, and toothed lock washers which secure six conduit clamps to angle support. Place regulator and relay

ENGINE — REMOVAL AND INSTALLATION

Figure 39 — Sling Attached to Engine

panel cover in position. Swing hinged bolts and washers into position and tighten bolts securely (fig. 34).

g. **Install Electrical Lines Felt Gasket** (fig. 33). Place in position two halves of gasket felt pads, and hold them in position with gasket cover. Install four cap screws, flat washers, and toothed lock washers securing gasket cover to channel support between left front and rear baffles.

h. **Install Upper Left Rear Baffle** (fig. 33). Place baffle in position, and rotate clamps to a horizontal position. Tighten stud nuts which lock two clamps.

i. **Connect Generator Lead and Ground Wire** (fig. 31). Place conduit in position and install two washers and stud nuts which secure generator lead and ground wire to studs in generator shielded connection. Tighten knurled nut which secures conduit. Install two cap screws which secure shielded cover plate to side of generator. Install safety wire.

j. **Connect Oil Lines** (fig. 31). Remove cloth from lines. Apply

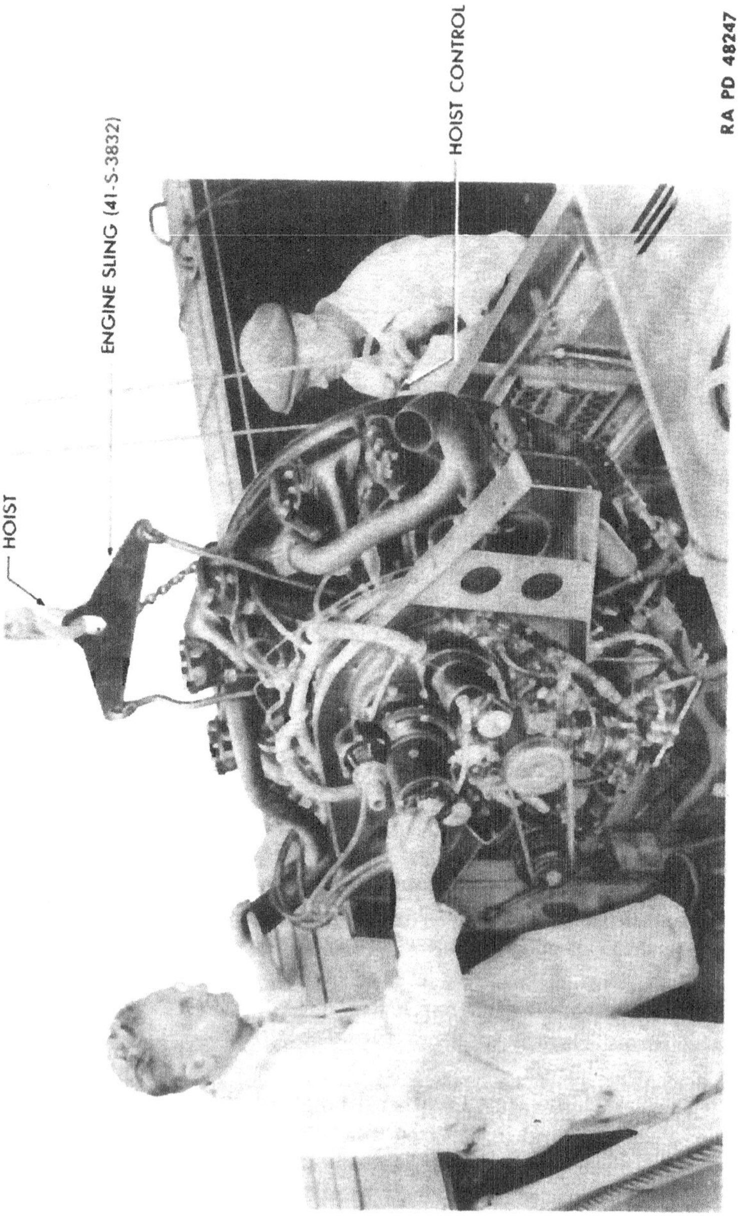

TM 9-775

LANDING VEHICLE TRACKED MK. I AND MK. II

Figure 40 — Lifting Engine from Vehicle

TM 9-775
62

ENGINE — REMOVAL AND INSTALLATION

Figure 41 — Engine Removed from Vehicle

white lead base antiseize compound to male threads. Connect oil inlet and oil outlet lines at oil pump. Connect oil scavenger line at accessory housing.

k. **Connect Engine Breather Line** (fig. 31). Connect engine breather line at accessory housing, above right magneto.

l. **Connect Tachometer Cable** (fig. 31). Slide tachometer cable into tachometer cable adapter, and tighten knurled nut which secures cable housing. Install nut and tighten clamp which secures tachometer cable to lower cylinder.

m. **Connect Oil Temperature Indicator Line** (fig. 31). Apply white lead base antiseize compound to threads of knurled nut. Plug oil temperature indicator line and attached bulb into its position below tachometer. Tighten knurled nut.

n. **Connect Primer Line** (fig. 31). Connect primer line, located on top right side of engine support beam, at union between flexible tubing and copper tubing. Install nut, bolt, and toothed lock washer securing clamp which holds primer line flexible tubing to engine support beam.

147

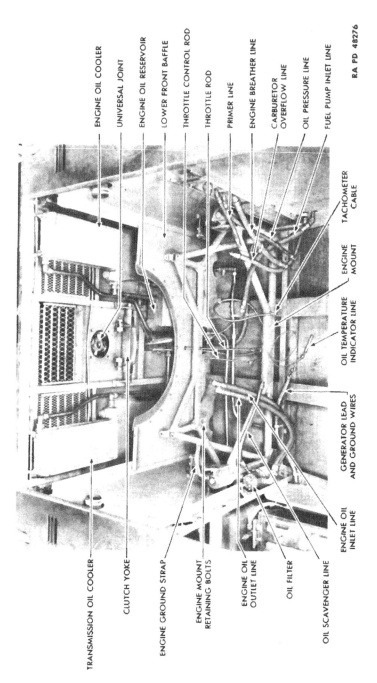

Figure 42 — Engine Room — Engine Removed

TM 9-775
62

ENGINE — REMOVAL AND INSTALLATION

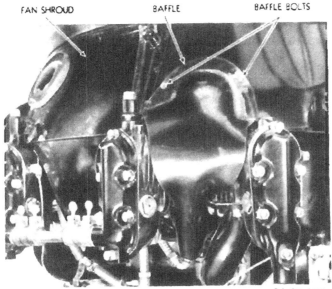

Figure 43 — Removing Spark Plug Baffle

o. **Connect Fuel Lines** (fig. 31). Remove cloth from lines. Connect fuel return line (from carburetor) at carburetor. Connect fuel inlet line (to fuel pump) at fuel pump.

p. **Connect Oil Pressure Line** (fig. 31). Remove cloth from line. Connect oil pressure line at fitting beneath right magneto.

q. **Install Magneto Guard.** Place magneto guard in position, and install screws which secure magneto guard to engine support beam.

r. Connect universal joints (par. 115).

s. **Install Mufflers and Flexible Tubing** The left and right mufflers are installed in same manner. Work one end of flexible tubing onto exhaust manifold and other end onto muffler. Insert tail pipe attached to muffler through hole in rear of vehicle. Install four bolts, nuts, and toothed lock washers which secure muffler to hull.

t. **Install Angle Support and Upper Front Baffle** (fig. 30). Install two nuts, bolts, and toothed lock washers (one each end) which secure angle support. Install two nuts, bolts, and toothed lock washers which secure upper front baffle to lower half.

u. **Install Stern Covers** (fig. 30). Install cap screws, bolts, nuts, and toothed lock washers holding bolted stern cover in position. Close hinged stern cover. NOTE: *If hinged stern cover has been removed, install it on vehicle by installing hinge retaining cap screws (two each side) and supporting arm cross recessed screws.*

LANDING VEHICLE TRACKED MK. I AND MK. II

Figure 44 — Removing Steady Bar

63. ENGINE SUPPORT BEAM — REMOVAL AND INSTALLATION.

a. Removal (Engine Out of Vehicle).

(1) PRELIMINARY OPERATIONS. Remove generator (par. 97), cranking motor (par. 95), governor, and magnetos (par. 91).

(2) DISCONNECT OIL AND FUEL LINES. Remove oil inlet and oil outlet elbow fittings from oil pressure and scavenger pump (fig. 26). Disconnect fuel lines from fuel pump to carburetor, and remove fittings on fuel pump (figs. 26 and 27).

(3) REMOVE ENGINE SUPPORT BEAM. Remove bolt and nut which secure primer line bracket to engine support beam. Remove eight bolts, nuts, flat washers, lock washers, and cotter pins securing engine support beam to crankcase. Lift off engine support beam (two men).

b. Installation.

(1) BOLT ENGINE SUPPORT BEAM IN POSITION. Place engine support in position on crankcase and install eight retaining bolts, nuts, flat washers, lock washers, and cotter pins. Install bolt and nut which secure primer line bracket to engine support beam.

(2) CONNECT FUEL AND OIL LINES. Install fittings in fuel pump, and connect fuel lines from fuel pump to carburetor. Install oil inlet and oil outlet elbow fittings.

ENGINE — REMOVAL AND INSTALLATION

TM 9-775
63

Figure 45 — Engine, Fan Shroud, and Steady Bar Removed

TM 9-775
63-64

LANDING VEHICLE TRACKED MK. I AND MK. II

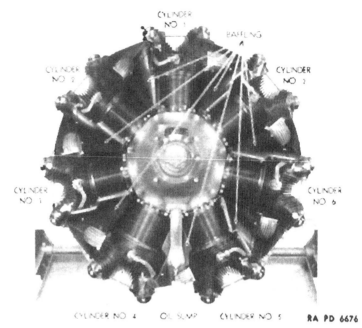

Figure 46 — Intercylinder Baffling — Front View

(3) CONCLUDING OPERATIONS. Install governor, magnetos (par. 91), cranking motor (par. 95), and generator (par. 97).

64. BAFFLES AND COWLING — REMOVAL AND INSTALLATION.

a. Maintenance.

(1) Keep all baffles and cowling clean and unobstructed so that air entering the front louvers will flow as directed to cool various parts of engine. Replace baffles damaged in service.

(2) Adjust fan shroud for correct clearance. There must be 0.125-inch clearance between fan blades and fan shroud at all points. If more clearance is needed at any point, slightly loosen all palnuts and nuts which secure fan shroud to cylinder rocker boxes. Drive a small wedge between fan blade and fan shroud, at point where more clearance is necessary. Check with a feeler gage. When 0.125-inch clearance is obtained, tighten all nuts and palnuts which secure fan shroud to cylinder rocker boxes. Remove screwdriver.

b. Removal.

(1) Remove engine from vehicle (par. 61).

(2) REMOVE SPARK PLUG BAFFLES (fig. 43). Remove two bolts, nuts, and toothed lock washers which secure each baffle to fan shroud.

TM 9-775
ENGINE — REMOVAL AND INSTALLATION

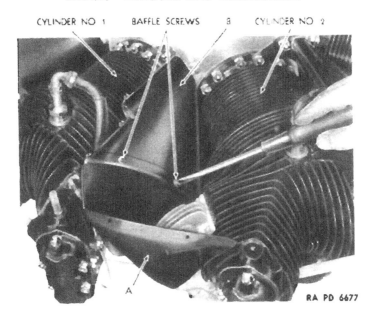

Figure 47 — Removing Intercylinder Baffling "A"

Remove one screw which attaches each baffle to cylinder head and lift off baffle. Remove remaining six baffles in same manner.

(3) REMOVE FAN SHROUD (figs. 44 and 45). Remove 12 bolts, nuts, and lock washers which hold fan shroud to intercylinder baffles. NOTE: *There are six intercylinder baffles, and they are installed between all cylinders except Nos. 4 and 5.* Remove seven palnuts and nuts which hold fan shroud to studs on cylinder rocker boxes. Cut safety wire, and loosen four cap screws which secure steady bar in clamps. Slide out steady bar. Pry fan shroud off studs on rocker boxes. Lift off two spacers between fan shroud and engine. (One is on No. 3 cylinder, and the other on No. 6 cylinder.)

(4) REMOVE INTERCYLINDER BAFFLING (A, fig. 47). CAUTION: *Before starting to remove intercylinder baffling, tag each individual baffle to show its proper location for assembling. The pieces in any one group of baffles are different from those in any other group. The pieces in each individual group of baffles should be designated as "A" and "B," or "A," "B," and "C," depending on the numbers of individual baffles in group* (figs. 47, 48, 49, and 50). Beginning with baffle "A" between No. 1 and No. 2 cylinders, remove the two screws which hold baffle "A" to baffle "B." Lift off baffle "A," and tag it to identify for assembly. In the same manner remove baffle "A" located between No. 2 and No. 3, No. 3 and No. 4, No. 5 and No. 6, No. 6 and No. 7,

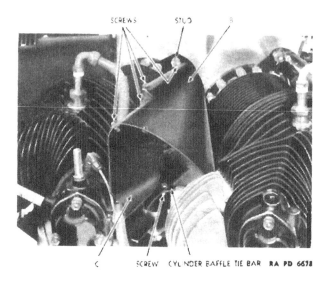

Figure 48 — Removing Intercylinder Baffling "B" and "C"

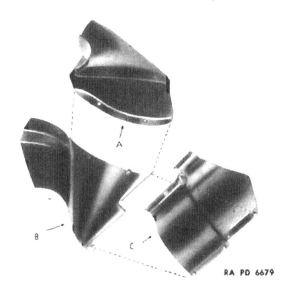

Figure 49 — Intercylinder Baffling — Exploded View

TM 9-775
ENGINE — REMOVAL AND INSTALLATION

No. 7 and No. 1 cylinders. NOTE: *There is a two-piece baffle between No. 4 and No. 5 cylinders, which is removed later.*

(5) REMOVE INTERCYLINDER BAFFLING (B and C, figs. 50 and 51). CAUTION: *Do not attempt to remove cylinder baffle tie bars at this time. Merely remove one screw from each.*

(6) DISCONNECT CYLINDER BAFFLE TIE BARS. Remove screw that holds cylinder baffle tie bar to baffle "C" between No. 1 and No. 2 cylinders. In same manner, remove screws that hold cylinder baffle tie bar to baffle "C" between No. 2 and No. 3, No. 3 and No. 4, No. 5 and No. 6, No. 6 and No. 7, No. 7 and No. 1 cylinders.

(7) DISCONNECT BAFFLE "B" FROM BAFFLE "C" (NO. 1 AND NO. 2 CYLINDERS). Beginning with baffling between cylinders 1 and 2, remove four screws which hold baffle "B" to baffle "C." Follow this same procedure with baffling "B" and "C" between No. 2 and No. 3, No. 3 and No. 4, No. 5 and No. 6, No. 6 and No. 7, No. 7 and No. 1 cylinders. Omit between No. 4 and No. 5 cylinders.

(8) REMOVE BAFFLES "B" AND "C" (NO. 3 AND NO. 4 CYLINDERS). Remove palnuts and nuts on two cylinder base studs which hold baffle "B" to base of No. 3 cylinder. If baffle "C" is still in place, remove it after lifting out baffle "B," together with baffle brace which is still attached to "B." Wire baffles "A," "B," and "C" of this group together and tag the group. Mark tag to indicate its position is between No. 3 and No. 4 cylinders.

(9) REMOVE BAFFLES "B" AND "C" (NO. 1 AND NO. 2 CYLINDERS). Return to baffling between No. 1 and No. 2 cylinders, and remove cotter pin and castle nut holding base of baffle "B" to the baffle to cylinder support. If baffle "C" has not fallen out previously and been tagged for position, lift it out after removing baffle "B," along with baffle brace which is still attached to "B." Remove stud from baffle to cylinder support.

(10) REMOVE BAFFLES "B" AND "C" (OTHER CYLINDERS). Carry out operation in step (9), above on baffling between No. 2 and No. 3, No. 5 and No. 6, No. 6 and No. 7, No. 7 and No. 1 cylinders. CAUTION: *Be sure to wire baffles in each group together and tag the group, marking on the tag the proper position of the group for assembly.*

(11) REMOVE TWO-PIECE SUMP BAFFLE (fig. 50). Remove two screws which hold small (sump) baffle "A" to cylinder (sump) baffle "B" between No. 4 and No. 5 cylinders. When lifting out baffle "A," rotate it to right, following curvature of cylinders. When lifting out baffle "B," rotate it to left, following curvature of cylinders.

(12) REMOVE ENGINE ROOM BAFFLES. Engine room baffles are removed as steps under the removal of engine (par. 61), fuel tanks (par. 71), fixed fire extinguishers (par. 146), and batteries (par. 100).

LANDING VEHICLE TRACKED MK. I AND MK. II

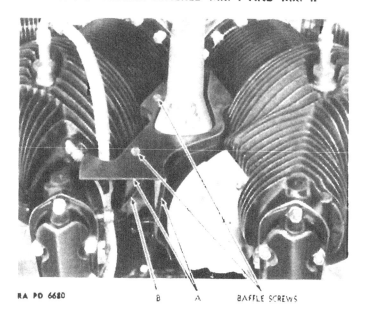

Figure 50 — Removing Two-piece Sump Baffle

c. Installation.

(1) INSTALL TWO-PIECE SUMP BAFFLE (fig. 50). Insert baffles "A" and "B" from right and left respectively, using a curving motion that follows curvature of each cylinder. Press two baffles into position, and fasten them together with two screws and lock washers.

(2) PLACE BAFFLES IN POSITION. Beginning with No. 1 cylinder and continuing in regular sequence, place baffles "B" and "C" from each group (except one located between No. 4 and No. 5 cylinders) between each pair of cylinders. CAUTION: *In each case, baffle "C" is put in place first, and baffle "B" is set on top of it.* The bracket on end of baffle "B" (except one attaching to base of No. 3 cylinder) will meet cylinder baffle to cylinder support, and is held to it by a drilled hexagon-head screw and a castle nut. Insert hexagon-head screw with its threaded end toward cylinders, and put on castle nut and cotter pin. As each pair of baffles ("B" and "C") is set in place, reach from underneath and insert screw and lock washer that hold cylinder baffle tie bar to baffle "C," but do not tighten screw. Following insertion of screw that holds tie bar to baffle "C," fasten baffles "B" and "C" together with four screws and lock washers. Reach from underneath and tighten screw holding each cylinder baffle tie bar to baffle "C" in its group. This baffle is attached to baffle "B" by two screws and lock washers. Tighten the two screws. Beginning

ENGINE – REMOVAL AND INSTALLATION

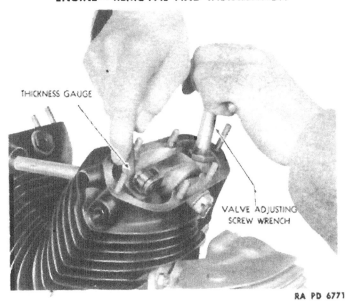

Figure 51 — Adjusting Valve Clearance

with No. 1 cylinder and continuing in regular order (except between No. 4 and No. 5 cylinders), install remaining baffle "A" in each group.

(3) INSTALL FAN SHROUD (figs. 44 and 45). Put a spacer on No. 3 intake valve rocker box to fan shroud stud, also one on No. 6 exhaust valve to rocker box fan shroud stud (two spacers). Place fan shroud on engine in such a position that one hole for steady bar is opposite intake valve rocker box of No. 6 cylinder. Install 7 nuts and palnuts which hold fan shroud to studs on rocker boxes. Tighten nuts enough to seat fan shroud. Slide steady bar in place so there is same length of bar on each side of shroud. Install 4 cap screws and safety wire which secure steady bar. Install 12 bolts, nuts, and lock washers which hold fan shroud to intercylinder baffles. (There is no baffle between No. 4 and No. 5 cylinders.) Install 14 bolts, nuts, and toothed lock washers which secure 7 spark plug baffles to fan shroud. Check for 0.125-inch clearance at all points between fan blades and fan shroud. If there is not 0.125-inch clearance at all points, adjust for proper clearance (par. 82 b (2)).

(4) INSTALL SPARK PLUG BAFFLES (fig. 43). Beginning with No. 1 cylinder, and proceeding at regular sequence, install each of 7 fan shroud to rear spark plug baffles. On each baffle install 2 screws with hexagonal nuts and lock washers which hold baffle to fan shroud. Then install screw and lock washer which attach end of baffle to cylinder head.

TM 9-775
LANDING VEHICLE TRACKED MK. I AND MK. II

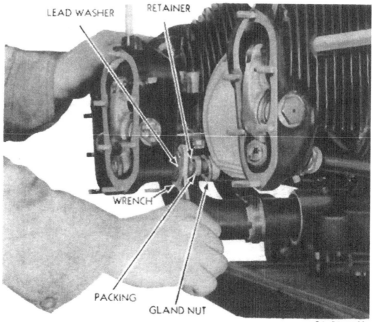

Figure 52 — Loosening Push Rod Housing Gland Nuts and Housing Retainers

(5) INSTALL ENGINE ROOM BAFFLES. Engine room baffles are installed as steps under installation of engine (par. 62), fuel tanks (par. 71), fixed fire extinguishers (par. 146), and batteries (par. 100).

65. VALVES AND VALVE PUSH RODS.

a. **Maintenance and Adjustment.**

(1) GENERAL. Inspect frequently to ensure that rocker box covers are tight, and that cover gaskets are in good condition. If an oil leak appears, replace cover gasket. Replace rocker arms and push rods when these parts are defective or damaged.

(2) VALVE ADJUSTMENT (ENGINE OUT OF VEHICLE) (fig. 51). Remove rocker box cover by removing six castle nuts and flat washers. Rotate crankshaft until top dead center has been reached (41-1-73-100, piston top dead center indicator). With a 0.010-inch feeler gage (41-G-404), check clearance between rocker roller and valve stem on both intake and exhaust valves. The feeler gage should slide between roller and valve stem with only a slight drag when clearance is correct. If clearance is correct, no further adjustment is necessary.

TM 9-775
65

ENGINE — REMOVAL AND INSTALLATION

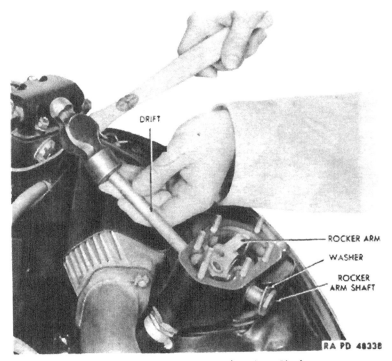

Figure 53 — Removing Rocker Arm Shaft

If not correct, loosen adjusting screw lock nut (41-W-1059 and 41-W-3812-440 wrenches), and turn adjusting screw until correct clearance is obtained. Tighten adjusting screw lock nut, and check valve clearance. Install new rocker box cover gaskets, and install rocker box covers.

(3) REPLACING PUSH ROD HOUSING GASKETS AND PACKING (ENGINE OUT OF VEHICLE).

(a) *Removal.* Remove push rod (subpar. b, below). Using push rod housing gland nut wrench (41-W-3812-530), unscrew push rod housing gland nut (fig. 52); using push rod housing retainer wrench (41-W-1986-100), unscrew housing retainer nut. Remove two palnuts which hold push rod housing to crankcase studs, then remove two hexagonal nuts. Slip the gland nut, gland nut packing, retainer nut, and lead washer down on the push rod housing far enough so the outer end of the housing clears the rocker box, and remove the push rod housing and base gasket.

(b) *Installation.* Install push rod (subpar. b, below). Place a push rod housing to crankcase gasket in position on the base studs, and slip base of push rod housing into position. Use a $7/16$-inch exten-

159

TM 9-775
LANDING VEHICLE TRACKED MK. I AND MK. II

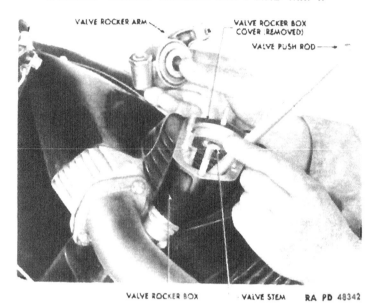

Figure 54 — Removing Valve Push Rod

sion speed socket wrench to put a plain hexagonal nut on each of the two push rod housing base studs. CAUTION: *Do not tighten or palnut these nuts.* Slip a gland nut, a gland nut packing, a retainer nut, and a lead washer onto the push rod housing. NOTE: *There are two ends to the retainer nut, one of which matches the gland nut, and the other the threads in the cylinder head.* Do not try to screw the wrong end into the cylinder head. Move the outer end of the push rod housing sideways into position, and enter the retainer nut in the threads in the cylinder head. Be careful not to cross-thread this nut. The lead washer is between this nut and the cylinder head. Use push rod housing retainer wrench (41-W-1986-100) to tighten the retainer moderately. Position the gland packing, and screw the gland nut in place, using a gland nut wrench (41-W-3812-530). CAUTION: *Tighten the gland nut carefully, not too tight. Excessive tightening will squeeze the gland packing out of position.* When the 13 push rod housings (all except exhaust rod housing on No. 5 cylinder) have been installed, tighten each of the base nuts to specified torque limits. Install palnuts in place on the base studs for the installed 13 push rod housings.

b. **Removal of Valve Rocker Arms and Push Rod** (figs. 53 and 54). Remove six castle nuts and flat washers which secure rocker box cover, and lift off cover. Turn crankshaft until valve is closed,

FUEL SYSTEM

and rocker arm is free (top dead center). Remove cotter pin, castle nut, and flat washer which secure rocker arm shaft. Tap out shaft and outside washer. Pry out rocker arm assembly. Lift out push rod. Remove all 14 rocker arms and push rods in similar manner.

 c. Installation (figs. 53 and 54). Place valve push rod in push rod housing. Press rocker arm assembly in, to position. Install rocker arm shaft and outside washer. Install castle nut, flat washer, and cotter pin which secure rocker arm shaft. Check valve adjustment (subpar. a (2), above). Place rocker box cover in position, and install six castle nuts and flat washers.

Section XVI

FUEL SYSTEM

	Paragraph
Description	66
Carburetor	67
Accelerator pedal linkage	68
Fuel pump	69
Fuel filter	70
Fuel tanks	71
Air cleaners	72
Priming pump	73
Lines, valves, and fittings	74

66. DESCRIPTION.

 a. General (fig. 55). Two main fuel tanks, mounted in the areas between the engine room and the hull, are used in the landing vehicles tracked. An additional small fuel tank containing 1 gallon of fuel is mounted above the fixed fire extinguishers for the auxiliary generator (used only on the vehicle with turret (LVT) (A) (1)).

 b. Fuel Flow System (fig. 55). From either of the two main fuel tanks, fuel goes into a collector or main fuel line. From there it flows through the fuel filter, and up into the fuel pump located above and to the right of the carburetor. From the fuel pump, fuel is forced down to the carburetor. A bypass line at the carburetor fuel inlet permits fuel in excess of carburetor requirements to return through a bypass pressure regulator valve which regulates the pressure of the fuel fed to the carburetor. On the armored cargo carrier (LVT (A) (2)), and on the unarmored cargo carrier (LVT (2)), the bypass line from the carburetor runs up and taps into the right fuel filter pipe just beneath the drip shield. On the armored tank (LVT (A) (1)),

TM 9-775

LANDING VEHICLE TRACKED MK. I AND MK. II

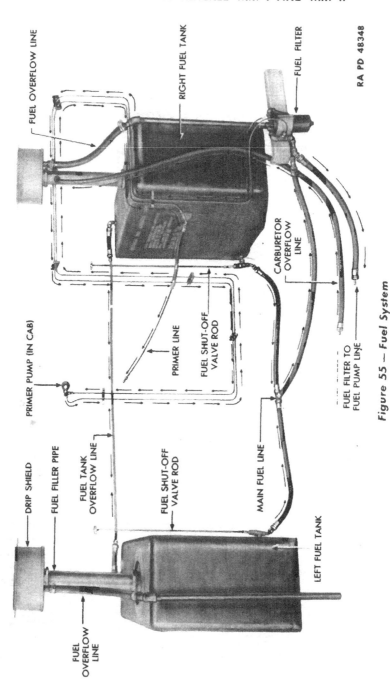

Figure 55 — Fuel System

FUEL SYSTEM

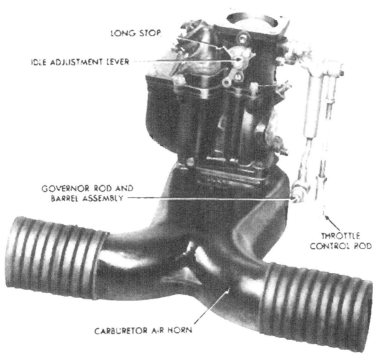

Figure 56 — Carburetor Removed — Front View

the bypass line runs to the auxiliary generator fuel tank, providing a source of fuel for the tank. No other method of filling the auxiliary generator tank is used. An overflow line runs from the auxiliary generator tank to the right fuel filler pipe.

c. **Primer Lines** (fig. 55). A fuel line, connected at the fuel filter, runs to the priming pump located in the cab. Upon operation of the pump plunger, the fuel is drawn into the pump and forced back to the priming distributor located on the engine, and then through the priming lines from the distributor to each of the 7 cylinders.

d. **Fuel Shut-off Valve** (fig. 55). A fuel shut-off valve is provided for each of the two main fuel tanks. The valve control rods are mounted in a vertical position at the side of each fuel tank. Access to the fuel shut-off valve rod handles is provided through the louvers in the stern bulkhead. To operate the valves, remove the outer stern bulkhead louver, and turn the handles in a clockwise direction to shut off the flow of fuel; and in a counterclockwise direction to turn on the flow of fuel.

TM 9-775
67

LANDING VEHICLE TRACKED MK. I AND MK. II

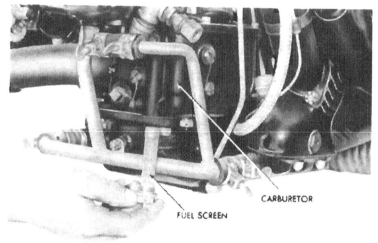

Figure 57 — Removing Fuel Screen

67. CARBURETOR.

a. Description. A single-barrel type NA-R6B (BS) carburetor is attached at the lowest point of the crankcase (fig. 26). The fuel mixture from the carburetor passes into the induction system of the crankcase rear section, from which it is passed to the cylinder intake valve ports through individual intake pipes.

b. Idle Adjustment (fig. 56). An idle adjustment lever is located on the carburetor body. Moving this lever laterally on its quadrant enriches or thins the idle mixture. When the upper end of the lever is against the long stop, the leanest point has been reached. A throttle stop on the throttle shaft makes it possible to adjust the engine idle speed to the correct revolutions per minute. To obtain a smooth idle, make both adjustments when the engine is hot.

c. Inspection and Cleaning (fig. 57). Remove fuel screen below fuel inlet of carburetor by cutting safety wire and unscrewing screen. Inspect screen and clean out any dirt or water which may have accumulated. Inspect entire carburetor to see that all parts are tight and properly locked. Apply a small quantity of oil in pump mechanism.

d. Removal (fig. 58). Disconnect and remove carburetor air horn scavenger line. Disconnect throttle control rod and governor rod at arm on governor (both ends of arm). Disconnect oil line at rocker box covers, and remove nut from bracket that secures oil line to carburetor air horn. Disconnect fuel pump to carburetor fuel line,

FUEL SYSTEM

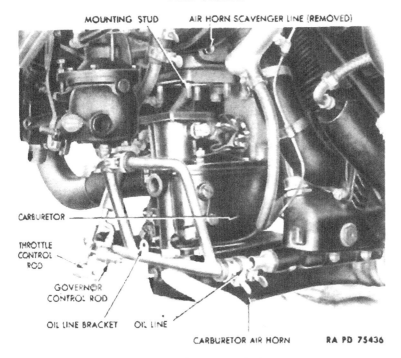

Figure 58 — Removing Carburetor

and unscrew brass fuel line fitting at carburetor. Remove safety wire from four castle nuts which secure carburetor to rear crankcase section. Unscrew and remove four castle nuts and four plain washers, and lift off carburetor. Strip off carburetor to crankcase gasket.

c. **Installation** (fig. 58). Place a gasket on carburetor flange of crankcase rear section. Place carburetor in position, and install four castle nuts and plain washers which secure carburetor to crankcase. Secure nuts with safety wire. Apply white lead base antiseize compound to threads, and install brass fuel line fitting at carburetor. Connect fuel pump to carburetor fuel line. Connect oil line at rocker arm box covers, and install nut on stud that secures carburetor air horn to oil line. Connect throttle control rod and governor rod at arm on governor. Install and connect carburetor air horn scavenger line.

68. ACCELERATOR PEDAL LINKAGE.

a. **Accelerator Pedal Adjustment.** With accelerator pedal fully depressed and against stop, loosen jam nuts and adjust turnbuckle (under power take-off support case cover) so that carburetor will be wide open without extending governor rod and barrel assembly

TM 9-775
68—69

LANDING VEHICLE TRACKED MK. I AND MK. II

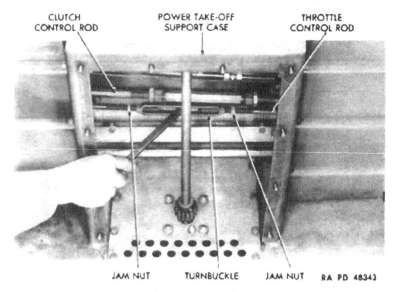

Figure 59 — Adjusting Throttle Control Rod

(fig. 59). The governor rod is spring-loaded to cushion carburetor butterfly valve arm against possible damage caused by sudden and severe opening of accelerator pedal. When accelerator pedal is fully depressed, spring on governor rod must not be completely extended.

69. FUEL PUMP.

a. **Description** (fig. 60). The fuel pump is mounted on the flange of the rocker scavenger oil pump, located on the accessory case to the right of the generator drive pulley. Fuel is pumped from the fuel tank through the fuel filter and to the carburetor.

b. **Maintenance.** No maintenance of the fuel pump should be attempted by using arm personnel. If the pump becomes inoperative, replace it with a serviceable unit.

c. **Removal.** Disconnect the fuel line (from fuel pump to carburetor) at the fuel pump. Disconnect the fuel line (from the fuel pump to fuel filter) at the fuel pump. Unscrew the palnuts on the four studs which hold the flange of the fuel pump to the mounting flange on the rocker scavenger oil pump. Unscrew the plain hexagonal nuts on the four studs, and remove nuts and washers. Lift off fuel pump and strip off the gasket.

d. **Installation.** Put a new gasket on the studs on the rocker scavenger oil pump. Place the fuel pump in position on the flange of the rocker scavenger oil pump. Place the clamp, which secures the

166

TM 9-775
69-70

FUEL SYSTEM

Figure 60 — Removing Fuel Pump

governor line, over the lower right stud. Install four plain washers and hexagonal nuts on the studs. Install four palnuts and tighten slightly. Connect the fuel lines.

70. FUEL FILTER.

a. Description. Fuel is cleaned by means of a fuel filter (CU) mounted in a bracket just inside the lower right rear baffle in the engine room. Fuel is drained into the filter from the main fuel line, and is forced through the filter element or cartridge. The cartridge consists of a series of fine, closely spaced, metal disks. The accumulated dirt is removed from the edges of the disks when the cartridge is turned by means of an external cleaning handle (fig. 64).

b. Maintenance. Turn the handle on top of the filter one complete turn daily to clean the disks. Access to the fuel filter drain is provided by a pipe plug located in the hull beneath the fuel filter (fig. 65). Remove the pipe plug, using a piece of 1-inch square stock 3 inches long, and an open-end wrench. Remove the fuel filter drain plug and copper washer, and drain the filter of accumulated dirt.

c. Removal of Fuel Filter. Open hinged rear cover. Disconnect the main fuel line at the elbow attached to the fittings which extend through the lower right rear baffle (from the fuel filter) (fig. 61). Disconnect the fuel line (fuel pump to fuel filter) at the fuel pump,

TM 9-775
LANDING VEHICLE TRACKED MK. I AND MK. II

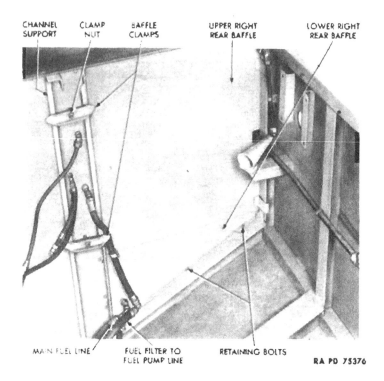

Figure 61 — Fuel Lines to Fuel Filter Through Baffles

then disconnect the same fuel line at the elbow fitting outside the lower right rear baffle. It is necessary to disconnect first at the fuel pump because the swivel nut is used only at the fuel pump. The entire line must be turned in order to disconnect it at the opposite end. Loosen the clamps (upper and lower) which hold the upper right rear baffle firmly to the channel support between the right front and right rear baffles. Rotate the clamps to a vertical position. Pry the upper right rear baffle out, and remove. Remove the three nuts, bolts, flat washers, and toothed lock washers which secure the bottom edge of the baffle to the hull. Remove the two nuts, bolts and toothed lock washers which secure the baffle to the channel support between the right front and right rear baffles. Work the baffle off the fittings attached to the fuel filter, and remove the baffle. Disconnect the primer line at the top of the fuel filter (fig. 62). Remove the four nuts, bolts, and toothed lock washers which secure the filter and assembled clamp bracket to the fuel filter bracket extending back from the hull. Lift the assembled filter and bracket from the vehicle

FUEL SYSTEM

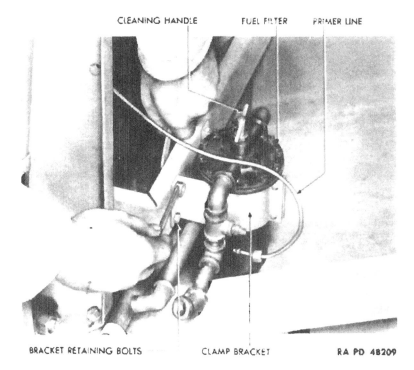

Figure 62 — Removing Fuel Filter

(fig. 62). Unscrew and remove the pipe fittings which extend from one side of the fuel filter across and beneath the bracket. Loosen the two screws which clamp the fuel filter in the bracket, and slide the fuel filter up and out of the bracket.

d. **Installation of Fuel Filter.** Place the filter in the fuel filter bracket (fig. 64). Apply white lead base antiseize compound to threads, and connect the 8-inch pipe fittings and elbow to the lower elbow on the fuel filter (figs. 63 and 64). When installed, the fittings extend forward from the filter through the cut-out hole in the baffle. Secure the filter assembly and bracket to the angle support bolted to the hull (fig. 62). Secure the fuel filter in its bracket by tightening the two cross-recess type screws and nuts (fig. 64). Manipulate the baffle into position by sliding the cut-out holes over the pipe fittings on the fuel filter. Secure the bottom edge of the baffle to the hull. Secure the baffle to the channel support (fig. 61). Connect the main fuel line (flexible hose) to the elbow on the fittings which extends through the baffle. Secure the line to the baffle with metal clamp. Apply white lead base antiseize compound to threads, and

LANDING VEHICLE TRACKED MK. I AND MK. II

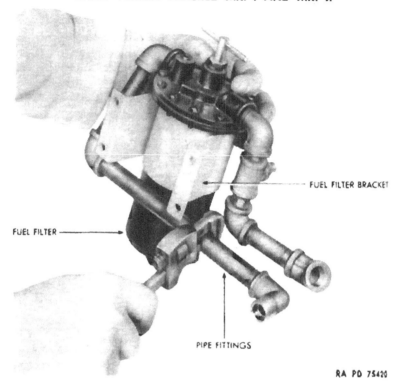

Figure 63 — Removing Fuel Filter Pipe Fittings

connect fuel return line (flexible hose to fuel pump) to the fittings extending through the rear hole in the baffle. Connect the $\frac{3}{16}$-inch copper primer line at the top of the fuel filter. Place the baffle in position with the two permanently installed brackets hooking over the hull support. Turn the two clamps on the channel support to a horizontal position, and tighten the two nuts (fig. 61).

71. FUEL TANKS.

a. **Description.** Two bullet-sealing fuel tanks are used in the armored models of the landing vehicles tracked (LVT (A) (1) and LVT (A) (2)). Steel fuel tanks are used on the unarmored landing vehicle (LVT (2)). A 1-gallon armored plate fuel tank (for the auxiliary generator) is used only in landing vehicle tracked with turret (LVT (A) (1)). The left fuel tank is located in the bulkheaded area, between the engine room and the hull, just in front of the batteries. The right fuel tank is located in the bulkheaded area, between the engine room and the hull, just in front of the fixed fire extinguishers.

FUEL SYSTEM

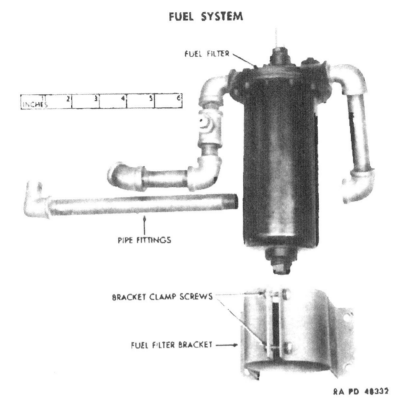

Figure 64 — Fuel Filter and Bracket

The auxiliary generator fuel tank is located above the fixed fire extinguishers.

b. **Data.** Make, type, and capacity of main fuel tanks are as follows:

	LVT (2)	LVT (A) (2)	LVT (A) (1)
Make	Terne Plate	Goodrich	Goodrich
Type	Steel	Self-sealing	Self-sealing
Capacity	110 gal	104 gal	104 gal

c. **Drains.** A sump drain plug is located at the bottom of each main fuel tank. Drain the tanks by removing pipe plugs in the hull beneath the sump drain plugs on the fuel tanks. The procedure of draining the tanks is covered under the removal of fuel tanks (par. 71).

d. **Removal.**

(1) GENERAL. The removal procedures for the left and right fuel tanks are identical, with the exception that the oil filter must also

TM 9-775
71

LANDING VEHICLE TRACKED MK. I AND MK. II

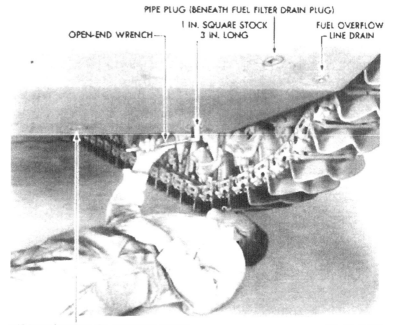

Figure 65 — Removing Hull Pipe Plug Beneath Fuel Tank-Sump Drain

be removed in order to remove the left fuel tank. Before proceeding with removal of either tank, close both fuel shut-off valves. Pump or siphon most of fuel from tanks, then remove hull pipe plugs from beneath each fuel tank (fig. 65). Remove fuel tank sump drain plugs, and drain remainder of fuel from both tanks.

(2) Remove engine (par. 61).

(3) REMOVE OIL FILTER (par. 84). It is necessary to remove the oil filter only when the left fuel tank is to be removed.

(4) REMOVE FIRE EXTINGUISHER LINE AND SHIELDED NOZZLE (fig. 66). Remove the two cap screws and toothed lock washers which secure the fire extinguisher shielded nozzle to the bracket on the left front baffle. Disconnect the fire extinguisher line (to the nozzle) at the fitting in the channel support between the upper left front and rear baffles.

(5) REMOVE ENGINE MOUNT. Remove the lacing wire and eight cap screws and flat washers (two each corner) which secure the engine mount to brackets on the bottom of the hull. Remove the cotter pin and rod end pin which secure throttle rod to the throttle arm attached to the engine mount. Loosen the jam nut on the throttle rod, and

TM 9-775
71

FUEL SYSTEM

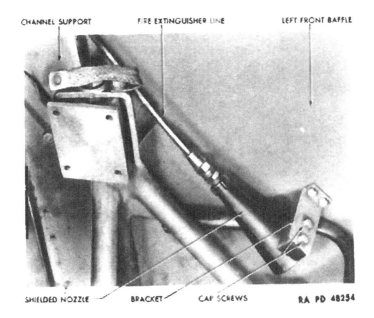

Figure 66 — Fire Extinguisher Shielded Nozzle

unscrew and remove the adjustable rod end and jam nut on the throttle rod (fig. 68). Loosen the jam nut at the opposite end of the throttle rod, then unscrew the throttle rod, and pull it out from its bushing in the engine mount (fig. 68). Attach a sling to the engine mount, then attach a hoist to the sling. Carefully lift the engine mount up and out of the vehicle (fig. 67).

(6) REMOVE LOWER FRONT BAFFLE (fig. 69). Remove the two bolts, nuts, flat washers, and toothed lock washers which secure the center bottom of the left front baffle. Slide the baffle up and out of the vehicle.

(7) REMOVE LEFT FRONT BAFFLE. Remove the six bolts, nuts, flat washers, and toothed lock washers which secure the left front baffle. Lift out the baffle.

(8) REMOVE CHANNEL SUPPORT (fig. 70). Disconnect fire extinguisher line that leads from the hull through the channel support. Remove the four nuts, bolts, and toothed lock washers (two at top and two at bottom), and the two cap screws and toothed lock washers (at the center), which secure the channel support between the front and rear baffles. Lift out the channel support.

(9) REMOVE AIR CLEANER (fig. 71). Remove the two stove bolts and lock washers from the two clamps which hold the air cleaner in position. Spread the clamps, and lift out the air cleaner.

173

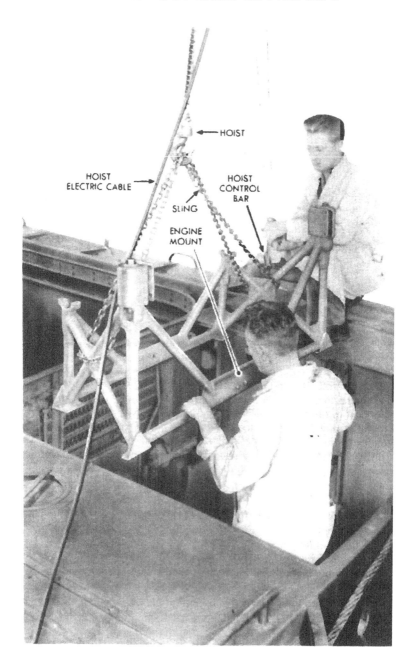

Figure 67 — Removing Engine Mount

FUEL SYSTEM

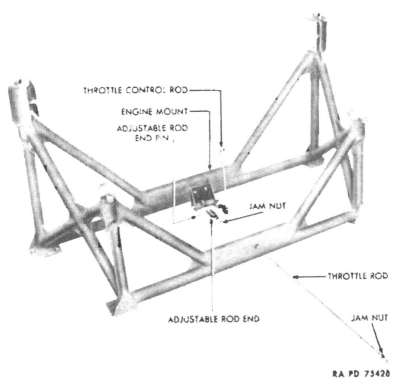

Figure 68 — Engine Mount

(10) REMOVE FUEL TANK FILLER PIPE ASSEMBLY. Remove the two bolts, nuts, and toothed lock washers which secure the channel support bracket to the hull (fig. 71). Lift out the bracket. Remove the eight nuts, bolts, and toothed lock washers which secure the fuel tank filler pipe packing gland flange to the hull (two men) (fig. 72). Disconnect the carburetor bypass line to the side of the pipe (on right tank only; pipe plug is used in left tank fuel filler pipe.) Loosen the lock nut at base of fuel tank filler pipe (fig. 73). Unscrew the fuel tank filler pipe assembly, and lift the assembly up and out of the fuel tank and from the vehicle (fig. 74).

(11) REMOVE FUEL TANK CLAMP (fig. 73). Remove the four fuel tank clamp bolts, nuts, and toothed lock washers. The two bolts in a vertical position, up and down the outer side of the fuel tank, must be removed completely. The two horizontal clamp bolts swivel up at their outer ends and, after being loosened, may remain in the clamp. Lift the clamp from the fuel tank.

(12) DISCONNECT FUEL SHUT-OFF VALVE (fig. 75). Remove the cotter pin which secures the universal joint of the fuel shut-off

LANDING VEHICLE TRACKED MK. I AND MK. II

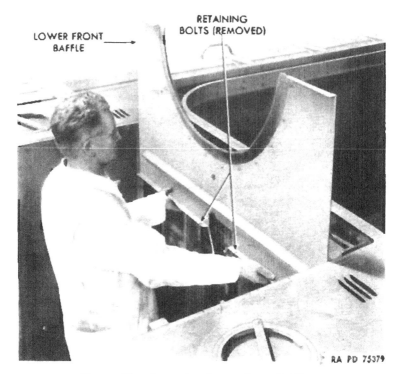

Figure 69 — Removing Lower Front Baffle

valve rod to the universal joint on the fuel shut-off valve. Lift the fuel shut-off valve handle and rod up and out of the way, and wire it in position. Disconnect the flexible tubing (main fuel outlet line) at base of fuel shut-off valve.

(13) DISCONNECT FUEL TANK OVERFLOW LINE (fig. 75). Disconnect the copper fuel tank overflow line at the fitting on top of the fuel tank.

(14) DISCONNECT THE FUEL OVERFLOW LINE (fig. 75). Loosen the hose clamp which secures the top of the fuel overflow line to the drip shield, and work the flexible tubing free. Rotate the elbow fitting at the lower end of the flexible tubing to one side if it interferes with the removal of the fuel tank.

(15) REMOVE OIL FILTER RETURN LINE (figs. 70 and 92). It is necessary to remove the oil filter return line (to the oil cooler) in order to provide room for the removal of the fuel tank. Disconnect the oil filter return line (oil tank inlet line) at fitting adjacent to bottom of oil reservoir. Remove the two halves of the cable and tube block which hold the oil filter outlet line, main fuel line, and oil tem-

TM 9-775
71

FUEL SYSTEM

Figure 70 — Left Fuel Tank Compartment

TM 9-775

LANDING VEHICLE TRACKED MK. I AND MK. II

Figure 71 — Removing Air Cleaner

perature indicator line in position. Remove the one clamp which also secures the return line to the bottom of the hull.

(16) REMOVE FUEL TANK. Lift the left fuel tank straight up approximately 8 inches. This will free the sump drain fitting from the bottom of the fuel tank compartment (fig. 75). Tilt the fuel tank upward, and pull the rear end of the fuel tank outward, in order to remove the fuel tank from the vehicle. Remove the two masonite sheet spacers that are beneath the fuel tank (fig. 76).

e. Installation.

(1) GENERAL. The two fuel tanks are installed in the same general manner. A few slight differences in the installation procedures are covered where they may occur in the text. In the following procedure the left fuel tank is installed.

(2) INSTALL COTTON BELTING (fig. 76). If cotton belting has been removed previously, install new belting to prevent the tank from rubbing or chafing. Cement the belting to brackets and panels.

(3) INSTALL TWO MASONITE SHEETS (fig. 76). Place the two masonite sheets in position. The hole in the forward sheet must be in the forward position beneath the tank sump drain fitting.

TM 9-775
71

FUEL SYSTEM

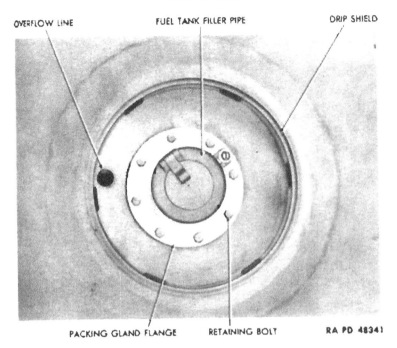

Figure 72 — Fuel Tank Filler Pipe Installed — Top View

(4) PLACE FUEL TANK IN POSITION (REQUIRES TWO MEN). Place fuel tank in position so the sump drain fitting extends down through the opening in the masonite sheet (fig. 76).

(5) CONNECT FUEL SHUT-OFF VALVE (fig. 75). Connect flexible tubing (main fuel outlet line) at base of fuel shut-off valve. Remove wire securing fuel shut-off valve rod. Drop rod down into position on fuel shut-off valve, and install cotter pin.

(6) CONNECT FUEL TANK OVERFLOW RETURN LINE (fig. 75). Connect the copper fuel tank overflow return line to the fitting on top of forward end of the fuel tank.

(7) CONNECT THE FUEL OVERFLOW LINE (fig. 75). Install the flexible rubber hose, and tighten the clamp at the top of the line. Adjust bottom elbow fitting to position.

(8) INSTALL FUEL TANK CLAMP (fig. 73). Place the L-shaped steel clamp in position. Install the two top horizontal bolts and the two vertical bolts. Install nuts and toothed lock washers loosely.
CAUTION: *Do not tighten bolts until filler pipe assembly has been installed.*

(9) INSTALL FUEL TANK FILLER PIPE ASSEMBLY. Cement a new filler pipe packing gland flange gasket in position in the drip shield.

179

TM 9-775
71

LANDING VEHICLE TRACKED MK. I AND MK. II

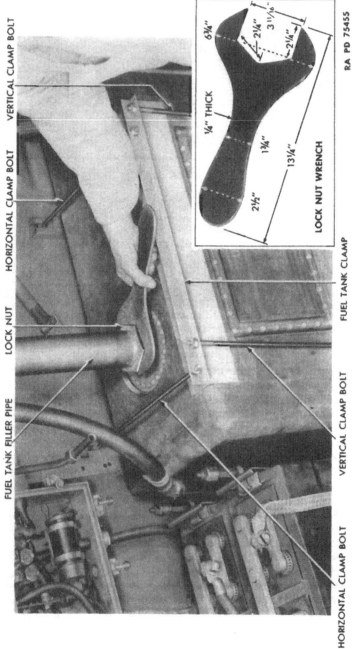

Figure 73 — Loosening Fuel Tank Filler Pipe Lock Nut

FUEL SYSTEM

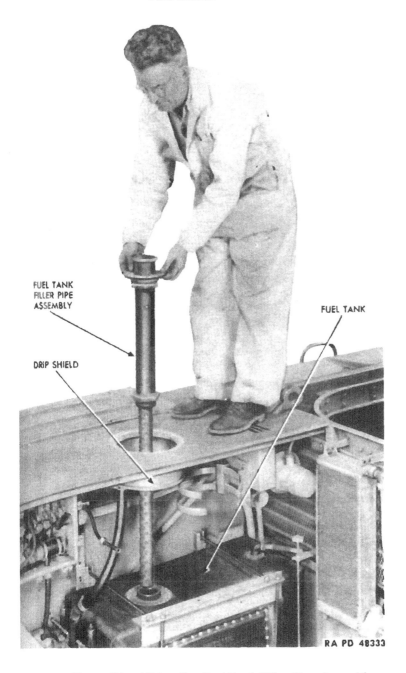

Figure 74 — Lifting Out Fuel Tank Filler Pipe Assembly

TM 9-775

LANDING VEHICLE TRACKED MK. I AND MK. II

Figure 75 — Left Fuel Tank Ready for Removal — Disconnections Made

FUEL SYSTEM

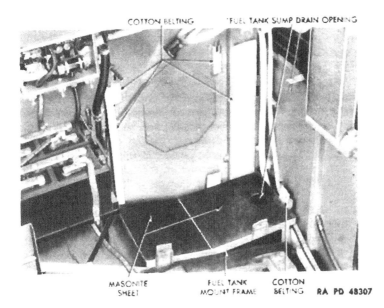

Figure 76 — Left Fuel Tank Compartment — Fuel Tank Removed

Apply white lead base antiseize compound to threads, and insert the screen end of the fuel filler pipe assembly into the fuel tank (fig. 74). Screw the pipe into the brass fitting on the fuel tank. Line up the eight holes of the filler pipe packing gland flange with the holes in the drip shield (fig. 74). Install eight nuts, bolts, and toothed lock washers. Tighten the lock nut at the fitting (fig. 73). Tighten the fuel tank clamp bolts alternately so the clamp fits snugly against the tank. Install filler pipe cover on drip shield. Connect carburetor overflow line to side of filler pipe (right fuel tank filler pipe only).

(10) INSTALL CHANNEL SUPPORT (fig. 70). Place the channel support in position and install the four securing nuts, bolts, and toothed lock washers (two at top and two at bottom).

(11) INSTALL CHANNEL SUPPORT BRACKET (fig. 71). Place bracket in position and install the two nuts, bolts, and toothed lock washers which secure the angle support to the hull. Then install the two cap screws and toothed lock washers which secure the channel support bracket to the channel support.

(12) INSTALL AIR CLEANER (fig. 71). Spread the two clamps, and place air cleaner in position. Install the two stove bolts and lock washers in the two clamps, and tighten.

(13) Install oil filter (par. 84).

LANDING VEHICLE TRACKED MK. I AND MK. II

(14) INSTALL LEFT FRONT BAFFLE. Slide the baffle into place behind the oil filter line, allowing the air cleaner intake duct to extend through the hole in the baffle. Install the six studs, nuts, and toothed lock washers, and the three flat washers which secure the baffle.

(15) INSTALL FIRE EXTINGUISHER LINE AND SHIELDED NOZZLE (fig. 66). Place nozzle in position, and install two cap screws and toothed lock washers which secure the nozzle to its bracket on the left front baffle. Connect the fire extinguisher line to the fitting on the channel support. Connect the fire extinguisher line at fitting on inside of channel support (fig. 71).

(16) INSTALL LOWER FRONT BAFFLE (fig. 69). Slide the baffle into place, pushing down so it rests on the hull support. Install the two nuts, bolts, flat washers, and toothed lock washers which secure the bottom of the baffle to the hull.

(17) INSTALL ENGINE MOUNT (figs. 67 and 68). Attach chain hoist to engine mount, and lower it so it rests on the four brackets on the bottom of the hull in the engine compartment. Install the eight cap screws and flat washers (two each corner) which secure the engine mount to brackets on bottom of the hull. Safety-wire the eight cap screws. Install jam nut on front (standard thread) end of rod (fig. 68). Slide the throttle rod through the holes in the lower front baffle and the engine mount. Install the jam nut on the rear (SAE thread) end of the rod as far as it will go. Install the adjustable rod end on the rear end of the rod by screwing it on as far as it will go. Connect the front end of the rod to the turnbuckle just forward of the baffle. Install the rod end pin which secures the throttle rod to the throttle arm on the engine mount. Install cotter pin in the rod end pin.

72. AIR CLEANERS.

a. Description. Two oil-bath air cleaners (VR models 3520D and 3530D) are used, one mounted above each of the two fuel tanks. Air, drawn through the rear bulkhead louvers, enters the open intake in the upper section of the air cleaner. Dust-laden air then passes down through the air cleaner, against the pool of oil in the lower section, and drops most of the dirt into the oil. Air is then drawn up through the filter section, and out into the lower intake tube which is attached by a flexible hose to the carburetor air intake tube. Dirt is collected below the oil level disk in the bottom of the lower section.

b. Maintenance.

(1) GENERAL. Change the oil in the air cleaners daily, unless the vehicle has been operated in wet weather, snow, or under unusually dust-free conditions. If the vehicle has been operated under particularly dusty conditions, it may be necessary to service air cleaners

TM 9-775
72

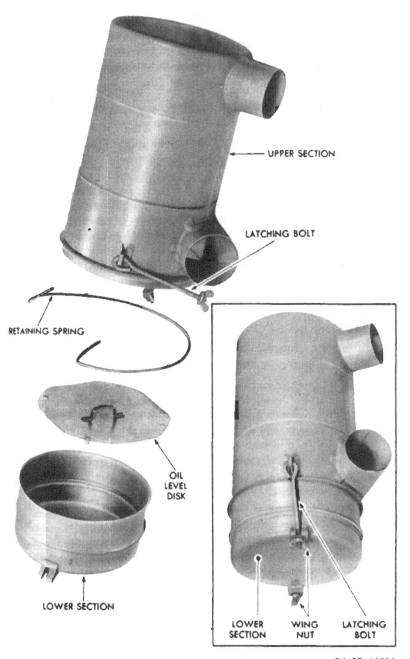

Figure 77 — Air Cleaner Assembly

LANDING VEHICLE TRACKED MK. I AND MK. II

even more frequently. In addition to the daily service given the air cleaners, make a regular and careful inspection of the entire air intake system. Check all flexible connections between the air cleaners and the carburetor air horns. NOTE: *A small hole or leak permits a large amount of dirt or dust to enter the engine.*

(2) CHANGING OIL IN AIR CLEANERS (fig. 77). To change the oil, lift out the rear bulkhead louver in front of the air cleaner, and loosen the two wing nuts on the latching bolts sufficiently to allow the bolts to be pulled free of the latches. This releases the lower section of the air cleaner. Remove the lower section, taking care not to tilt the section and spill oil in the compartment. Compress the retaining spring slightly and remove spring and oil level disk. Pour oil from the oil reservoir. Scrape out dirt. Clean the reservoir with dry-cleaning solvent. Fill oil reservoir up to the groove which holds the oil level disk, using engine oil SAE 30, for temperatures about 32° F, and engine oil SAE 10, for temperatures below 32° F. Snap the oil level disk into position, then install the retaining spring. Lift the lower section into place on the filter, hook the latching bolts and wing nuts beneath the latches, and tighten the two wing nuts. Repeat the operation on the other air cleaner.

c. *Removal.* Removal of the air cleaners is covered under the removal of the fuel tanks (par. 71).

d. *Installation.* Installation of the air cleaners is covered under the installation of the fuel tanks (par. 71).

73. PRIMING PUMP.

a. *Description.* A priming pump (DV) is used to inject a spray of fuel into the engine cylinders to facilitate starting. When the plunger is pulled out, it draws a charge of gasoline into the priming pump cylinder. When the pump plunger is pushed in, this gasoline is forced through the priming lines to the priming distributor on the engine. The priming distributor lines run to each engine cylinder.

b. *Maintenance.* If more than a few strokes are required to prime the engine, the leather packing on the end of the plunger should be checked for leakage. To stop leakage, compress packing by half turns on the packing nut located behind the priming plunger handle until the leakage stops (fig. 78). If the primer pump no longer delivers gasoline to the engine, as evidenced by lack of resistance to pump-handle operation, replace the primer pump.

c. *Removal* (fig. 78). Disconnect the two primer lines at the fittings on the priming pump. Mark or tag the lines to assure correct installation. At rear of priming pump and hand throttle bracket, unscrew the retaining nut that holds the pump to the bracket. Remove the packing nut on front end of pump barrel, and pull out plunger. Pull pump out of panel from rear.

FUEL SYSTEM

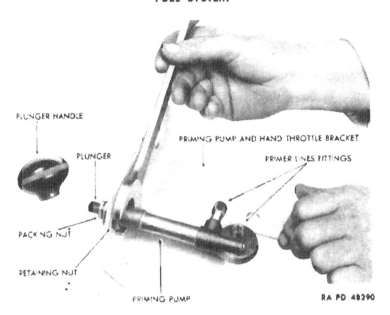

Figure 78 — Removing Priming Pump

d. **Installation** (fig. 78). Remove plunger and retaining nut from new or replacement priming pump, and install threaded end of barrel from rear of priming pump and hand throttle bracket. Install plunger and special nut on bracket end of pump and tighten the locking nut at the back. Connect the inlet and outlet lines at the fittings on the pump.

74. LINES, VALVES, AND FITTINGS.

a. **Fuel Lines.** Flexible fuel lines are used in all cases except for the fuel tank overflow line and the priming lines to the primer pump. These lines are copper. Inspect all lines frequently to make sure they are not leaking. When replacing a defective fuel line, cut new line the same length as the one which was removed. Make sure when replacing copper fuel lines that the new line conforms exactly to the shape of the original.

b. **Fuel Shut-off Valves.** A fuel shut-off valve (CR) is provided for each fuel tank. Fuel shut-off valve handles are located just behind the outer left and right stern bulkhead louvers. The valves themselves are located at the front bottom of each fuel tank where the main fuel line reaches the fuel tank.

LANDING VEHICLE TRACKED MK. I AND MK. II

Section XVII

INTAKE AND EXHAUST SYSTEMS

	Paragraph
Description	75
Mufflers, flexible tubing, and connections	76
Engine exhaust system	77
Removal of engine room blower	78
Installation of engine room blower	79
Intake pipe gland nut	80

75. DESCRIPTION.

a. The exhaust system includes the engine exhaust system, mufflers, flexible tubing and connections, and the two engine room blowers. The function of the exhaust system is to provide an efficient means of disposing of exhaust gases created by the combustion of the gasoline in the engine.

76. MUFFLERS, FLEXIBLE TUBING, AND CONNECTIONS.

a. The mufflers, on one each side, are mounted on the rear of the hull, inside the engine compartment. Flexible tubing connects the exhaust manifolds to the mufflers, and an exhaust pipe extends through the rear of the hull to expel the exhaust gas to the atmosphere.

b. The removal and installation of mufflers, flexible tubing, and connections are covered under the removal of the engine (par. 61).

77. ENGINE EXHAUST SYSTEM.

a. Description (figs. 79 and 80). The engine exhaust system is composed of two groups of exhaust pipes. The right group has three branches that are attached to the exhaust elbow of No. 2, No. 3, and No. 4 cylinders. The left group has four branches that are attached to the exhaust elbow of No. 1, No. 7, No. 6, and No. 5 cylinders. The branches of each group merge into a single stack (one stack for each branch) that attaches to flexible tubing, which is connected to the muffler.

b. Removal.

(1) REMOVE RIGHT EXHAUST PIPES (figs. 79 and 80). Remove safety wire, brass castle nut, and plain washer from each of the six studs that hold the right exhaust pipes in place on the exhaust elbows of No. 2, No. 3, and No. 4 cylinders. Lift the upper exhaust pipe off the elbow of No. 2 cylinder. Lift the lower exhaust pipe off the

INTAKE AND EXHAUST SYSTEMS

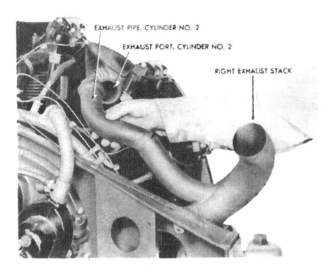

Figure 79 — Removing Exhaust Pipe — No. 2 Cylinder

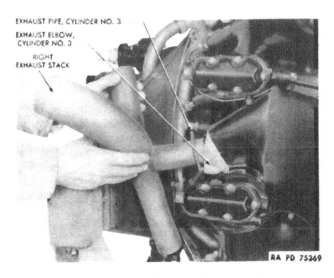

Figure 80 — Removing Exhaust Pipe — No. 3 Cylinder

elbow of No. 4 cylinder, and manipulate the exhaust pipe (and stack) off the elbow of No. 3 cylinder.

(2) REMOVE LEFT EXHAUST PIPES. Remove safety wire, brass castle nut, and plain washer from each of the eight studs that hold the left exhaust pipe assembly in place on the exhaust elbows of No. 1, No. 7, No. 6, and No. 5 cylinders. Lift the upper exhaust pipe off the elbow of No. 1 cylinder. Lift the exhaust pipe (and stack) off the elbow of No. 7 cylinder. Lift the exhaust pipes off the elbows of No. 6 and No. 5 cylinders.

(3) INSTALL PORT COVERS. Install port covers on all exhaust elbows to prevent dirt from entering the cylinders.

c. Installation.

(1) REMOVE PORT COVERS. Remove the nuts which hold the exhaust elbow port covers in place on the studs, and remove the covers.

(2) INSTALL RIGHT EXHAUST PIPES (figs. 79 and 80). Place the right exhaust pipe assembly (three pieces) in position on the exhaust elbow mounting pads of No. 4, No. 3, and No. 2 cylinders, in that order. Install a plain washer and brass castle nut on each of the six studs that secure the flanges of the right exhaust pipe assembly. Safety-wire the nuts.

(3) INSTALL LEFT EXHAUST PIPES. Place the left exhaust pipe assembly (four pieces) in position on the exhaust elbow mounting pads of No. 5, No. 6, No. 7, and No. 1 cylinders, in that order. Install a plain washer and brass castle nut on each of the eight studs that secure the flanges of the left exhaust pipe assembly. Safety-wire the nuts.

78. REMOVAL OF ENGINE ROOM BLOWER.

a. General. Two identical engine room blowers (TA-361-S) are used, one on each side of the engine room. Because the blowers are identical, one must necessarily be installed upside down (in relation to the other) in order to perform the same function. However, the blowers are removed in the same manner.

b. Remove bolted stern cover (par. 61).

c. Remove the Bulkhead Louver. Lift out the bulkhead louver in front of the blower to be removed.

d. Disconnect Electrical Lead. Remove two screws securing the blower motor cover. Lift off the cover (fig. 81). Lift up the paper insulation cover which protects the terminal stud nut at the base of the blower motor (fig. 81). Remove the outer terminal stud nut, thus disconnecting the blower electrical lead (fig. 82). Unscrew the

TM 9-775
78

INTAKE AND EXHAUST SYSTEMS

Figure 81 — Engine Room Blower Installed

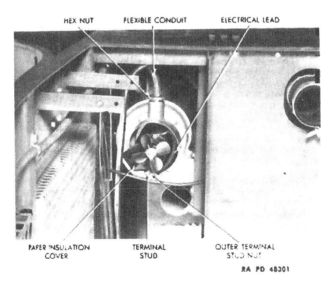

Figure 82 — Engine Room Blower — Cover Removed

191

TM 9-775
78-79

LANDING VEHICLE TRACKED MK. I AND MK. II

Figure 83 — Removing Gas Exhaust Duct Bottom

flexible conduit retaining hexagonal nut, and pull the lead out of the blower (fig. 81).

e. **Remove the Gas Exhaust Duct Bottom** (fig. 83). Remove the four nuts, bolts, and toothed lock washers that secure the gas exhaust duct bottom. It is necessary to remove the gas exhaust duct bottom in order to remove the four nuts and bolts securing the blower to gas exhaust duct.

f. **Removing Engine Room Blower** (fig. 83). While using one wrench and working up in the gas exhaust duct to hold the nuts, use another wrench to remove the four nuts, bolts, and toothed lock washers which secure the blower to the gas exhaust duct. Work the blower in toward the engine slightly to free it from the horizontal intake duct, then remove the blower and horizontal intake duct.

79. INSTALLATION OF ENGINE ROOM BLOWER.

a. **Install the Engine Room Blower and Horizontal Intake Duct.** Place the blower with the horizontal intake duct attached in position,

192

INTAKE AND EXHAUST SYSTEMS

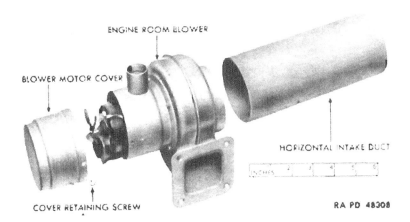

Figure 84 — Engine Room Blower

and secure to gas exhaust duct with four bolts, nuts, and toothed lock washers (fig. 83). It is necessary that this operation be done before the gas exhaust duct bottom is installed, in order to have access to the four nuts securing the blower to the gas exhaust duct.

b. **Install the Gas Exhaust Duct Bottom.** Using four bolts, nuts, and toothed lock washers, secure the gas exhaust duct bottom to duct (fig. 83).

c. **Install Electrical Lead.** If the blower motor cover is installed, remove the two screws securing it, and lift off the cover. Slide the conduit retaining hexagonal nut up the conduit, and force the electrical lead through the opening on top of the blower. Place lead on terminal stud and secure with outer terminal stud nut (fig. 82). Place paper insulation cover over terminal stud. Install the blower motor cover using two screws (fig. 81). Tighten the conduit retaining hexagonal nut.

d. **Install the Bulkhead Louver.** Slide the upper edge of the louver up and behind the bulkhead. Pull louver down so that the lower edge is resting under the two retaining clips.

e. **Install the Bolted Stern Cover.** Secure the bolted stern cover with 16 cap screws and toothed lock washers.

80. **INTAKE PIPE GLAND NUT.**

a. **Tighten Intake Pipe Gland Nut** (fig. 85). Use intake pipe gland nut wrench (41-W-1536-500) to tighten intake pipe gland nut which secures base of intake pipe to crankcase.

Figure 85 — Tightening Intake Pipe Gland Nuts

Section XVIII
COOLING SYSTEM

	Paragraph
Description	81
Flywheel and fan assembly	82

81. DESCRIPTION.

a. The engine is cooled by an air blast produced by a fan mounted on the engine flywheel. When the engine is running, the fan draws air through the louvers in the stern bulkhead, and forces it between and around the finned cylinders of the engine. Warm air then passes out through the grilled, hinged stern cover. Air ducts are formed on the engine by baffles bolted around and between each cylinder and cylinder head. A fan shroud on the engine forms a further duct for the inlet of air through the stern bulkhead louvers. Upper and lower baffles which meet around the fan shroud force all air drawn in by the fan to pass around the engine cylinders.

82. FLYWHEEL AND FAN ASSEMBLY.

a. Description. A flywheel and fan assembly (one unit) is attached to the crankshaft on the forward side of the engine. The flywheel and fan unit fits inside the fan shroud, and there must be 0.125-inch clearance between the fan blades and the fan shroud. The clutch is attached to the forward side of the flywheel.

b. Maintenance. Keep fan blades clean. Check clearance between fan blades and fan shroud (subpar. a, above).

Section XIX
LUBRICATION SYSTEM

	Paragraph
Description	83
Oil filter	84
Oil coolers	85
Oil reservoir	86
Engine and transmission oil lines	87
Oil pressure and scavenger pump	88
Oil screens	89

83. DESCRIPTION.

a. Lubrication of the engine depends upon the forced circulation of oil from a remote oil reservoir, through the engine to the oil pump,

TM 9-775
83

LANDING VEHICLE TRACKED MK. I AND MK. II

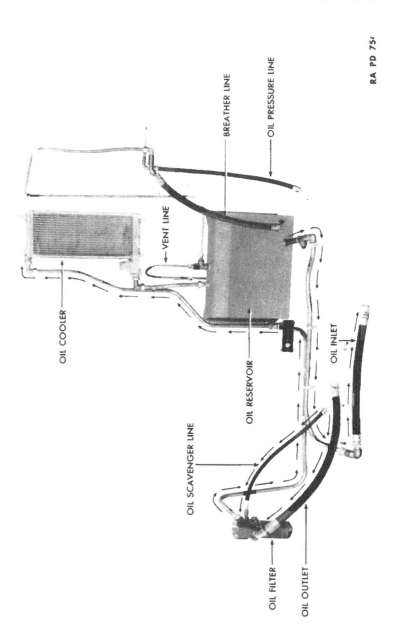

Figure 86 — Engine Oil System

TM 9-775
83-84

LUBRICATION SYSTEM

out to the oil filter, through the oil cooler, then back to the oil reservoir (fig. 86). The amount of pressure built up in the system is determined by an oil pressure regulator valve located in the oil pump. Connected to the accessory housing is a breather line which runs to a fitting on top of the oil reservoir. To this fitting is attached a vent line which vents the reservoir to the atmosphere.

84. OIL FILTER.

a. Description. An automatic, disk-type oil filter (CU) is used. When the engine is operating, the pressure of the oil automatically turns the cleaning plates within the filter.

b. Maintenance. Remove the sump drain plug at the bottom of the oil filter, and drain and clean the pump of the filter (fig. 89). Check operation of filter by removing cap at top of filter, and turning the plates by hand. Clogging of the filter is indicated by stiff action, and should be reported to ordnance personnel.

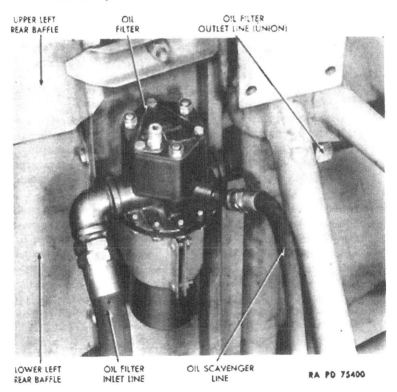

Figure 87 — Oil Filter Installed

197

TM 9-775

LANDING VEHICLE TRACKED MK. I AND MK. II

Figure 88 — Removing Oil Filter and Bracket

c. Removal.

(1) DISCONNECT OIL LINES (fig. 87). Open hinged rear cover. Disconnect the oil filter inlet line at the engine, then disconnect the line at the filter. This procedure is necessary, as a swivel fitting is used only at the engine end of the line. Disconnect the oil scavenger line at the engine, then disconnect the line at the filter. Disconnect the oil filter outlet line at the oil filter.

(2) REMOVE OIL FILTER. Open hinged rear cover. Remove upper left rear baffle (par. 64). Remove screws which secure the lower left rear baffle and lift off the baffle. Remove four bolts, nuts, and toothed lock washers which secure the oil filter bracket to the channel support between the front and rear baffle (fig. 88). Lift off the assembled oil filter and bracket.

(3) REMOVE OIL FILTER BRACKET (fig. 89). Loosen the two clamp screws which tighten the oil filter bracket on the oil filter. Slide the oil filter out of the oil filter bracket.

d. Installation.

(1) INSTALL OIL FILTER BRACKET (fig. 89). Tap the oil filter bracket to its correct position on the body of the filter (nearly flush against the base). CAUTION: *Do not tighten the two clamp screws until lines are connected.*

TM 9-775
84-85

LUBRICATION SYSTEM

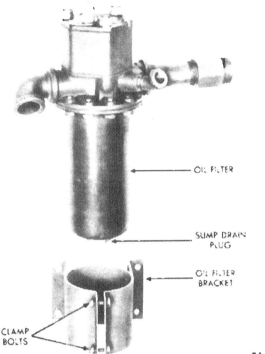

Figure 89 — Oil Filter and Bracket

(2) ATTACH ASSEMBLED OIL FILTER AND BRACKET (fig. 87). Place the assembled oil filter and bracket in position and install the four bolts, nuts, and toothed lock washers which secure the oil filter bracket to the channel support.

(3) CONNECT OIL LINES (fig. 87). Connect oil filter outlet line (to oil cooler) at fitting on forward side of the oil filter. Secure the line (oil filter to oil cooler) by clamp bolted to the hull support. Connect oil scavenger line at oil filter, then at engine. Connect oil filter inlet line at oil filter, then at engine. After the oil filter is installed and lines are connected, tighten the oil filter clamp screws. Place lower left rear baffle in position, and install retaining screws. Install upper left rear baffle (par. 64). Close hinged rear cover.

85. OIL COOLERS.

a. **Description.** Individual oil coolers are provided to cool engine oil and transmission oil. Oil enters at the top of the cooler, is forced

Figure 90 — Removing Oil Cooler

LUBRICATION SYSTEM

downward through finned passages, and returned to the system through the oil return line at the bottom. A spring-loaded relief valve allows cold or thick oil to pass directly to the oil return line without passing through the oil cooler passages. This occurs whenever the resistance of the oil is passing through the cooler is greater than the tension of the spring of the relief valve. Both oil coolers are located on the engine room side of the stern bulkhead (fig. 42).

b. Maintenance. Regularly remove and clean oil coolers. Clean out interior oil passages with steam, or flush with dry-cleaning solvent. Blow out fins and air passages with compressed air. Remove the bypass valve by turning out plug and lifting off spring and valve assembly. Clean all parts and check for free action before assembling. Use new parts if operation is faulty.

c. Removal.

(1) GENERAL. Both the transmission and engine oil coolers are identical in design and function. The removal of either cooler is accomplished in the same manner.

(2) DRAIN OIL. Remove the pipe plug in the hull beneath the oil reservoir, using a piece of 1-inch square stock about 3 inches long (fig. 65). Remove the drain plug in the bottom of the oil reservoir, and drain oil into a suitable container.

(3) REMOVE OIL COOLER (fig. 90). Remove bolted rear cover (par. 61). Disconnect the oil cooler inlet line at the elbow in the upper side of the cooler. Disconnect the oil cooler outlet line (to the oil reservoir) at the bottom of the cooler. Remove the eight bolts, nuts, and toothed lock washers holding the oil cooler to the four brackets welded to the engine room bulkhead. Lift off the oil cooler.

d. Installation. The transmission and engine oil coolers are identical and are installed in the same manner. Install the drain plug, and place cooler in position. Install eight nuts, bolts, and toothed lock washers holding cooler to the four brackets welded to the engine room bulkhead. Connect oil cooler outlet line (to oil reservoir) at bottom of cooler. Connect oil cooler inlet line to elbow fitting at top of the cooler. Install hull pipe plug. Install bolted rear cover (par. 62).

86. OIL RESERVOIR.

a. Description. An engine oil reservoir, of 23-quart capacity, is mounted on brackets on the floor of the engine room, just behind the stern bulkhead. Access to the oil reservoir is provided through an opening in the cargo compartment side of the stern bulkhead, covered by a plate secured by two wing nuts (fig. 9).

TM 9-775

LANDING VEHICLE TRACKED MK. I AND MK. II

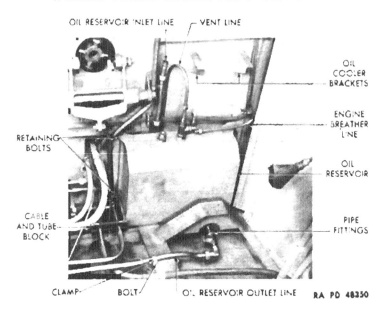

Figure 91 — Engine Oil Reservoir Ready for Removal

b. Removal.

(1) GENERAL. Drain the oil from the engine oil reservoir.

(2) Remove engine (par. 61).

(3) Remove engine mount (par. 63).

(4) Remove engine oil cooler (par. 85). NOTE: *Although the engine oil cooler is remote from the engine oil reservoir, it is necessary to remove the engine oil cooler in order to provide room to remove the oil reservoir inlet line from the oil cooler.*

(5) DISCONNECT OIL LINES (fig. 91). Disconnect the oil reservoir inlet line (from oil cooler) at top of oil reservoir. Disconnect the oil reservoir outlet line (to the engine) at the elbow fitting at the rear bottom of the oil reservoir, and at flexible oil line passing through channel support on hull bottom. Remove two clamp bolts, nuts, and toothed lock washers securing the two clamps which hold the line to the hull. Remove the line. Disconnect and remove the oil filter return line to the oil cooler (par. 84).

(6) REMOVE VENT LINE (fig. 91). Disconnect and remove the curved vent line at the top of the oil reservoir.

(7) DISCONNECT BREATHER LINE (fig. 91). Disconnect, from the engine, the breather line which taps into the fitting on top of the oil reservoir.

LUBRICATION SYSTEM

(8) REMOVE OIL RESERVOIR (fig. 91). Unscrew and remove the pipe fittings at the rear bottom of the oil reservoir. It is necessary to remove the fittings to provide room for the removal of the oil reservoir. Remove the four bolts, nuts, and toothed lock washers which secure the oil reservoir to the stern bulkhead. Lift out the oil reservoir.

c. **Installation.**

(1) PLACE RESERVOIR IN POSITION. Place reservoir in position and install four bolts, nuts, and toothed lock washers which secure reservoir to stern bulkhead (fig. 95). Apply white lead base antiseize compound to threads; and install pipe fitting assembly at rear bottom of reservoir (fig. 91).

(2) CONNECT OIL LINES (fig. 91). Connect oil reservoir outlet line (to engine) at pipe fittings at rear bottom of oil reservoir, and at flexible oil line passing through hull support. Secure the line by two metal clamps bolted to hull supports. Connect oil reservoir inlet line (from the oil cooler) to the fitting at top of oil reservoir.

(3) CONNECT BREATHER LINE. Connect the breather line to the T-fitting on top of oil reservoir.

(4) CONNECT VENT LINE. Connect the vent line to the top of T-fitting on top of oil reservoir.

(5) Install oil cooler (par. 85).

(6) Install engine mount (par. 63).

(7) Install engine (par. 62).

(8) FILL OIL RESERVOIR. If the pipe plug in the hull beneath the oil tank has been removed, install the plug, using a piece of 1-inch square stock about 3 inches long (fig. 65). Fill the oil reservoir to prescribed level (sec. VIII).

87. ENGINE AND TRANSMISSION OIL LINES.

a. **Description.** In almost all cases copper tubing is used for the engine and transmission oil lines. Flexible lines are used from the oil filter to the engine, and from the transmission copper tubing line in the control tunnel up to and around the transmission.

b. **Maintenance.** Clean by removing all lines and blowing out with compressed air. NOTE: *Inspection of lines should be made, if possible, while the lines are installed, since leaks will be easily discovered due to the oil or dust deposits.*

c. **Replacements.** When a defective copper tubing oil line is to be replaced, it is of extreme importance that the new line be made to conform exactly to the shape of the original. Cut all flexible tubing oil lines to exact length of the original lines.

LANDING VEHICLE TRACKED MK. I AND MK. II

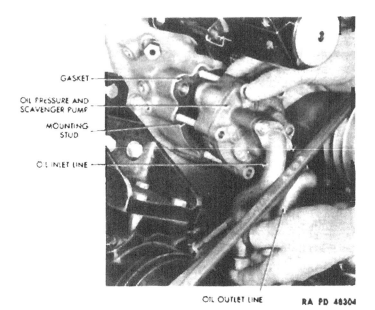

Figure 92 — Removing Oil Pressure and Scavenger Pump

88. OIL PRESSURE AND SCAVENGER PUMP.

a. **Description.** A dual-purpose, gear-type oil pump supplies oil under pressure to the various engine parts, and incorporates a scavenger pump which draws oil from the engine sump, and returns it to the supply tank. The pump is located on the lower left side of the accessory case. Pressure section of the duplex pump picks up oil from the oil supply tank of the vehicle, and forces it under pressure through the oil screen into the arterial lines of the accessory case. Oil from all sections of the crankcase drains into the oil sump, and is pumped from the sump by the scavenger section of the duplex oil pump, through the scavenger oil screen, through an oil filter, an oil cooler, and then to the oil supply tank. Thus a continuous circulation of the lubricating oil is obtained.

b. **Maintenance.** If the pump becomes inoperative, replace it with a new unit.

c. **Removal** (fig. 92). Disconnect oil inlet line and outlet line at the unions with the flexible tubing. Plug the end of each line. Remove safety wire, and unscrew the five castle nuts on the flange of the duplex pressure and scavenger oil pump. Remove the five flat washers, and lift off the pump. Strip off the gasket.

LUBRICATION SYSTEM

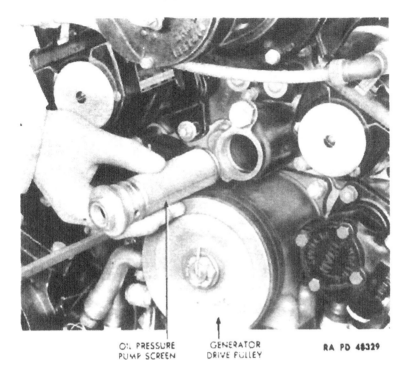

Figure 93 — Removing Oil Pressure Pump Screen

d. **Installation** (fig. 92). Install a new gasket on the mounting studs. Place the duplex pump on the studs. Install the five flat washers and five castle nuts. Tighten the castle nuts, and secure with safety wire. Connect oil inlet line and outlet line at the unions with the flexible tubing.

89. OIL SCREENS.

a. **Description.** An oil pressure pump screen to remove foreign matter from the engine oil is located on the accessory case above the generator drive pulley, and provides a receptacle for the oil temperature indicator. A scavenger oil screen to remove foreign matter from the scavenger lines is located in the accessory case to the left of the oil pressure pump screen and to the right of the duplex pressure and scavenger oil pump.

b. **Maintenance.** After each 25 hours of engine operation, remove both oil screens. Inspect screens for foreign matter and presence of metal particles. Clean with dry-cleaning solvent and dry with com-

TM 9-775
89

LANDING VEHICLE TRACKED MK. I AND MK. II

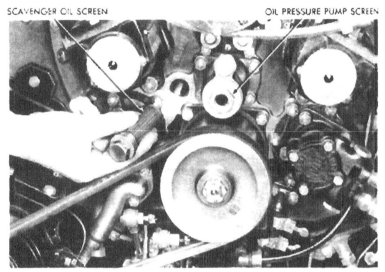

Figure 94 — Removing Scavenger Oil Screen

pressed air. If either screen becomes unserviceable, replace the assembly as a unit.

c. **Removal of Oil Pressure Pump Screen** (fig. 93). Loosen retaining nut, and remove the oil temperature indicator. Remove safety wire from oil pressure screen nut. Unscrew the screen assembly from the accessory case.

d. **Installation of Oil Pressure Pump Screen** (fig. 93). Screw the screen assembly into the accessory case. Install safety wire in the oil pressure pump screen nut. Install the oil temperature indicator, and tighten the retaining nut.

e. **Removal of Scavenger Oil Screen** (fig. 94). Remove safety wire, and unscrew the screen assembly from the accessory case.

f. **Installation of Scavenger Oil Screen** (fig. 94). Screw the scavenger oil screen into the accessory case, and install safety wire through the hexagonal head.

TM 9-775
90-91

Section XX

IGNITION SYSTEM

	Paragraph
Description	90
Magnetos	91
Spark plugs	92
Booster coil	93

90. DESCRIPTION.

a. The ignition system consists of the magneto switch, the two magnetos, the booster coil, and the spark plugs. The system supplies current for igniting the gas air mixture in the combustion chambers of the engine.

91. MAGNETOS.

a. **Description and Data.** Dual ignition is furnished by two radio shielded VMN7-DFA magnetos (SCI) (fig. 26) which are flanged-mounted on the accessory case. One is mounted at the upper right side, and the other at the upper left side. The right magneto fires the front set of spark plugs. The left magneto fires the rear set of spark plugs.

b. **Maintenance.**

(1) BREAKER POINT ADJUSTMENT (fig. 97). Release safety ring on both sides of the magneto, and remove breaker cover. Hand-crank the engine until the breaker points reach their maximum gap. The correct gap is 0.012 inch. If the gap is less than 0.010 inch or more than 0.014 inch, loosen the lock nut, on the stationary contact screw and reset the screw, using a 0.012-inch feeler gage. When the correct gap is obtained, tighten the lock nut and install breaker cover. If the magneto points are burned (ash-colored), do not adjust the gap, but replace the magneto.

(2) LUBRICATION. Examine the felt wick at the bottom of the contact breaker to make sure it is saturated with oil. Squeeze the felt with the fingers, if oil appears on the surface of the felt, no additional lubricant is needed. If it is dry, add a few drops of engine oil, grade SAE 30.

c. **Removal.** NOTE: *The left and right magnetos are removed in the same manner.*

(1) REMOVE MAGNETO LEAD WIRES. Loosen knurled nut, and pull out magneto wire from both magnetos.

(2) REMOVE MAGNETO DISTRIBUTOR BLOCK CASES (fig. 98). Remove the safety pin from the latch at the lower end of the case half. Remove the two cover clamp bolts at the top of the magneto

LANDING VEHICLE TRACKED MK. I AND MK. II

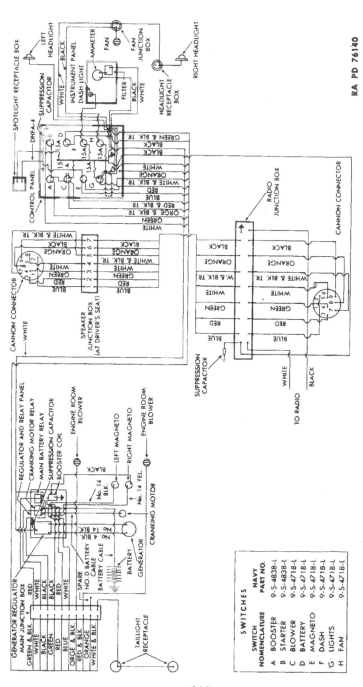

Figure 95 — Wiring Diagram — LVT (2) and LVT (A) (2)

IGNITION SYSTEM

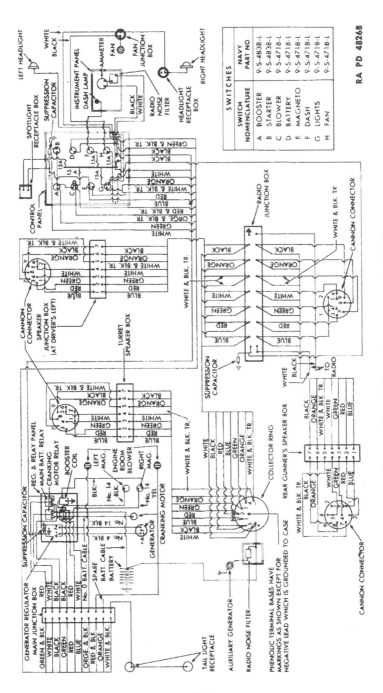

Figure 96 — Wiring Diagram — LVT (A) (1)

TM 9-775
91

LANDING VEHICLE TRACKED MK. I AND MK. II

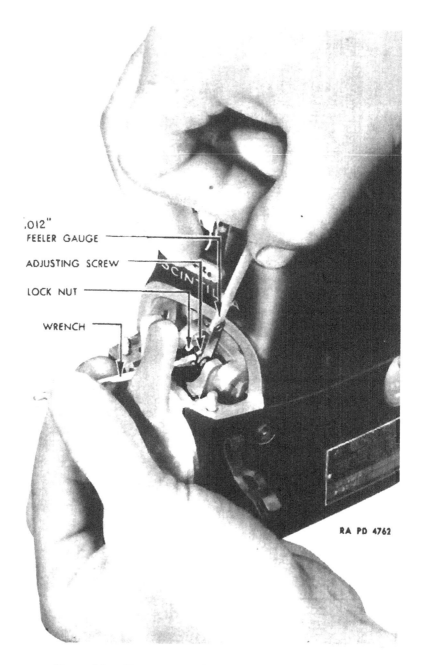

Figure 97 — Magneto Breaker Point Gap Adjustment

TM 9-775
91

IGNITION SYSTEM

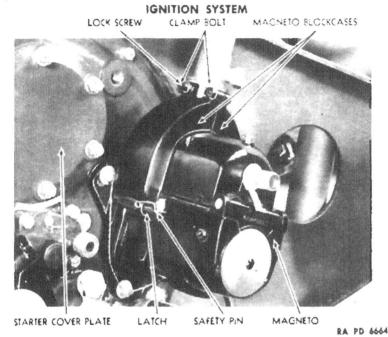

Figure 98 — Removing Magneto Block Cases

case. Loosen the lock screw on the side of the cover to wiring conduit clamp, and lift off the two sections of the magneto block case.

(3) REMOVE MAGNETO DISTRIBUTOR BLOCKS. Remove the safety pin from the distributor block top clamp. Loosen the distributor block top clamp set screw a distance of 1/8 inch. Loosen the distributor block clamp on the magneto. Carefully pry up the base of the distributor block, and lift out the block. (There are two blocks for each magneto.) CAUTION: *Do not chip the blocks while removing them.*

(4) REMOVE MAGNETOS FROM ACCESSORY. Remove safety wire from three castle nuts on the flange of the magneto, and unscrew the castle nuts. Remove three flat washers, and lift off the magneto.

d. **Installation and Timing of Right Magneto — Engine Removed.**

(1) LOCATE TOP DEAD CENTER. Remove spark plugs and rocker box covers from No. 1 cylinder. Install top dead center indicator (41-1-73-100) in the rear spark plug opening. Install timing disk (41-D-1266) and pointer on front end of crankshaft. Rotate the crankshaft clockwise until pointer comes up to one of the lines on dial of top dead center indicator during compression stroke. Mark this line. Mark the location of pointer on timing disk at this point. Continue rotation of crankshaft until pointer of top dead center in-

211

LANDING VEHICLE TRACKED MK. I AND MK. II

Figure 99 — Removing Magneto Distributor Blocks

dicator moves to its maximum point and returns to the line marked. Mark location of pointer on timing disk at this point. Move crankshaft so that needle points to the line midway between the two pencil marks on timing disk. No. 1 cylinder is now at top dead center. Move the pointer to zero on the disk. Turn the engine crankshaft in its proper direction of rotation until the piston in No. 1 cylinder is exactly 20 degrees before top center on its compression stroke, as indicated on the timing disk. Place a new gasket on the right magneto mounting studs.

(2) FIX MAGNETO GEAR POSITION. With the distributor block shielding covers and the distributor blocks removed, the large gear inside the case can be examined easily. There are several marks on this gear. Among them are the letters "L" and "R," both of which may be disregarded at this time. The important marks to observe are two parallel lines near the gear teeth of the large bronze gear. Revolve the large bronze gear inside the case of the right magneto until the two parallel marks on the gear edge line up with two similar marks on the left side of the magneto case itself (fig. 100). Turn magneto in same direction until the points are ready to open. Now the marks on the magneto will be approximately $\frac{3}{8}$ inch past (above) the scribe marks on the case.

(3) PLACE MAGNETO IN POSITION (fig. 100). Hold the magneto in this position, and engage the spline shaft on the magneto

TM 9-775
91

IGNITION SYSTEM

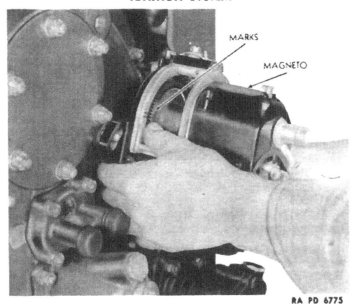

RA PD 6775

Figure 100 — Installing Magneto

with the drive coupling in the accessory case in such manner that the mounting studs will be located in right-hand end of slotted holes. Put a plain washer and castle nut on each of the three studs, and screw the nuts down lightly. The piston in No. 1 cylinder should still be in a position 20 degrees before top center.

(4) ADJUST BREAKER POINT GAP. Remove the magneto breaker point cover, and insert a 0.0015-inch feeler gage between the breaker points. Move the magneto in a clockwise direction a very little at a time by tapping on the side of its case with a fiber hammer. NOTE: *The three holes in the magneto mounting flange are oblong in shape, and tapping sideways on the magneto case causes it to turn slightly, with the shaft as an axis.* The instant the breaker points come apart sufficiently to offer only a very slight resistance to the withdrawal of the 0.0015-inch feeler gage, stop moving the magneto.

(5) CHECK MAGNETO TIMING. Turn crankshaft in direction opposite normal rotation ¼ turn to remove the backlash. Turn in normal direction of rotation until a few degrees before firing position (original crankshaft setting). Again insert 0.0015-inch feeler gage between breaker points, and turn crankshaft slowly until breaker points offer just a slight resistance to withdrawal of the feeler gage. Look at the position of the timing disk with relation to the pointer. If it shows that No. 1 piston is 20 degrees before top center, the right

213

LANDING VEHICLE TRACKED MK. I AND MK. II

magneto is timed correctly. Tighten and safety-wire the castle nuts on magneto flange. CAUTION: *If the right magneto is not timed correctly, take off the retainer castle nuts and washers and remove the magneto, then install it to the engine with the serrations correctly engaged.*

(6) TIME LEFT MAGNETO. Use the timing disk and bar to turn the engine crankshaft in its proper direction of rotation until the piston in No. 1 cylinder is exactly 17 degrees before top center on its compression stroke. CAUTION: *If the crankshaft is inadvertently turned past 17 degrees (for example, 16 or 15 degrees), be sure to turn the shaft backward, and again turn it up to just 17 degrees. This is necessary to take up all the lash in the direction of rotation.* Time the left magneto in the same manner as the right magneto (steps (2), (3), (4), and (5), above).

(7) INSTALL DISTRIBUTOR BLOCKS. Install distributor blocks, and tighten distributor block clamps. Install safety pins.

(8) INSTALL DISTRIBUTOR BLOCK CASES (fig. 98). Place two sections of magneto block case in position. Tighten lock screws on side to cover magneto conduit clamp. Install two cover clamp bolts at top of magneto case. Install safety pin in latch at lower end of case half.

(9) CONNECT MAGNETO LEAD WIRES. Plug magneto lead wires in position, and tighten knurled retaining nuts.

e. Timing Magnetos — Engine in Vehicle.

(1) GENERAL. Procedure given in subpar. d above is for timing magnetos with the engine removed. It is necessary to remove the engine to use a timing disk (41-D-1266) and pointer (41-1-73-100) to locate top dead center. Then the number of degrees before top dead center at which magnetos fire (17 degrees for left magneto and 20 degrees for right magneto) may be ascertained. The following procedures, steps (2) and (3) give a quick, temporary method of setting the crankshaft in the proper position to fire either magneto (if the other magneto has not been removed and is timed correctly) with the engine installed. After fixing crankshaft position, continue with regular timing procedure given under subparagraph d (2), (3), (4), and (5), above.

(2) TIME LEFT MAGNETO. If right magneto is timed correctly, left magneto can be timed on vehicle by turning flywheel ½ inch in direction of rotation after breaker points on right magneto break. This ½ inch is equal to 3 degrees, and puts the crankshaft 17 degrees before top center, the correct position for firing the left magneto.

(3) TIME RIGHT MAGNETO. If the left magneto is timed correctly, the right magneto can be timed on vehicle by setting the flywheel ½ inch ahead of the position at which left magneto breaker

TM 9-775
91–92

IGNITION SYSTEM

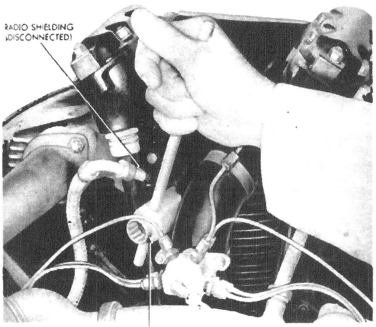

Figure 101 — Removing Spark Plugs

points open. This ½ inch equals 3 degrees, and places the crankshaft 20 degrees before top center, the correct position for firing the right magneto.

92. **SPARK PLUGS.**

a. **Description and Data.** Either BG 417S or CP 63S spark plugs are supplied as standard equipment with the W670-GA engine.

b. **Maintenance.** Clean carbon from spark plugs with dry-cleaning solvent. Check spark plug gap with a feeler gage, and adjust to 0.018 inch. Replace worn or damaged spark plug cables. NOTE: *Spark plug cables usually are replaced in sets. In emergencies, however, a single cable may be replaced.*

c. Removal (fig. 101). Disconnect the locking nut which fastens the conduit elbow of the radio shielding to the sparkplug. NOTE: *It also may be necessary to loosen the nut which fastens the conduit elbow to the flexible radio shielding on the cable.* Place a deep socket wrench over the hexagonal nut on the base of the spark plug, and unscrew the spark plug. CAUTION: *Do not twist the wrench*

TM 9-775
92–93

LANDING VEHICLE TRACKED MK. I AND MK. II

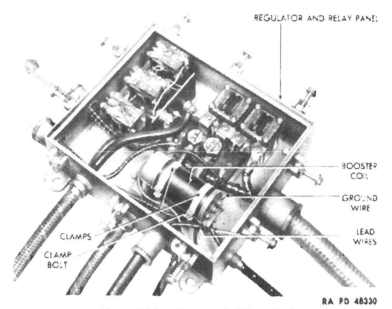

Figure 102 — Booster Coil Installed

against the top of the spark plug, or break the cooling fins of the cylinder head. Install dummy plugs to prevent dirt from entering the engine.

d. **Installation** (fig. 101). Remove dummy plugs from the 14 spark plug holes. Install the spark plugs, placing a new copper spark plug gasket between the base flange of the spark plug and the cylinder head. Use a deep socket or spark plug wrench to tighten the plug. Tighten the locking nuts which fasten the conduit elbows of the radio shielding to the spark plugs. If necessary, tighten the nuts which fasten the conduit elbows to the flexible radio shielding on the cable.

93. BOOSTER COIL.

a. **Description and Data.** A battery-operated booster coil (DR 1115481), located in the regulator and relay panel in the battery compartment, serves to supply ignition spark during the starting period of the engine.

b. **Maintenance.** No maintenance of the booster coil will be attempted by using arm personnel. If, when the booster switch is pressed, no buzzing sound is heard at the engine compartment, remove the defective coil and install a new coil.

STARTING SYSTEM

c. Removal.

(1) Open hinged rear cover.

(3) Remove upper left rear baffle (par. 64).

(3) REMOVE REGULATOR AND RELAY PANEL COVER. Loosen the 10 hinged bolts which secure the regulator and relay panel cover. Swing the bolts out of the way, and lift off the cover.

(4) REMOVE BOOSTER COIL (fig. 102). Disconnect the two lead wires and one ground wire attached to terminal studs in front of the booster coil. Tag the wires and terminals as wires are removed in order to assure correct assembly. Remove the nut and bolt from each of the two clamps securing the booster coil in the regulator relay panel. Spread the clamps, and lift out the booster coil.

d. Installation.

(1) Place the booster coil in position in the regulator and relay panel. Install the one nut and bolt in each of the two clamps securing the booster coil (fig. 102). Connect the two lead wires and the one ground wire to their proper terminals in the front of the booster coil.

(2) INSTALL THE REGULATOR AND RELAY PANEL COVER. Place the regulator and relay panel cover on the regulator and relay panel. Swing the hinged bolts back into the cover brackets, and tighten bolts.

(3) Install upper rear baffle (par. 64).

(4) Close hinged rear cover.

Section XXI

STARTING SYSTEM

	Paragraph
Description	94
Cranking motor	95

94. DESCRIPTION.

a. The starting system consists of the starter switch, cranking motor, and bendix drive. The system is designed to supply electrical energy for rotating the crankshaft to start the engine. Provision is also made for hand-cranking the engine if the electrical system fails.

95. CRANKING MOTOR.

a. Description and Data. A DR 12-volt electric cranking motor (1108682), with provision for hand cranking, is flange-mounted at the top of the accessory case.

TM 9-775

LANDING VEHICLE TRACKED MK. I AND MK. II

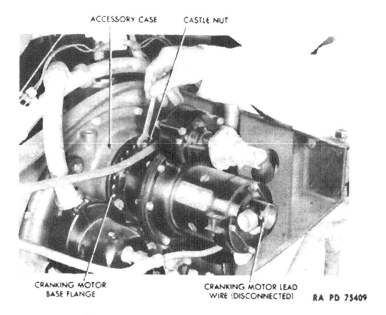

Figure 103 — Removing Cranking Motor

b. **Removal** (fig. 103). Remove safety wire and unscrew cap on rear face of cranking motor. Remove nut and lock washer and disconnect lead wire. Unscrew knurled nut that secures conduit to motor housing. Remove safety wire from castle nuts on the six studs which hold the cranking motor base flange to the accessory case. Using crowfoot wrench (41-W-871-45) to remove inaccessible nuts, unscrew and remove the six castle nuts and the six plain washers, supporting the cranking motor with one hand as the last nut is removed. Pull the cranking motor, with gasket, straight out from the engine.

c. **Installation** (fig. 103). Inspect flange gasket. If worn or torn, install a new gasket. NOTE: *There are three parts to the cranking motor flange gasket; a fiber gasket next to the accessory case, a lead or aluminum gasket, and a fiber gasket next to the base of the cranking motor.* Push the cranking motor straight into position, supporting it with one hand and twisting to secure proper alinement. Install plain washers and castle nuts on the six studs. Install safety wire on the six castle nuts. Secure the lead wire on the terminal stud with a nut and lock washer. Install the cap on rear face of cranking motor. Secure with safety wire. Tighten knurled nut that secures the conduit to cranking motor housing.

Section XXII
GENERATING SYSTEM

	Paragraph
Description	96
Generator	97
Generator regulator	98

96. DESCRIPTION.

a. The generating system consists of the 12-volt generator and the generator regulator. It is designed to supply current for keeping the battery in a fully-charged condition.

97. GENERATOR.

a. Description and Data (fig. 104). A 12-volt generator (DR 1106458), is bracket-mounted on the web of the lower left side of the engine support beam. The generator is driven by two V-belts from a generator drive pulley on the accessory case.

b. Maintenance.

(1) GENERAL. No maintenance, other than belt adjustment and replacement, should be attempted by the using arm personnel. The generator requires no lubrication between engine overhaul periods. Every 25 hours of engine operation, inspect the generator to make sure that terminals are clean and tight.

(2) ADJUST DRIVE BELTS (fig. 104). Remove locking wire, and loosen the four cap screws in the adjusting slots so the mounting bracket can be moved by tapping. Move generator mount as needed to provide the correct belt tension. (Easy thumb pressure on the belts at the center point between pulleys should depress the belts approximately ½ inch.) Tighten cap screws, and install locking wire.

(3) REPLACE DRIVE BELTS (fig. 104). Remove three of the four generator bracket mounting cap screws and loosen the fourth one (the left forward one). Tilt the generator, and lift the belts off the pulley. Install new belts and adjust for correct tension.

c. Removal (fig. 104). Disconnect generator lead and ground wires (par. 61). Cut safety wire, and remove four cap screws which secure generator to generator mounting bracket. Tilt the pulley end of the generator upward, and remove the drive belts. Lift off generator.

d. Installation (fig. 104). Place the generator in position and install, loosely, the four cap screws which secure the generator to the generator mounting bracket. Place the generator drive belts on the generator and on the generator drive pulley, and adjust belt tension

TM 9-775
97–98

LANDING VEHICLE TRACKED MK. I AND MK. II

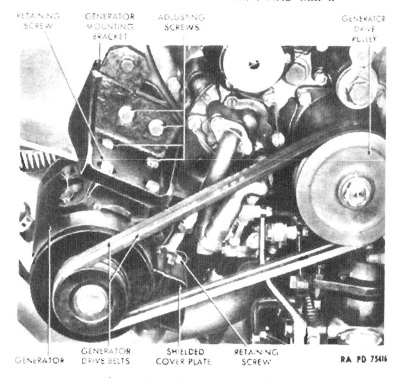

Figure 104 — Generator Installed

(subpar. b (2), above). Tighten the retaining screws, and lock with safety wire. Connect generator lead and ground wires (par. 62).

98. GENERATOR REGULATOR.

a. **Description and Data.** A generator regulator, part number DR 1118473, is used in the landing vehicle tracked. Generator regulator is located in the regulator and relay panel within the battery compartment. Function of the generator regulator is to control generator output so that under all conditions of operation, battery charge will be maintained. There are three separate units in the regulator, consisting of the voltage regulator and cut-out relay. The voltage regulator permits the output of the generator to reach the battery in proportion to its state of charge. The current regulator limits the maximum output of the generator to a safe value. The cut-out relay prevents the battery from discharging through the generator when the generator is not producing enough voltage to overcome the voltage of the battery.

b. **Removal.**

(1) Open hinged rear cover.

TM 9-775
98

GENERATING SYSTEM

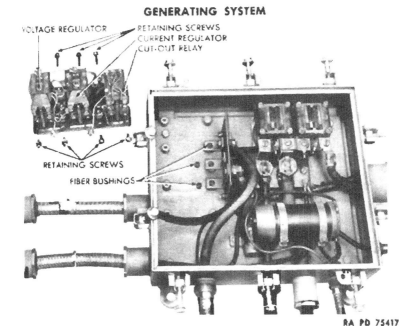

Figure 105 — Generator Regulator Removed

(2) Remove upper left rear baffle (par. 61).

(3) REMOVE REGULATOR AND RELAY PANEL COVER. Loosen the 10 hinged bolts which secure the regulator and relay panel cover. Swing the bolts out of the way, and lift off the cover.

(4) REMOVE GENERATOR REGULATOR (fig. 105). Remove the seven screws, with flat washers and lock washers, which secure the generator regulator to the base of the regulator and relay panel, and lift out the generator regulator. Take care not to lose the three small fibre bushings through which the three retaining screws on front side of the generator regulator pass.

c. Installation.

(1) INSTALL GENERATOR REGULATOR. Place the generator regulator in position in the regulator and relay panel. Be sure the three small fibre bushings are in place in the regulator and relay panel. Install the seven screws, with flat washers and lock washers, which secure the generator regulator to the base of the regulator and relay panel (fig. 105).

(2) INSTALL REGULATOR AND RELAY PANEL COVER. Place the cover on the regulator and relay panel. Swivel the 10 hinged bolts into position on the cover brackets, and tighten bolts securely.

(3) Install upper left rear baffle (par. 62).

(4) Close hinged rear cover.

221

TM 9-775
99-100

LANDING VEHICLE TRACKED MK. I AND MK. II

Section XXIII

BATTERY AND LIGHTING SYSTEM

	Paragraph
Description	99
Batteries and cables	100
Headlights	101
Taillights	102
Spotlight receptacle box	103
Fan junction box	104
Headlight receptacle box	105
Main junction box	106
Circuit breaker	107
Flexible conduits and wiring	108
Solenoids	109
Radio junction box	110
Speaker junction box	111

99. DESCRIPTION.

a. The battery and lighting system consists of two 6-volt storage batteries mounted in the compartment between the engine room and the left side of the hull, two taillights and two headlights, with the necessary wiring and junction boxes to complete the circuits. Provision is made for connecting the cab fan by means of a junction box located on the hull between the driver and the assistant driver. On the forward bulkhead behind the driver is a receptacle for connecting a spotlight.

100. BATTERIES AND CABLES.

a. Description and Data. Two 6-volt storage batteries (ES) are connected in series to maintain the voltage of the system at 12 volts. Batteries are installed in the battery rack, located in the compartment between the engine room and the left side of the hull (fig. 34). Access to the batteries is provided by removing the upper and lower left rear engine room baffles.

b. Maintenance.

(1) CAPACITY AND TEMPERATURE DATA. At temperatures below freezing, the load on the battery becomes greater, and the relative capacity of the battery is reduced. For this reason, when low temperatures prevail, it will be necessary to maintain the specific gravity of the battery electrolyte at 1.250 or higher, and to replace the battery when its gravity reading is below that point. The following data shows the capacity of the batteries and the relative freeezing point of the electrolyte.

BATTERY AND LIGHTING SYSTEM

RA PD 43402

Figure 106 — Hydrometer Correction Chart

Specific Gravity	Freezing Temperature F.	Specific Gravity	Freezing Temperature F.
1.100	18	1.220	− 31
1.120	+ 14	1.240	− 51
1.140	+ 8	1.260	− 75
1.160	+ 2	1.280	− 92
1.180	− 6	1.300	− 95
1.200	− 17		

(2) DETERMINING SPECIFIC GRAVITY (fig. 106). To determine the actual specific gravity of the electrolyte, it is necessary to check the temperature of the solution with a thermometer. If the temperature is normal (80° F.), the specific gravity reading will be correct. However, if the temperature is above or below 80° F., it will be necessary to make an allowance to determine the actual specific gravity. This is due to the fact that the liquid expands when warm, and the same volume weighs less than when it is at normal temperature. The reverse is also true, and when the temperature is below normal, or 80° F., the liquid has contracted and the same volume weighs more than it does when normal. The correction chart (fig. 106) shows the figures to be used to make these corrections. For example, when the specific gravity (as shown by the hydrometer

TM 9-775
100

LANDING VEHICLE TRACKED MK. I AND MK. II

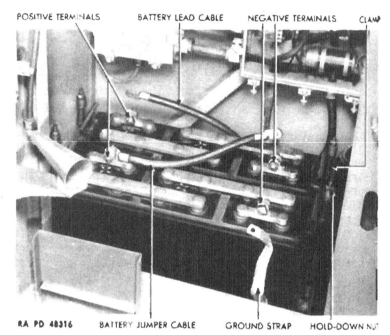

Figure 107 — Battery Installed — Cables Disconnected

reading) is 1.290, and the temperature of the electrolyte is 60° F., it will be necessary to subtract 8 points or 0.008 from 1.290, which gives 1.282 as the actual specific gravity. If the hydrometer reading shows 1.270, at a temperature of 110° F., it will be necessary to add 12 points or 0.012 to the reading, which gives 1.282 as the actual specific gravity.

c. **Removal** (fig. 107). Open the hinged rear cover. Remove left muffler and flexible tubing (par. 61). Remove the upper left rear baffle (par. 61). Disconnect the battery ground strap (negative) and battery lead (positive) cables, then disconnect and remove the battery jumper that connects the two batteries together. Loosen the two hold-down nuts which secure each battery clamp, and slide off the four clamps. Lift the batteries up and out of their compartments.

d. **Installation** (fig. 107). Place the batteries in position so that the two negative terminals are forward, and the two positive terminals are toward the rear. Slide the four clamps in place, and tighten the nuts that hold the clamps against the batteries. Connect the battery ground to the negative terminal of the inner battery. Connect the battery lead to the positive terminal of the outer battery.

224

TM 9-775
100—101

BATTERY AND LIGHTING SYSTEM

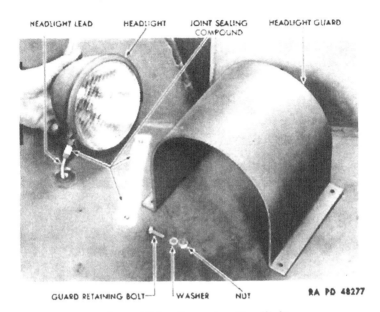

Figure 108 — Removing Headlight

Connect the battery jumper to the positive terminal of one battery, and the negative terminal of the other. Install the upper left rear baffle (par. 62). Install the left muffler and flexible tubing (par. 62). Close the hinged rear cover.

101. **HEADLIGHTS.**

a. **Description.** Headlights are mounted on the left and right sides of the cab cover at the front of the vehicle. They are double-filament, sealed-beam type. Semicircular steel guards bolted to the cab cover surround each light, and protect it from damage. Headlights are controlled by the light switch on the control panel.

b. **Removal of Headlight** (fig. 108). Remove the four bolts, nuts, and toothed lock washers which secure the headlight guard to the cab cover. This will require two men: one to hold the bolts on the outside of the vehicle, and another to turn off the nuts within the vehicle. Lift off the guard. Remove headlight receptacle box cover and disconnect lead. From within the cab, remove the headlight retaining nut and washer from the bottom of the headlight support bracket. It is necessary for one man to hold light to keep it from turning while another man removes the nut. Pull the headlight, with lead wire attached, from the vehicle.

c. **Installation of Headlight** (fig. 108). Coat base of headlight

225

TM 9-775
101

LANDING VEHICLE TRACKED MK. I AND MK. II

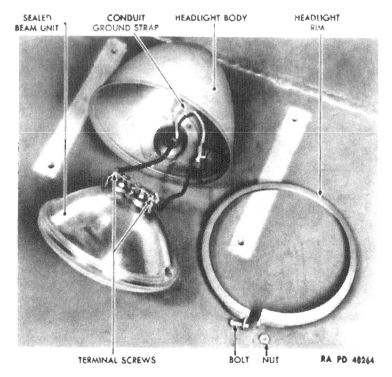

Figure 109 — Installation of Sealed Beam Lamp-unit

with joint sealing compound. Place the headlight in position, and insert headlight lead wire through the opening in the cab cover. Install the retaining nut and washer. Connect headlight lead wire to the proper terminal within the headlight receptacle box. Place the headlight guard in position, and install the four retaining bolts, nuts, and toothed lock washers.

d. **Removal of Sealed Beam Lamp-unit** (fig. 109). Remove headlight guard (subpar. b, above). Remove the headlight rim bolt and nut, and lift off rim. Slide the sealed beam lamp-unit part way out of the headlight body. Disconnect the two wires from the terminals at the back of the sealed beam lamp-unit. Remove sealed beam lamp-unit.

e. **Installation of Sealed Beam Lamp-unit** (fig. 109). Place the sealed beam lamp-unit part way into the headlight body, and connect the two wires to the proper terminals at the back of the sealed beam lamp-unit. Push the sealed beam lamp-unit the remainder of the way into the body. Place the rim against the sealed beam lamp-unit and

TM 9-775
101—102

BATTERY AND LIGHTING SYSTEM

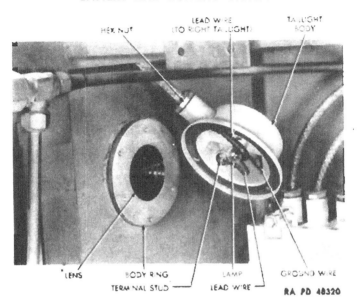

Figure 110 — Taillight Removed

on the headlight body, and install the rim bolt and nut. Install headlight guard (subpar. e, above).

102. **TAILLIGHTS.**

a. **Description.** Taillights, left and right, are used in the landing vehicles tracked. Taillights are controlled by the light switch on the control panel.

b. **Maintenance.** If taillight lenses are broken, install new lenses. If the lamps within the taillight assembly are burned out, install new lamps. Should a complete taillight assembly be damaged beyond repair, install a new taillight assembly.

c. **Removal.**

(1) GENERAL. A left and right taillight are provided. Access to the taillights is provided by removable baffles at the left and right of the engine room.

(2) Open hinged rear cover.

(3) Remove left (or right) upper rear baffles (par. 61).

(4) REMOVE TAILLIGHT (fig. 110). From inside the engine room, remove three screws that secure the taillight to the hull. Swing the taillight away from the hull, allowing the assembly to be supported by the electrical wiring (fig. 110). Disconnect ground and lead wires at

227

LANDING VEHICLE TRACKED MK. I AND MK. II

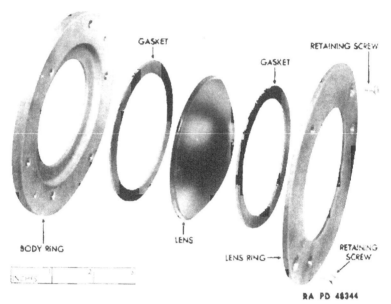

Figure 111 — Taillight Lens — Disassembled

terminal stud. Loosen the two flexible conduit retaining hexagonal nuts, and pull wires out of body. If a new lamp is to be installed, twist out the defective lamp.

(5) REMOVE TAILLIGHT LENS. On the exterior rear of the vehicle remove the three cap screws holding the taillight body ring to the hull. From the engine room, pull the body ring and lens off the inside of the hull.

(6) DISASSEMBLE TAILLIGHT LENS. Remove two screws from the lens ring of the taillight lens. Lift off the lens ring, glass lens, and two rubber gaskets (fig. 111).

d. Installation.

(1) GENERAL. The left and right taillights are identical, and are installed in the same manner.

(2) ASSEMBLE TAILLIGHT LENS. Place the rubber gaskets and glass lens in position on the body ring (fig. 111). Then place the lens ring on so it holds the lens in place. Be sure all holes in the two rings are alined. Install two screws that hold the assembly together.

(3) INSTALL TAILLIGHT LENS. Cover rear face of lens ring with joint sealing compound, and place the taillight lens assembly in position on the inside of the engine room. While it is held in position, install the three cap screws which secure the body ring to the hull (fig. 110).

BATTERY AND LIGHTING SYSTEM

(4) INSTALL TAILLIGHT. Insert the two flexible conduits into the body of the taillight. Connect ground and lead wires to terminal stud. Tighten conduit retaining hexagonal nuts. Install a new lamp if the old one was removed. NOTE: *If the right taillight is being installed, only one conduit need be installed. From inside the engine room, install three cap screws and toothed lock washers which secure the taillight to the hull.*

(5) Install left (or right) upper rear baffle (par. 62).

(6) Close the hinged rear cover.

103. SPOTLIGHT RECEPTACLE BOX.

a. **Description.** The spotlight receptacle is located on the forward bulkhead, directly behind the driver. A cap fastened to the box by means of a small chain covers the receptacle plug opening in the front of the box. To use the spotlight furnished with the vehicle, unscrew the cap from the front of the spotlight receptacle box and plug in the spotlight.

b. **Maintenance.** If the spotlight receptacle box is damaged, install a new box. If lead wires in the box are loose, broken, or otherwise defective, tighten connections or install new wires. Only one conduit connects to spotlight receptacle box, connecting the receptacle box to the control panel.

c. **Removal** (fig. 112). Remove the four bolts, nuts, and flat washers which secure the cover to the spotlight receptacle box, and lift off the cover. Remove the two screws which secure the lead wires within the spotlight receptacle box. Unscrew conduit retaining hexagonal nut. Pull the conduit, with wires, out of the spotlight receptacle box. Pull out the bolts, nuts, and toothed lock washers which secure the spotlight receptacle box to the forward bulkhead. Lift off the box.

d. **Installation** (fig. 112). Place the spotlight receptacle box in position, and install the two retaining bolts, nuts, and toothed lock washers. Insert the flexible conduit, with two lead wires, within the conduit opening in the spotlight receptacle box. Connect lead wires to the proper terminal within the box. Place cover on the spotlight receptacle box. Then install the four retaining nuts, bolts, and flat washers. Screw the knurled cap over the opening in the front of the cover.

104. FAN JUNCTION BOX.

a. **Description** (fig. 206). Mounted on the hull, directly in front of and between the driver and the assistant driver, is the fan junction box. The function of the fan junction box is to provide an electrical connection from the control panel for the cap fan and the right headlight.

TM 9-775
LANDING VEHICLE TRACKED MK. I AND MK. II

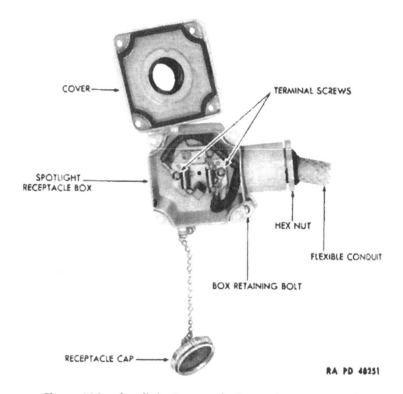

Figure 112 — Spotlight Receptacle Box — Cover Removed

b. **Maintenance.** If the cab fan junction box is damaged or if wires are loose, broken, or otherwise damaged, install a new cab junction box, or tighten or replace defective wiring.

c. **Removal.** Remove the four nuts and bolts which secure the junction box cover to the junction box. Lift off the cover. Disconnect the lead wires at the terminals within the fan junction box. CAUTION: *Tag all wires and terminals as wires are removed, in order to assure correct assembly.* Loosen the conduit hexagonal nuts securing the conduits within the junction box, and pull out the conduits with wires. Remove the two stud nuts and toothed lock washers which hold the junction box in position. Lift the junction box off the studs.

d. **Installation.** Place the junction box in position, and install the two stud nuts with toothed lock washers which secure the box. Insert the conduits in the proper conduit opening in the junction box, and connect wires to the proper terminals. These terminals and wires were tagged at removal to assure correct assembly. Place the junction box cover on the junction box, and install the four retaining bolts and nuts.

BATTERY AND LIGHTING SYSTEM

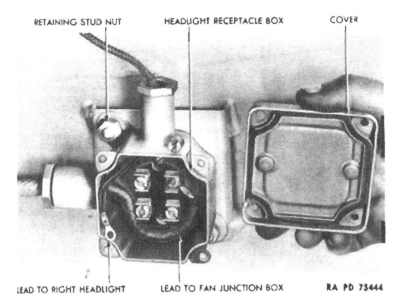

Figure 113 — Headlight Receptacle Box — Cover Removed

105. HEADLIGHT RECEPTACLE BOX.

a. **Description.** The headlight receptacle box is mounted on the hull to the right of the assistant driver. It provides a junction for electrical current supplied to the right headlight. Two conduits lead from the headlight receptacle box; one to the cab fan junction box, and one up to the right headlight.

b. **Maintenance.** If the headlight receptacle box is damaged, install a new box. If wiring is loose, broken, or otherwise defective, tighten connections or install new conduits and wires.

c. **Removal** (fig. 113). Remove the four nuts and bolts which secure the headlight receptacle box cover to the box, and lift off the cover. Remove the stud nuts which secure the two lead wires within the receptacle box. Identify terminals and wires, to assure correct assembly. Loosen hexagonal nuts, and pull the two conduits with wires out of the receptacle box. Remove the two stud nuts and toothed lock washers which secure receptacle box in position. Lift the box from the studs.

d. **Installation** (fig. 113). Place the headlight receptacle box on its mounting studs, and install the two retaining stud nuts and toothed lock washers. Insert the two conduits, with wires, into the proper conduit openings in the receptacle box. Connect the wires to their proper

LANDING VEHICLE TRACKED MK. I AND MK. II

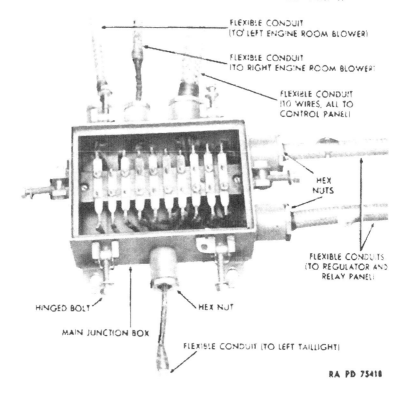

Figure 114 — Main Junction Box — Cover Removed

terminals within the box. Tighten hexagonal nuts which secure the conduits within the box. Place the cover on the receptacle box, and install the four retaining bolts and nuts.

106. MAIN JUNCTION BOX.

a. **Description** (fig. 114). The main junction box is located in the battery compartment at the rear of the regulator and relay panel. All electrical connections at the rear of the vehicle, such as the taillights, regulator, and relay panel, engine room blowers, and booster coil, lead into the main junction box, and are connected by terminals to wiring running to switches on the control panel.

b. **Maintenance.** Maintenance consists of replacing defective wiring leading into the main junction box, and, in case of injury to the case, cover, or terminal block, the replacement of the complete main junction box.

TM 9-775
106

BATTERY AND LIGHTING SYSTEM

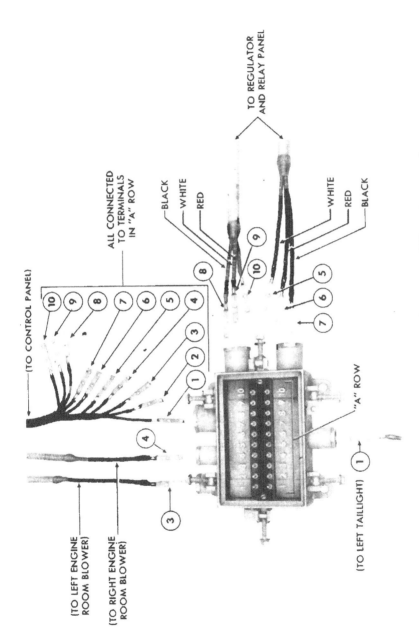

Figure 115 — Main Junction Box — Wires Removed

TM 9-775
106
LANDING VEHICLE TRACKED MK. I AND MK. II

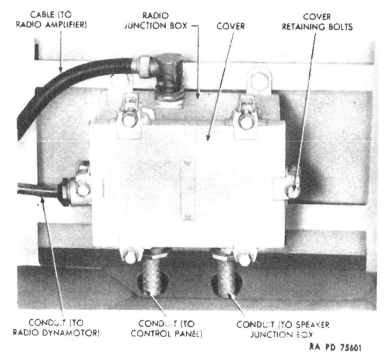

Figure 116 — Radio Junction Box — Installed

c. Removal.

(1) Open hinged rear cover.

(2) Remove upper left rear baffle (par. 61).

(3) REMOVE MAIN JUNCTION BOX COVER (fig. 114). Loosen the six hinged cover bolts. Swing bolts to one side and lift off cover.

(4) DISCONNECT MAIN JUNCTION BOX (fig. 115). Disconnect the wires leading into the main junction box. There are 19 wires, all equipped with terminals, which mount on the terminal studs in the box. Loosen the hexagonal nuts on the conduits through which the wires lead into the main junction box, and pull out the conduits, with wires. NOTE: *The method of attaching terminals, which consists of alternating terminal stud nuts between upper and lower terminals, should be carefully studied before disconnecting the wires so that wires may be connected at assembly in the same manner* (fig. 114).

(5) REMOVE MAIN JUNCTION BOX. Remove the four stud nuts and toothed lock washers which secure the main junction box to brackets on the inside of the hull. Lift out the main junction box.

234

BATTERY AND LIGHTING SYSTEM

TM 9-775
106–108

d. Installation.

(1) INSTALL MAIN JUNCTION BOX. Place the main junction box in position, and install four stud nuts and toothed lock washers which secure the main junction box to the inside of the hull.

(2) CONNECT MAIN JUNCTION BOX (figs. 114 and 115). Insert the wires into their proper conduit openings in the main junction box. Place the wires with terminals on the proper terminal studs in the main junction box (fig. 115). Install the terminal stud nuts, alternating them between upper and lower terminals as they were installed before disconnection of main junction box.

(3) INSTALL MAIN JUNCTION BOX COVER. Place the main junction box cover on the main junction box. Rotate the six hinged cover bolts into position. Tighten bolts securely.

(4) Install upper left rear baffle (par. 62).

(5) Close hinged rear cover.

107. CIRCUIT BREAKER.

a. Description. On earlier models of the landing vehicle tracked, a 75-ampere fuse, controlling the generating circuit, is mounted on a fuse block within the control panel (fig. 185). In later models of the landing vehicles tracked, this 75-ampere fuse is removed, and in its place a circuit breaker has been installed. The circuit breaker is mounted in the lower rear corner of the control panel (fig. 186), and is operated from within the panel by means of a rubber push-type button. To close the circuit after it has been broken by the circuit breaker, push in on this button.

b. Maintenance. If circuit breaker has proved to be defective, either through continually breaking the circuit for no apparent reason, or refusing to break the circuit under an excessive overload, install a new circuit breaker.

c. Removal (fig. 186). Remove the two cap screws with flat washers which secure the circuit breaker to brackets within the control panel. Disconnect the two lead wires at the terminals on the circuit breaker. Lift out the circuit breaker.

d. Installation. Place the circuit breaker in position, and install the two retaining cap screws and flat washers. Connect the two circuit breaker lead wires to their respective terminals on the circuit breaker.

108. FLEXIBLE CONDUITS AND WIRING.

a. Diagnosis. The only service which using arm personnel should attempt with respect to conduits or wiring is diagnosis of trouble (par. 49), and installation of new conduits or wiring.

TM 9-775
108—109

LANDING VEHICLE TRACKED MK. I AND MK. II

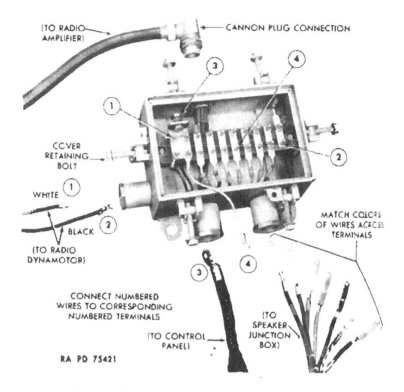

Figure 117 — Radio Junction Box — Disconnected

b. **Installation of New Conduits and Wiring.** When removing or installing conduits, make sure the conduits do not twist or kink. Make sure conduits are connected to the proper terminals. Follow the appropriate wiring diagram for reference on connections. Make certain that all conduit retaining bolts are installed, and that cables are held securely in their proper position.

109. SOLENOIDS.

a. Two solenoids are used to electrically fire 37-mm gun and cal. .30 machine gun in the landing vehicle tracked, with turret (LVT (A) (1)). Solenoids are not used on any other model of the landing vehicles tracked. If solenoids will not operate to fire the guns when switches are turned on, and electric triggers are depressed, assuming wiring has been tested and found in good condition, install new solenoids.

236

TM 9-775
110

BATTERY AND LIGHTING SYSTEM

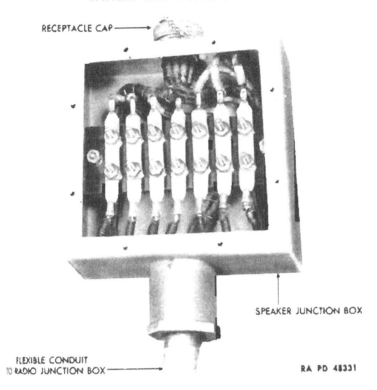

Figure 118 — Speaker Junction Box — Cover Removed

110. **RADIO JUNCTION BOX.**

a. **Description.** Located directly behind the assistant driver's seat, the radio junction box is used as a junction between electrical current from the control panel and radio equipment, and the remote control units. The radio junction box is connected to the control panel, speaker junction box, radio amplifier, and radio dynamotor.

b. **Maintenance.** If the radio junction box case or cover is damaged through accidental or combat use, replace the junction box as a unit. If wiring within the junction box is loose, broken, or otherwise defective, tighten terminals or replace the broken or defective wires.

c. **Removal** (figs. 116 and 117). Loosen the six hinged bolts which secure the radio junction box cover. Swing the hinged bolts out of the way, and lift off the cover. Remove terminal stud nuts, and disconnect wires connected to terminal posts within the junction box. Tag all wires and terminal posts as wires are removed, in order

LANDING VEHICLE TRACKED MK. I AND MK. II

to assure correct assembly. Loosen hexagonal nuts which secure the conduits within the junction box and pull the conduits, with wiring, out of the box. Remove the four bolts, nuts, and toothed lock washers which secure the radio junction box to front bulkhead. Lift out the radio junction box.

d. **Installation** (figs. 116 and 117). Place the radio junction box in position and install the four retaining bolts, nuts, and toothed lock washers. Insert lead wires and conduits into the proper conduit openings in the radio junction box. Connect the wires to the proper terminal posts within the junction box. When connecting the conduit wires from the speaker junction box within the radio junction box, match up the colors of the wires in the conduit with the corresponding colored wires on opposite ends of terminals within the junction box. Wires in the junction box connect to the cannon plug in the top of the box. Place the radio junction box cover on the junction box. Swing the six hinged bolts up into the brackets of the cover, and tighten the bolts.

111. SPEAKER JUNCTION BOX.

a. **Description.** Located on the side of the hull, to the rear of the control panel, the speaker junction box provides a means of electrically connecting driver's radio remote control unit. One conduit enters the speaker junction box at the bottom. This conduit comes from the radio junction box located on the floor behind the assistant driver's seat. A cannon-type plug connection is located on top of the speaker junction box, and is connected to the driver's remote control unit.

b. **Maintenance.** If the speaker junction box is defective, install a new one. If the driver's remote control unit will not work due to loose, broken, or otherwise defective wiring, tighten the terminals within the box, or replace defective wiring.

c. **Removal** (figs. 118 and 119). Remove the speaker junction box cover and gasket. Remove the 14 terminal stud nuts, which secure the 7 terminal wires, with terminals, to the terminal posts within the speaker junction box. Loosen the conduit retaining hexagonal nut, and pull the conduit with wires out of the speaker junction box. Remove the 4 cross-recessed screws and washers located within the speaker junction box, which secure the box to the bracket on the side of the hull. Lift off the box.

d. **Installation** (figs. 118 and 119). Place the new speaker junction box in position, and install the 4 retaining cross-recessed screws and lock washers. Insert the conduit from the radio junction box into the speaker junction box. Match up the colors of the wiring running through the conduit to the colors of the wiring attached to

TM 9-775

BATTERY AND LIGHTING SYSTEM

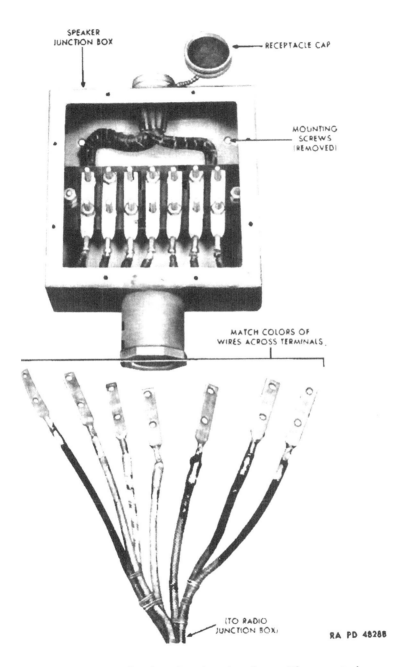

Figure 119 — Speaker Junction Box — Disconnected

TM 9-775
111–112

LANDING VEHICLE TRACKED MK. I AND MK. II

the terminals within the junction box, and which lead to the cannon plug. Place the 7 terminals or wires on their respective terminal posts, and install the 14 terminal stud nuts. Shellac a new gasket to the speaker junction box cover. Place the cover on the speaker junction box, and install the 8 retaining screws and toothed lock washers.

Section XXIV

RADIO INTERFERENCE SUPPRESSION SYSTEM

	Paragraph
Description	112
Maintenance	113

112. DESCRIPTION.

a. The radio interference suppression system consists of the complete shielding of wiring and spark plugs with the addition of filters, condensers, and bonding. It is designed to reduce interference to radio reception produced by operation of the electrical system of the vehicle. It is very essential that this system function perfectly at all times, since its failure may prevent the reception of some vital radio communication.

b. The interference produced by the generator-regulator and/or ignition system that would be carried by the wiring to points where radio reception might be affected is bypassed to ground (hull of vehicle).

(1) A 0.5 microfarad condenser is mounted in the regulator-relay panel and connected to the generator armature terminal at the regulator (fig. 36).

(2) An SPR-model FL-19 filter is mounted in the instrument panel and connected between the ammeter and B+ terminal of the regulator (fig. 188).

(3) A 0.25 microfarad condenser is mounted in control panel and connected to the load side of the generator fuse (fig. 185).

(4) A 0.01 microfarad condenser is mounted under the terminal strip in the radio junction box and connected to positive radio power supply terminal (fig. 117).

(5) Fourteen bonds (ground straps) are mounted as follows:

(a) Engine to engine support (two) (fig. 37).

(b) Throttle control rod to hull (three) (fig. 120).

(c) Clutch control rod to hull (three) (fig. 120).

(d) Tachometer line, oil pressure line, oil temperature line, priming pump line to hull (five).

RADIO INTERFERENCE SUPPRESSION SYSTEM

(e) Instrument panel box to hull (one) (fig. 188).

c. On LVT (A) (1) vehicles additional suppression components are used as follows:

(1) An FL-12 filter is mounted in a metal box located under the propeller shaft in control tunnel, between the two station uprights and connected between the auxiliary generator and the battery.

(2) One bond connects the auxiliary engine-generator unit to the hull (fig. 220).

113. MAINTENANCE.

a. Regulator Relay Condenser (fig. 36).

(1) SERVICE. Keep connections tight. If defective, replace.

(2) REMOVAL. Loosen the 10 cover retaining bolts and swing clear. Lift off panel cover. Disconnect lead from regulator panel. Remove mounting screw. Lift out condenser.

(3) INSTALLATION. Place condenser in position. Install mounting screw. Connect lead to terminal on regulator panel. Replace cover. Swing cover retaining bolts into position. Tighten nuts.

b. Instrument Panel Filter (fig. 188).

(1) SERVICE. Keep connections tight. Disconnect leads from terminals when testing for open or short circuit to ground. Replace if faulty.

(2) REMOVAL. Remove the 11 cross-recessed screws which secure the instrument panel cover to the panel. Lift off the panel cover. Disconnect the lead from the ammeter to the filter, at the filter. Disconnect the lead from the control panel, at the filter. Remove the two nuts, bolts, and copper washers which secure the brackets around the radio interference filter. Lift out the filter and bracket.

(3) INSTALLATION. Place the radio interference filter in position, then place the bracket around the filter. Install the two retaining nuts, bolts, and copper washers. Connect the lead from the ammeter to the filter. Connect the lead wire from the control panel to the filter. Place the instrument panel cover in position, and install the 11 cross-recessed retaining screws.

c. Control Panel Condenser (fig. 185).

(1) REMOVAL. Loosen the 12 cover retaining bolts and swing clear. Lift off cover. Disconnect lead from distribution panel. Remove retaining nut. Lift out condenser.

(2) INSTALLATION. Place condenser in position. Install lock washer and place lead over terminal; install lock washer and nut.

LANDING VEHICLE TRACKED MK. I AND MK. II

d. **Radio Junction Box Condenser.**

(1) REMOVAL. Loosen the six cover retaining bolts and swing clear. Lift off cover. Remove screw holding terminal block at battery terminal end (fig. 117). Disconnect lead. Lift out condenser.

(2) INSTALLATION. Place condenser with mounting strap between box clip and terminal block. Install and tighten terminal block screw. Connect lead and tighten nut.

e. **Auxiliary Engine Generator Filter (on LVT (A) (1) Models Only).**

(1) REMOVAL. Remove control tunnel side panels, making filter box available. Remove 10 round-head screws. Remove cover. Disconnect generator lead at filter. Disconnect terminal block strap at filter. Remove screws and washers holding bracket in place. Lift out filter and bracket.

(2) INSTALLATION. Place filter in position, then place bracket around the filter. Install mounting screws and lock washers. Connect lead from generator and strap from terminal block. Place cover in position, and install the 10 cover retaining screws. Replace control tunnel side panels.

Section XXV

CLUTCH ASSEMBLY

	Paragraph
Clutch assembly	114

114. CLUTCH ASSEMBLY.

a. *Description.* The clutch is located on the forward side of the flywheel (fig. 28), and provides a means of gradually applying the power of the engine to the power train of the vehicle. When the clutch is disengaged (clutch pedal depressed), the clutch drive plates (lined) and driven plates (unlined) are separated, and impart no motion to the power train. Gears can be changed only when the clutch is disengaged. Clutch engagement, especially when the vehicle is at rest, must be gradual to insure against stalling the engine, or damaging the clutch or power train.

b. *Maintenance.*

(1) INSPECTION AND REPLACEMENTS. Check the facings for wear, and inspect the unlined plates for cracks and abrasions. Check the condition of all bearings. Install new parts for worn or damaged parts. Check linkage for rust and corrosion.

(2) ADJUSTMENTS OF CLUTCH.

(a) *Adjustment of Linkage (Pedal Adjustment) (fig. 120).* As the clutch facing wears, the amount of pedal free play is reduced. In

CLUTCH ASSEMBLY

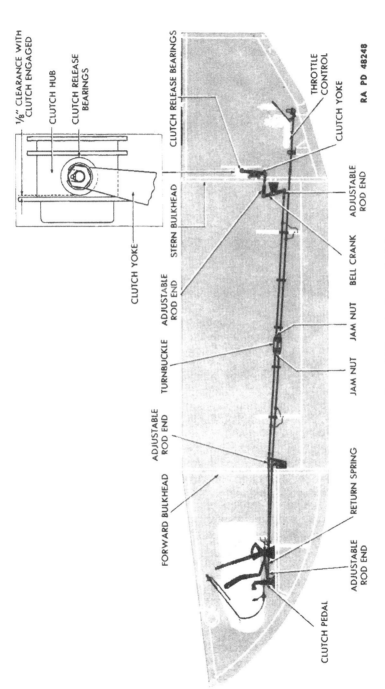

Figure 120 — Clutch and Throttle Control Linkage

TM 9-775
LANDING VEHICLE TRACKED MK. I AND MK. II

Figure 121 — Removing Clutch Spring Housing

time, this will result in slipping of the clutch due to lack of clearance at the release bearing. Unless this free travel is maintained by adjustment, the pedal will finally ride the stop and prevent full engagement. This condition allows the clutch plates to slip and burn, and makes early replacement necessary. Adjust the linkage as follows: Remove either the right or left control tunnel side plate. With the clutch engaged, measure from the center of the lower hole of the bell crank to the rear bulkhead. CAUTION: *Be sure this distance is* $3\frac{3}{4}$ *inches.* Remove the power take-off support case cover. Loosen the jam nuts on the turnbuckle, and adjust the clutch linkage so there is $\frac{1}{16}$-inch clearance between the clutch release bearing and the clutch forward flange. This will result in approximately $\frac{1}{2}$-inch free play in the clutch pedal, which is the desired amount of free travel. Total sleeve movement should be $1\frac{1}{16}$ inches, and never less than $1\frac{3}{16}$ inch.

(b) *Major Clutch Adjustment.* Although a worn clutch should be replaced, a new one may not be available in an emergency, and in

TM 9-775
114

CLUTCH ASSEMBLY

Figure 122 — Removing Clutch Spindle

this case, a major clutch adjustment will stop the clutch from slipping. This adjustment is made to restore 1 1/16-inches clearance from forward edge of clutch collar sleeve to the front flange (with the clutch engaged). The adjustment involves removing a shim from under each clutch release arm pivot pad. This is accomplished by removing the two elastic stop nuts from each of the pads, and pulling out one shim. Check the clearance between the clutch release arms (fingers) and the pivot pads on the shims. This clearance must be 0.004 inch at all three points. NOTE: *Adjusting for 0.004-inch clearance may slightly change the 1 1/16-inches measurement desired at the flange, but not enough to matter.*

c. **Removal.** Take out the cotter pin, and remove the clutch spindle nut and washer from the end of clutch spindle. Pull the clutch yoke off the spindle. Cut safety wire, and remove 12 cap screws which secure clutch spring housing. Lift off spring housing (fig. 121). Remove five clutch plates (three unlined and two lined). Screw slide hammer clutch spindle puller (41-P-2957-40) onto threaded end of clutch spindle and pull spindle; or remove clutch spindle by inserting 1/2-inch standard thread bolts in two

245

TM 9-775
114

LANDING VEHICLE TRACKED MK. I AND MK. II

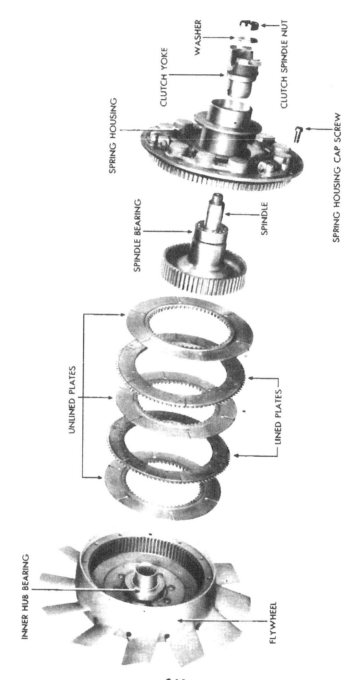

Figure 123 — Clutch Assembly — Disassembled

PROPELLER SHAFTS AND UNIVERSAL JOINTS

threaded holes of spindle and alternately turning bolts clockwise to force out spindle (fig. 122).

d. **Installation** (fig. 123). Place the clutch spindle on crankshaft, and tap spindle on the inner hub bearing with a soft hammer. Place the five clutch plates on the crankshaft, beginning with an unlined steel plate, and alternating with a lined plate. Place the spring housing assembly on the crankshaft so the mark (.) on the flywheel tooth fits between the two marks (o o) on the teeth of the spring housing. Install 12 cap screws which secure the spring housing. Lace the cap screws with safety wire. Place the clutch yoke on the crankshaft. Place the flat washer on the crankshaft, and install the clutch spindle nut. Secure the nut with a cotter pin.

Section XXVI
PROPELLER SHAFTS AND UNIVERSAL JOINTS

	Paragraph
Propeller shafts and universal joints	115
Power take-off	116

115. PROPELLER SHAFTS AND UNIVERSAL JOINTS.

a. **Description.** Two propeller shafts are used in the landing vehicle tracked. The rear propeller shaft is connected to the universal joint assembly on the engine, and runs to the center of the vehicle. The front propeller shaft is connected to the universal joint assembly on the transmission, and also runs to the center of the vehicle. Between the two propeller shafts is mounted a power take-off, to which is attached the bilge pump drive shaft. This drive shaft extends vertically down to the bilge pump, and drives the bilge pump whenever the propeller shafts are turning. Each propeller shaft is equipped with a universal joint at each end, and a slip joint which permits slight variations in the length of the shaft. Propeller shafts are enclosed in a propeller shaft housing, which extends the full length of the cargo compartment. This forms the top of the control tunnel, so-called because side plates, extending down from both sides of the housing to the floor, form a tunnel through which pass all engine controls leading into the cab.

b. **Maintenance.** Maintenance of propeller shafts and universal joints consist of frequent and proper lubrication (sec. VIII). No repair should be attempted on any part of the propeller shafts or universal joints. Shafts and joints are balanced at manufacture, and any attempt to repair the unit will destroy the balance of the propeller shafts and universal joints, and will result in noisy and inefficient operation.

TM 9-775
LANDING VEHICLE TRACKED MK. I AND MK. II

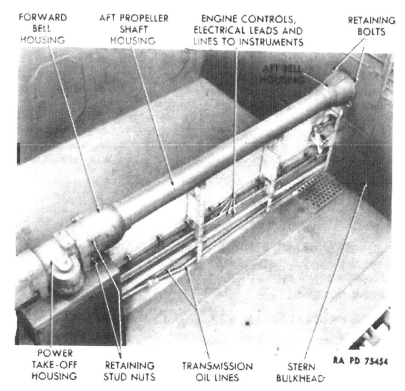

Figure 124 — Control Tunnel — Side Plate Removed

c. **Removal.**

(1) GENERAL. The removal of either the front or rear propeller shaft is accomplished in much the same manner. One end of each of the propeller shafts is connected to the power take-off at the center of the vehicle. All four universal joints are identical. In the following procedure the rear propeller shaft, with its two bell housings, is removed.

(2) REMOVE CONTROL TUNNEL SIDE PLATES (fig. 124). Remove the 16 cap screws and toothed lock washers which secure one of the control tunnel side plates. Lift out any ammunition boxes which are in the way, and the ammunition box sta-rod, and remove the side plate. Remove the opposite plate in the same manner.

(3) DISCONNECT BELL HOUSING CASTINGS (fig. 125). Remove the four nuts, bolts, and toothed lock washers which secure the rear bell housing to the rear bulkhead. Slide the housing back on the propeller shaft housing, thus providing access to the universal joints connecting the rear propeller shaft to the engine (fig. 32). Remove

TM 9-775
115

PROPELLER SHAFTS AND UNIVERSAL JOINTS

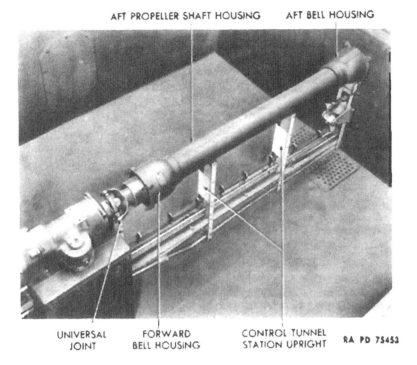

Figure 125 — Propeller Shaft Housing and Forward Bell Housing Casting in Position to Disconnect Forward Universal Joint

the four stud nuts and toothed lock washers which secure the forward bell housing to the power take-off housing (fig. 124). Slide the bell housing back on the propeller shaft housing (fig. 125).

(4) DISCONNECT UNIVERSAL JOINTS. Tap down the locking lugs which lock the universal joint retaining cap screws. Remove the cap screws (four each universal joint) with locking lugs (two each universal joint). Tap the universal joints apart, then lift the two needle bearing retainers off the prongs of each universal joint.

(5) DISCONNECT CONTROL TUNNEL STATION UPRIGHTS (fig. 125). Remove the two nuts, bolts, and toothed lock washers which secure each control tunnel station upright to the bracket welded on the propeller shaft housing.

(6) REMOVE PROPELLER SHAFT AND HOUSING (fig. 126). Lift the propeller shaft with assembled propeller shaft housing from the control tunnel station uprights, and remove.

(7) DISASSEMBLE PROPELLER SHAFT AND PROPELLER SHAFT HOUSING. Unscrew the dust collar on the rear end of the propeller

TM 9-775
LANDING VEHICLE TRACKED MK. I AND MK. II

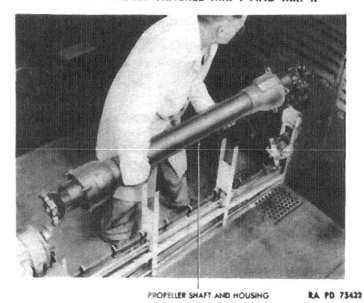

Figure 126 — Lifting Off Propeller Shaft and Housing

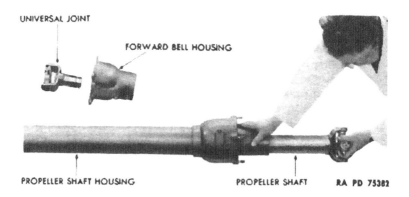

Figure 127 — Removing Propeller Shaft

shaft (fig. 127). Pull the splined universal joint off the shaft. From the opposite end of the housing, slide the propeller shaft out of the propeller shaft housing (fig. 127). Slide front bell housing and rear bell housing off the propeller shaft housing.

d. **Installation.**

(1) ASSEMBLE PROPELLER SHAFT TO PROPELLER SHAFT HOUSING. Slide the front bell housing and rear bell housing on the pro-

TM 9-775
115

PROPELLER SHAFTS AND UNIVERSAL JOINTS

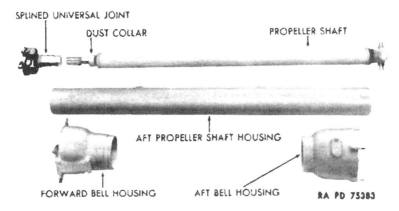

Figure 128 — Propeller Shaft and Propeller Shaft Housing

peller shaft housing (fig. 128). Slide the propeller shaft into the propeller shaft housing (fig. 127). Line up the arrows (one on the universal joint and one on the large part of the propeller shaft), and slide the universal joint on the splines of the propeller shaft. Tighten dust collar which secures splined universal joint on propeller shaft.

(2) PLACE PROPELLER SHAFT AND HOUSING IN POSITION. Lift the propeller shaft with the propeller shaft housing in position on the control tunnel station uprights (fig. 126).

(3) CONNECT CONTROL TUNNEL STATION UPRIGHTS. Install the two nuts, bolts, and toothed lock washers which secure each control tunnel upright to the bracket welded on the propeller shaft housing.

(4) CONNECT UNIVERSAL JOINTS (fig. 125). Place a needle bearing retainer on each prong of the universal joint. Work either universal joint into position against the other half of the joint. Slip two cap screws into a locking lug, then install the cap screws through the universal joint. Bend the locking lug up to lock the cap screws securely in position.

(5) CONNECT BELL HOUSINGS (fig. 124). Slide the bell housings up into position. Install the four stud nuts and toothed lock washers which secure the front bell housing to the power take-off housing. Install the four nuts, bolts, and toothed lock washers which secure the rear bell housing to the stern bulkhead.

(6) INSTALL CONTROL TUNNEL SIDE PLATES (fig. 129). Place the control tunnel side plate in position, and install the 16 retaining cap screws and toothed lock washers. Install the opposite side plate in the same manner. Stow ammunition boxes, and install ammunition box sta-rod.

TM 9-775
LANDING VEHICLE TRACKED MK. I AND MK. II

Figure 129 — Removing Power Take-off Retaining Cap Screws

116. POWER TAKE-OFF.

a. **Description.** Mounted between the front and rear propeller shafts is the power take-off. The universal joints of the power take-off are connected to the universal joints of the propeller shafts. The power take-off housing is connected to the propeller shaft housing. Function of the power take-off is to provide a gear-driven method of driving the bilge pump. This power is transmitted by a bilge pump drive shaft connected by chain-coupled sprockets (one each end) to the power take-off, and to the bilge pump.

b. **Maintenance.** Lubrication of the power take-off is accomplished through Zerk fittings. Method and frequency of lubrication are covered in the Lubrication Guide (sec. VIII).

c. **Removal.**

(1) DISCONNECT UNIVERSAL JOINTS. Remove the four cap screws and toothed lock washers securing the rear end of the front control

TM 9-775
116

PROPELLER SHAFTS AND UNIVERSAL JOINTS

POWER TAKE-OFF VERTICAL RETAINING TOOTHED
SUPPORT CASE   SHAFT   CAP SCREWS LOCK WASHERS   RA PD 75402

Figure 130 — Lifting Off Power Take-off

tunnel side plates. Remove the four cap screws and toothed lock washers securing the front end of the rear control tunnel side plates. Spread plates apart slightly. Disconnect propeller shaft bell housings at the power take-off, and slide the housings back on the propeller shaft housing (par. 115). Disconnect the universal joints at each end of the power take-off (par. 115).

(2) DISCONNECT BILGE PUMP DRIVE SHAFT. The complete removal of the bilge pump drive shaft is covered under the removal of the bilge pump (par. 156). However, to remove the power take-off, it is necessary only to disconnect the bilge pump drive shaft at the top, where it connects to the power take-off (par. 156).

(3) REMOVE COUPLING SPROCKET. Loosen the set screw which secures the sprocket to the power take-off vertical shaft from within the power take-off support case. Pry the sprocket off the shaft. This operation is identical to that of removing the sprocket from the bilge pump vertical shaft (par. 156). CAUTION: *Take care not to lose the key which locks the coupling sprocket on the shaft.*

253

LANDING VEHICLE TRACKED MK. I AND MK. II

(4) REMOVE POWER TAKE-OFF. Remove the four cap screws and toothed lock washers which secure the power take-off to the power take-off support case. These must be turned from within the power take-off support case (fig. 129). Lift off the power take-off (fig. 130).

d. Installation.

(1) PLACE POWER TAKE-OFF IN POSITION. Place the power take-off in position, and install the four cap screws and toothed lock washers which secure the power take-off to the power take-off support case (figs. 129 and 130). These cap screws must be installed upward from within the power take-off support case.

(2) INSTALL COUPLING SPROCKET. Slide a coupling sprocket on the vertical shaft of the power take-off. Tap the key in place to lock the sprocket on the shaft. Tighten the sprocket set screw.

(3) CONNECT BILGE PUMP DRIVE SHAFT. Line up the sprocket at the top of the bilge pump drive shaft with the coupling sprocket just installed on the vertical shaft of the power take-off. Wrap the coupling chain around the two sprockets. Slide the master link through the two ends of the chain, inserting the two center chain links into one end of the chain link. Install the master link lock.

(4) CONNECT UNIVERSAL JOINTS. Connect the propeller shaft universal joints to the universal joints on the power take-off (par. 115).

(5) CONNECT PROPELLER SHAFT HOUSING. Slide the propeller shaft bell housing into position against the power take-off housing. Connect the bell housing to the propeller shaft housing (par. 115).

(6) SECURE CONTROL TUNNEL SIDE PLATES. Install the four cap screws and toothed lock washers which secure the forward end of each of the four control tunnel side plates.

Section XXVII

FINAL DRIVE

	Paragraph
Final drive assembly	117
Driving sprockets and sprocket hubs	118

117. FINAL DRIVE ASSEMBLY.

a. Description. The final drive assembly is composed of two units; an interior, and an exterior unit. The interior unit is bolted to the transmission, while the exterior unit is coupled to the interior unit by a through bolt. Both units are supported by a bell housing bolted to the hull. A coupling, connected by a flange adapter to the interior final drive unit, forms a base for a through bolt which projects outward through the bell housing. The splined exterior final drive assembly

TM 9-775
117

FINAL DRIVE

Figure 131 — Lifting Off Driving Sprocket

passes through the bell housing, over the through bolt, and meshes with splines within the coupling. The driving sprocket is attached to studs in the outer end of the final drive assembly.

b. Maintenance. Maintenance of the final drive assembly consists solely of replacing those parts which, through wear or damage, become defective.

c. Removal.

(1) BREAK TRACK. (par. 128). Break the track, then roll the upper section of the track back from the driving sprocket.

(2) REMOVE DRIVING SPROCKET. Remove stud nuts and washers which secure the driving sprocket to the final drive assembly. Lift off the driving sprocket (fig. 131). CAUTION: *The driving sprocket weighs more than 100 pounds, and should be lifted off by two men. When doing so, protect hands with rags, since the teeth of the sprocket are very sharp.*

(3) REMOVE FINAL DRIVE ASSEMBLY. Remove cotter pin and hexagonal nut securing final drive assembly through bolt nut. Remove

255

TM 9-775
LANDING VEHICLE TRACKED MK. I AND MK. II

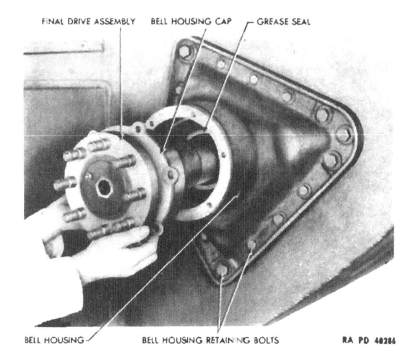

Figure 132 — Removing Final Drive Assembly

the six cap screws and toothed lock washers which secure the bell housing cap to the bell housing. Tap around the circumference of the final drive assembly with a soft hammer to free the bell housing cap, and loosen the final drive assembly in its seat. Pull the final drive assembly out of the bell housing (fig. 132). CAUTION: *Pull the final drive assembly straight out slowly, in order that the bell housing grease seal is not damaged. If the final drive assembly is tilted, or if the weight of the assembly is permitted to drop down, the grease seal within the bell housing will be ruined.*

(4) REMOVE BELL HOUSING. Remove the 16 bolts, nuts, and toothed lock washers that secure the bell housing to the hull. Support the bell housing while removing the bolts and nuts in order to prevent it from dropping free of the vehicle. Lift the bell housing from the vehicle (fig. 133). CAUTION: *Pull bell housing straight out, over the through bolt. Do not drop bell housing down on through bolt, or the grease seal within the bell housing will be damaged.*

(5) REMOVE FINAL DRIVE COUPLING (fig. 134). Remove the 12 bolts, nuts, and toothed lock washers which secure the coupling to the

TM 9-775
117

FINAL DRIVE

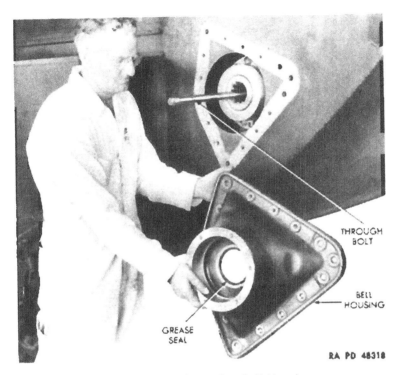

Figure 133 — Removing Bell Housing

flange adapter on the interior final drive. Tap the coupling free, and lift the coupling, with gasket attached, out of the vehicle.

(6) DISASSEMBLE COUPLING (figs. 135 and 136). Push the through bolt out through the male supporting ring, then pull out the hub from the supporting ring.

d. Installation.

(1) ASSEMBLE COUPLING (fig. 136). Place the hub within the male supporting ring. Push the through bolt into position in the supporting ring.

(2) INSTALL COUPLING. Shellac a new gasket on the male supporting ring. Lift the coupling hub into position against the flange adapter on the interior final drive. Install the 12 bolts, nuts, and toothed lock washers which secure the coupling to the flange adapter.

(3) INSTALL BELL HOUSING. Coat the mounting face of the bell housing with joint sealing compound. Carefully lift the bell housing up and slide it over the through bolt against the hull. CAUTION: *Do not drop bell housing down against through bolt, as this will damage*

257

TM 9-775
LANDING VEHICLE TRACKED MK. I AND MK. II

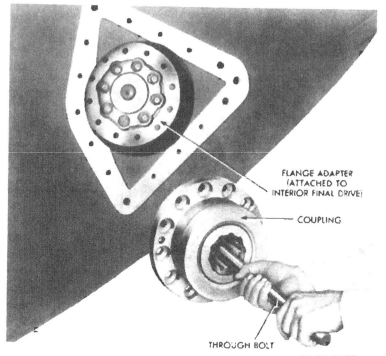

Figure 134 — Removing Final Drive Coupling

the grease seal within the bell housing. Install the 16 bolts, nuts, and toothed lock washers which secure the bell housing to the hull. This will require two men.

(4) INSTALL FINAL DRIVE ASSEMBLY (fig. 137). Insert the special final drive installing bar (41-B-19-540) firmly into the final drive assembly. Using the bar, lift the final drive assembly up, and carefully insert the final drive assembly over the through bolt which extends out from the hull. Use the bar as a guide so that, when inserting the final drive shaft through the bell housing, the bell housing grease seal is not damaged. Mesh the splines of the final drive shaft with the splined slots within the coupling. Line up the screw holes in the bell housing cap with screw holes in the bell housing. Install the six cap screws and toothed lock washers which secure the bell housing cap to the bell housing. Slip the through bolt washer over the end of the through bolt, and install the through bolt hexagonal nut and cotter pin.

(5) INSTALL DRIVING SPROCKET. Lift the driving sprocket hub into position over the bell housing and on the studs of the final drive assembly (two men). Install the stud nuts and washers which secure the sprocket to the final drive.

TM 9-775
117

FINAL DRIVE

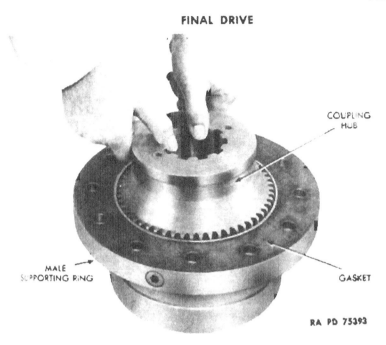

Figure 135 — Removing Coupling Hub

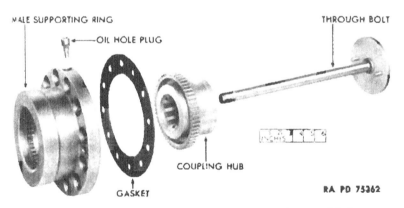

Figure 136 — Final Drive Coupling — Disassembled

(6) CONNECT TRACK. Roll the upper section of the track over the driving sprocket to facilitate meshing track with sprocket. Shift the transmission into third gear, and turn the sprocket over slowly with the cranking motor, to roll the track over the sprocket. Connect the track (par. 129).

259

TM 9-775
LANDING VEHICLE TRACKED MK. I AND MK. II

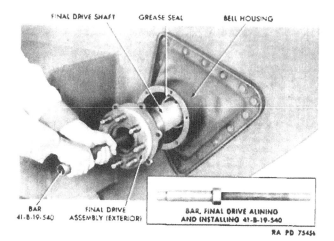

Figure 137 — Installing Final Drive Assembly

Figure 138 — Driving Sprocket Removed from Sprocket Hub

POWER TRAIN

118. DRIVING SPROCKETS AND SPROCKET HUBS.

a. **General.** Each driving sprocket consists of a sprocket hub on which is mounted two sprockets. Each sprocket is identical. Removal of any of the sprockets from either of the hubs is accomplished in the same manner.

b. **Maintenance.** If sprocket hubs are cracked or otherwise damaged, replace sprocket hubs. If sprockets are cracked, or if teeth have been broken off, install a new sprocket on the sprocket hub.

c. **Removal (fig. 138).** Remove the 14 cap screws and toothed lock washers which secure the sprocket to the sprocket hub. Lift off the sprocket. Remove the other sprockets in the same manner.

d. **Installation.** Place the sprocket on the sprocket hub. Aline the punched "o" mark on one sprocket with punched "o" mark on the opposite sprocket. Install the 14 retaining cap screws and toothed lock washers. Install remaining sprockets in the same manner.

Section XXVIII

POWER TRAIN

	Paragraph
Description	119
Transmission and differential lubrication system	120
Removal of transmission and differential	121
Installation of transmission and differential	122
Adjustment of brake shoes	123
Flange adapter	124
Controlled differential	125

119. DESCRIPTION.

a. **General.** Power is transmitted from the propeller shafts to the tracks by means of the power train, which consists of the transmission, controlled differential, final drives, and driving sprockets.

b. **Transmission.** The transmission is of the synchromesh type, having five forward speeds and one reverse. First and reverse speeds are through a sliding gear. The transmission and differential are enclosed in one case called the transmission and differential case (fig. 139).

c. **Differential.** The controlled differential transmits engine power to the final drive, and also incorporates a device for steering the tank. The differential contains two "halves," each half being separately controlled by a brake drum and brake shoe. When one brake is applied,

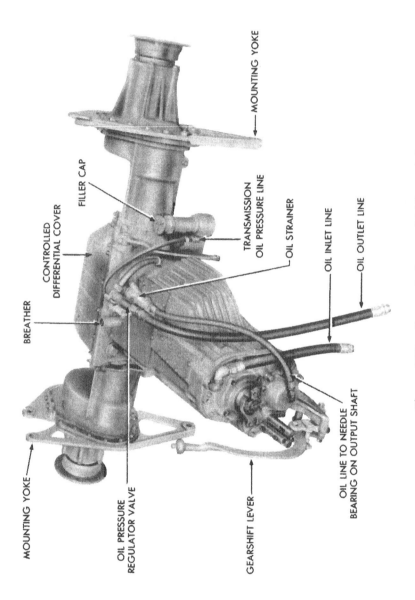

Figure 139 — Transmission and Differential

POWER TRAIN

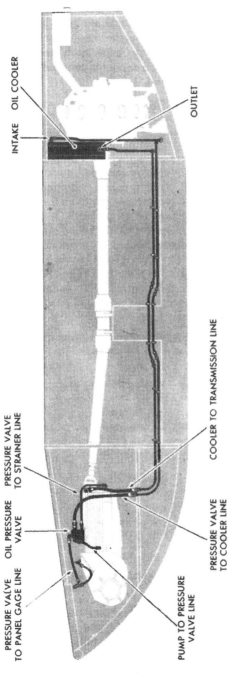

Figure 140 — Transmission and Differential Lubrication System

one-half of the differential slows down while the other half increases in speed. By this means the vehicle is steered. It is impossible to lock one final drive sprocket as long as the other sprocket is in motion. Stopping the vehicle is possible only when both steering levers are simultaneously pulled fully to the rear, thus contracting the brake shoes in each "half" of the differential.

120. TRANSMISSION AND DIFFERENTIAL LUBRICATION SYSTEM.

a. Lubrication of the transmission and differential is performed by the same oiling system (fig 140). An oil pump, mounted in the transmission case, draws oil from the screened pocket in the bottom of the differential case. From the pump, oil is pumped under pressure to the pressure regulator valve located on top of the transmission. This valve is for the purpose of controlling oil pressure in the transmission. Two main outlet lines lead from the valve: The main oil line leads to the oil cooler where oil is cooled, and is then returned to the transmission gears for lubrication and recirculation through the system; the other line leads to a disk-type oil strainer, and from the strainer to the needle bearing on the transmission output shaft. Oil then passes into the screened pocket in the bottom of the differential case, and is picked up by the oil pump for recirculation throughout the lubrication system.

b. A third outlet line from the valve runs to the instrument panel, indicating on the transmission oil pressure gage the pressure of oil circulating through the transmission.

c. Transmission oil cooler is mounted in the engine room, on the left side of the bulkhead. This unit is similar, in structure and function, to the engine oil cooler. A pressure relief valve in the cooler outlet provides an automatic means for bypassing oil, when cold, to avoid damaging the core of the cooler.

d. The entire lubrication system must be given regular periodic inspections to detect any leaks or damage to oil lines before the damage can cause loss of transmission oil pressure. It is particularly important to check proper seating of the oil temperature gage bulb, since air leaks may develop at this point. Regular inspections of the lubrication system are included in the preventive maintenance service sections (secs. VII and XI).

121. REMOVAL OF TRANSMISSION AND DIFFERENTIAL.

a. Disconnect Headlights. Remove control panel cover (par. 140). Disconnect left headlight lead, and pull lead out of control panel. Disconnect bolt, nut, and flat washer which secure the lead to the left front corner of the hull. Remove four bolts and nuts which secure the headlight receptacle box cover. Lift off cover. Disconnect the right

POWER TRAIN

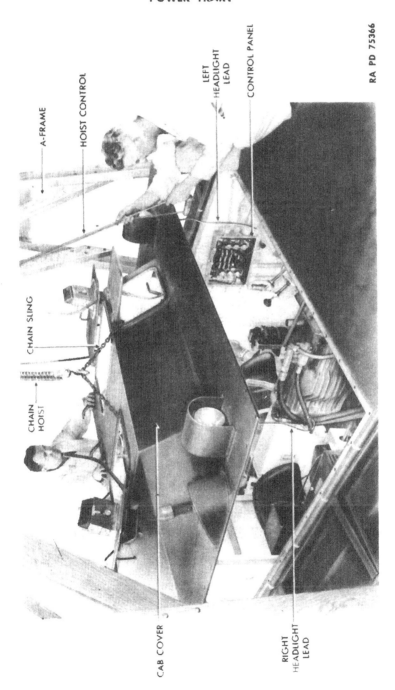

Figure 141 — Removing Cab Cover

TM 9-775
121
LANDING VEHICLE TRACKED MK. I AND MK. II

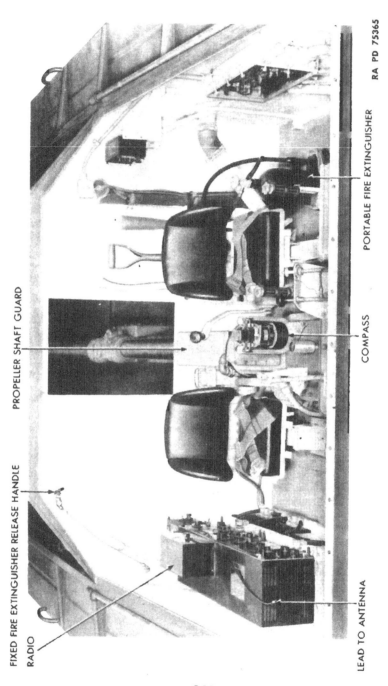

Figure 142 — Cab — Cover Removed

TM 9-775

POWER TRAIN

headlight lead, and pull out the lead. Remove bolt, nut, and flat washer from the clamp which holds the lead to the hull.

b. **Disconnect Radio Antenna.** Loosen the knurled nut, and pull out the wire lead to the radio antenna from the radio loading coil.

c. **Remove Cab Cover** (fig. 141). Remove cross-recessed screws, bolts, cap screws, and stud nuts which secure the cab cover to the hull. Minor differences will be noticed in the removal of the cab covers used on earlier and later models of the landing vehicles tracked, but in general all cabs are removed in the same manner. Install a chain sling on the cover. Hook a hoist to the sling, and carefully lift the cover from the vehicle.

d. **Remove Compass** (fig. 14). Remove the three cap screws and flat washers which secure the compass to the top of the transmission. Lift off the compass.

e. **Remove Radio and Miscellaneous Equipment** (figs. 10 and 142). Remove radio transmitter, receiver, dynamotor, and amplifier. Lift the water tank straight up out of its retaining bracket, then compress the tank wire bracket together and remove the bracket (fig. 10). Lift out the signal light case. Swivel the hand lantern to a horizontal position in its bracket in the right side of the transmission, and lift out the lantern (fig. 24). Remove ropes lying on the floor of the cab (fig. 11). Make sure all personal equipment is removed from the floor of the cab.

f. **Disconnect Instrument Panel.**

(1) GENERAL. In disconnecting the instrument panel (in order to remove the transmission) it is not necessary to disconnect all wires and tubing leading to the instrument panel. Only those which enter through the back of the panel must be disconnected. The panel may then be swung to one side out of the way.

(2) DISCONNECT INSTRUMENT PANEL COVER. Remove 11 cross-recessed screws which secure the instrument panel cover to the instrument panel. Tilt cover back and downward away from panel.

(3) DISCONNECT ENGINE AND TRANSMISSION OIL PRESSURE LINES (fig. 143). The engine and transmission oil pressure lines must be disconnected in order to remove the transmission. On earlier model vehicles the construction of the lines was such that they had to be disconnected at the fittings just in back of, and above, the instrument panel. The lines could not be disconnected at the gages. In later model vehicles, coiled sections of copper tubing have been connected to the gages so that the lines can be disconnected at the gages. CAUTION: *When disconnecting the lines at the junction of the flexible tubings with the copper tubings, always hold the nut on the flexible tubing while turning the nut on the copper tubing. This will prevent*

TM 9-775
121
LANDING VEHICLE TRACKED MK. I AND MK. II

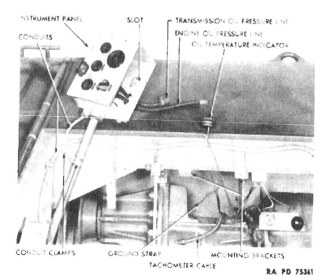

Figure 143 — Instrument Panel Removed

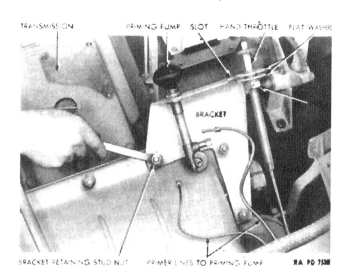

Figure 144 — Removing Priming Pump and Hand Throttle Bracket

POWER TRAIN

damage to the lines. Remove nut and screw on clamps which secure the lines to brackets mounted on the hull.

(4) DISCONNECT TACHOMETER CABLE (fig. 143). Unscrew knurled nut which secures tachometer cable in the back of the tachometer. Pull the cable from the tachometer, and push it out the back of the instrument panel.

(5) DISCONNECT OIL TEMPERATURE INDICATOR (fig. 143). On early model vehicles the oil temperature indicator line is fed through a hole in the back of the panel, through the control tunnel, and back to the engine. On later model vehicles, the oil temperature indicator line comes up under the bottom of the instrument panel. In order to simplify the removal of the indicator on early model vehicles, it is necessary to cut a slot from the opening in the back of the panel, through which passes the indicator line, down to the edge of the panel. In this manner, the line may be slid down out of the panel. Remove the two bolts, nuts, and washers securing the oil temperature indicator to the back face of the instrument panel cover. Drop the gage down toward the bottom of the panel, and slide the line out through the slot cut in the back of the panel.

(6) DISCONNECT INSTRUMENT PANEL GROUND STRAP (fig. 147). Remove cross-recessed screw which holds ground strap to rear of the instrument panel.

(7) REMOVE INSTRUMENT PANEL. Remove the four cap screws (two each side) securing the instrument panel brackets to the hull. Remove two bolts, nuts, and flat washers securing the instrument panel to the two brackets extending out from the hull to the top of the panel. Remove the two bolts, nuts, and flat washers securing the two clamps which hold the two conduits leading into the left side of the instrument panel. Carefully lift the instrument panel from its brackets, and lay it forward on top of the hull (fig. 147).

g. Disconnect Priming Pump and Hand Throttle Bracket (fig. 144). Disconnect the two primer lines to the priming pump. Loosen nut which secures the hand throttle cable in its slot in the bracket. Lift hand throttle and attached cable with nut and two flat washers, out of slot. Remove three stud nuts securing the bracket to the transmission, and remove the bracket with installed priming pump.

h. Disconnect Instrument Line Block. Remove the one bolt, nut, and toothed lock washer which secure the instrument lines block to the back of the transmission. This wood block, made in two halves, holds in position the two primer lines, engine oil pressure line, tachometer cable, and oil temperature indicator line.

i. Disconnect Transmission Oil Pressure Line (fig. 145). The transmission oil pressure line has already been disconnected behind the instrument panel. However, another section of flexible tubing runs

TM 9-775
LANDING VEHICLE TRACKED MK. I AND MK. II

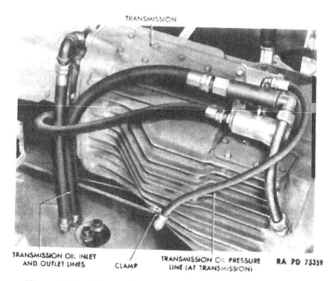

Figure 145 — Transmission Oil Lines — Disconnected

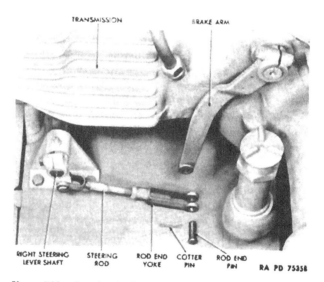

Figure 146 — Steering Brake Linkage (Right Steering Lever)

POWER TRAIN

PROPELLER SHAFT   UNIVERSAL JOINT   TRANSMISSION   RA PD 75357

Figure 147 — Propeller Shaft Disconnected at Transmission

from the copper tubing under the top of the hull to the transmission. In order to free the transmission, remove the cap screw from the clamp which holds this piece of flexible tubing to the hull, in front of the cab fan. Disconnect the line at the junction of the copper tubing and flexible tubing. Lay the flexible tubing on the transmission. There is no need to disconnect the opposite end of the tubing (at the transmission), unless a new line must be installed.

    j. **Disconnect Transmission Oil Lines** (fig. 145). Disconnect the transmission oil inlet and outlet lines at the fittings extending up through cab floor to the left of the assistant driver's seat.

    k. **Disconnect Steering Brake Linkage** (fig. 146). Remove cotter pin and rod end pin securing the steering brake linkage of the right steering lever to the brake arm mounted on transmission. Disconnect the steering brake linkage of the left steering lever in same manner.

    l. **Remove Propeller Shaft.** Remove the six cross-recessed screws (three each side) with flat washers and toothed lock washers which secure the propeller shaft guard at the rear of transmission (fig. 142). Lift off guard. Remove locking wire and four cap screws with lock washers which secure front propeller shaft bracket to transmission. Lift off the bracket. Remove front propeller shaft housing, disconnect front propeller shaft (fig. 147) and remove propeller shaft and housing (par. 115). NOTE: *The front propeller shaft bracket on the trans-*

271

TM 9-775
121
LANDING VEHICLE TRACKED MK. I AND MK. II

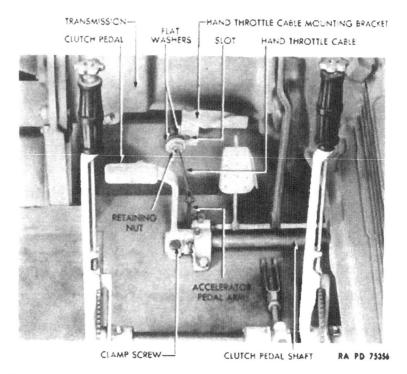

Figure 148 — Hand Throttle Cable Disconnected from Mounting Bracket

mission must be removed in order to provide room to disconnect the front propeller shaft to universal joint at the transmission.

m. Disconnect Hand Throttle Cable (fig. 148). Loosen the two nuts (front and rear) which secure the hand throttle cable to the bracket beneath transmission, directly behind clutch pedal. Slide cable, two flat washers (one each side of bracket), and retaining nuts out of slot at side of bracket. It is not necessary to disconnect cable from accelerator pedal arm.

n. Remove Clutch Pedal (fig. 148). Remove the one cap screw and washer which clamp the clutch pedal on the splined clutch pedal shaft. Tap the clutch pedal off the shaft.

o. Remove Floor Sections. Remove the four self-tapping screws with dome lock washers which secure the left front floor section. Lift out the floor section. Remove the five self-tapping screws with dome lock washers which secure the right front floor section.

p. Remove Transmission and Differential Mounting Bolts (fig. 149). Remove cotter pins and four bolts and nuts which secure each

POWER TRAIN

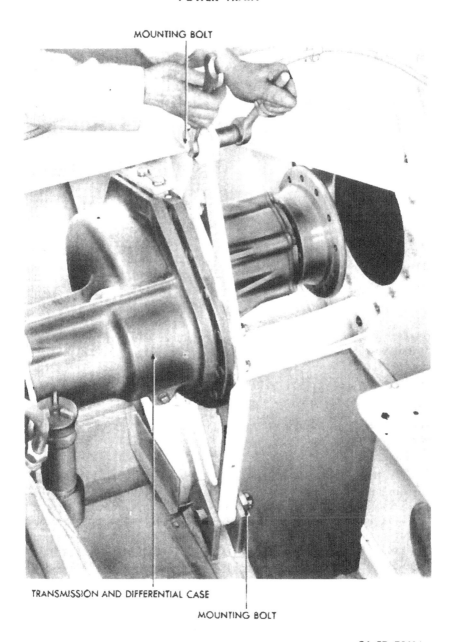

Figure 149 — Removing Transmission and Differential Mounting Bolts

TM 9-775
121—122

LANDING VEHICLE TRACKED MK. I AND MK. II

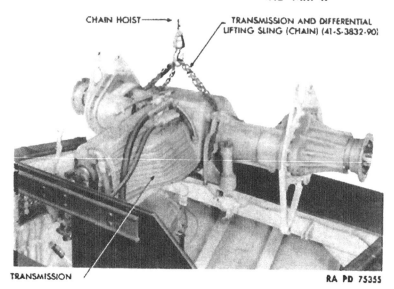

Figure 150 — Transmission and Differential Lifting Sling (41-S-3832-90) — Installed

transmission and differential mounting yoke (right and left) to the mounting brackets welded to the hull.

q. Remove Transmission and Differential Lifting Sling (fig. 150). Wrap chain lifting sling (41-S-3832-90) around the center of the transmission and differential just in front of the brake arms mounted on the splined shafts extending out from the controlled differential. At this point, the transmission and differential are perfectly balanced.

r. Remove Transmission and Differential (fig. 151). Connect chain hoist to the lifting sling. Carefully lift the transmission and differential assembly up and to the rear. In moving assembly back, it will be necessary to tilt the rear of assembly upward so that the gearshift lever will clear the forward bulkhead, and also so that the mounting yokes will slide out from under the hull. After assembly has been worked free of the mounting brackets, lift it straight up and out of the vehicle. Swing to one side, and lower to a previously prepared stand, or place on blocks.

122. **INSTALLATION OF TRANSMISSION AND DIFFERENTIAL.**

a. Install Transmission and Differential Lifting Sling (fig. 150). Wrap chain lifting sling (41-S-3832-90) around the center of the

274

TM 9-775
122

POWER TRAIN

Figure 151 — Lifting Transmission and Differential from Vehicle

transmission and differential assembly just in front of the brake arms mounted on the splined shafts extending out from the controlled differential.

b. **Lift Transmission into Vehicle** (fig. 151). Connect the chain hoist to the lifting sling. Lift transmission and differential assembly up over the cab, and carefully lower into the cab, tilting front slightly in order to permit mounting yokes to pass under floor of hull. CAUTION: *Make sure, when lowering assembly into the cab, not to strike against brackets and other fittings within the cab.*

c. **Install Transmission and Differential Mounting Bolts** (fig. 152). When the mounting yokes are in the welded brackets on the

Figure 152 — Alining Transmission and Differential Mounting Yokes

POWER TRAIN

hull, drive the special alining bars (41-P-560 and 41-P-560-10) through the brackets and through the mounting yokes to line up the transmission. Remove the alining bars one at a time. Install the four bolts and nuts which secure each transmission mounting yoke (right and left) to the brackets.

d. **Install Front Floor Section.** Position the left front floor section, and install the four retaining self-tapping screws with dome lock washers. Place the right front floor section in position, and install its five retaining self-tapping screws with dome lock washers.

e. **Install Clutch Pedal** (fig. 191). Slide the clutch pedal on the splines of the clutch pedal shaft so that the pedal is against the stop on the shaft mounting bracket. Install the one cap screw and washer which clamp the pedal on the shaft.

f. **Connect Hand Throttle Cable** (fig. 148). Slide the hand throttle cable into the slot on the bracket beneath the transmission, directly behind the clutch pedal. A flat washer and nut should be on each side of the bracket. Tighten the retaining nuts.

g. **Install Propeller Shaft** (figs. 127 and 147). Lift the front propeller shaft and housing up into position, and connect the propeller shafts to the power take-off and to the transmission (par. 115). Place the front propeller shaft bracket against the transmission and install the four cap screws with lock washers. Safety-wire the four cap screws. Place the propeller shaft guard over the front universal joint, and install the six retaining screws (three each side) with flat washers and toothed lock washers.

h. **Connect Steering Brake Linkage** (fig. 146). Line up the rod and yoke of the steering rod with the brake arm on the transmission, and install the rod end pin and cotter pin. Connect the steering brake linkage of the opposite steering lever in the same manner. If rod ends have not been turned on the steering rods while the transmission was out of the vehicle, steering brakes will probably not need adjusting. If an adjustment is needed, refer to the paragraph on the steering brake shoes (par. 123).

i. **Connect Transmission Oil Lines** (fig. 145). Connect the transmission inlet and outlet oil lines at the fittings extending up through the cab floor to the left of the assistant driver's seat.

j. **Connect Transmission Oil Pressure Line** (fig. 145). Connect the transmission oil pressure line at the junction of the copper tubing and flexible tubing underneath the hull. Install the cap screw and clamp which secure the flexible tubing to the hull just in front of the cab fan.

k. **Connect Instrument Line Block.** Install the bolt, nut, and toothed lock washer which secure the instrument line block in back of the transmission. This wood block, made in two halves, holds in position the two primer lines, engine oil pressure line, tachometer cable, and the oil temperature indicator line.

l. **Connect Priming Pump and Hand Throttle Bracket** (fig. 144). Place the priming pump and hand throttle bracket on the transmission. Install the three stud nuts which secure the bracket to the transmission. Place the hand throttle cable in its slot in the bracket. One of the two flat washers on the cable must be on each side of the bracket. Tighten cable retaining nut. Connect the two primer lines to the two primer fittings on the priming pump.

m. **Mount Instrumental Panel** (fig. 143). Lift the instrument panel from on top of the hull where it was placed out of the way. Install the two bolts, nuts, and flat washers which secure the instrument panel to the upper brackets extending out from the hull. Install the four cap screws (two each side) which secure the instrument panel lower brackets to the hull. Install the two bolts, nuts, and flat washers securing the two clamps which hold the two conduits entering the left side of the instrument panel.

n. **Connect Instrument Panel Ground Strap** (fig. 143). Install a cross-recessed screw which secures the ground strap to the rear of the instrument panel.

o. **Connect Oil Temperature Indicator** (fig. 143). Lift the oil temperature indicator up through the bottom of the instrument panel, and place it in position on the back of the instrument panel cover. Install the two bolts, nuts, and washers which secure the indicator to the cover.

p. **Connect Tachometer Cable** (fig. 143). Lead tachometer cable through opening in rear of instrument panel, and tighten knurled nuts which secure tachometer cable to the back of tachometer.

q. **Connect Engine and Transmission Oil Pressure Lines** (fig. 143). Connect engine and transmission oil pressure lines at the junction of the copper tubing and the flexible tubing. Turn the nut on the copper tubing while holding the nut on the flexible tubing, to prevent damage to the lines.

r. **Connect Instrument Panel Cover.** Tilt instrument panel cover into position against instrument panel, and install 11 retaining screws.

s. **Install Radio and Miscellaneous Equipment** (figs. 10 and 142). Install radio on its brackets in right side of cab. Stow all Pioneer equipment and other vehicular equipment in their proper places within cab.

t. **Install Compass** (fig. 14). Place compass in position on transmission, and install three cap screws and flat washers which secure compass.

POWER TRAIN

u. **Install Cab Cover** (fig. 141). Loop chain sling through cab window and around cab cover. Hook a chain hoist to sling and carefully lift cover up into position on cab. Install cross-recessed screws, bolts, cap screws, and stud nuts which secure cover to hull.

v. **Connect Radio Antenna.** Connect wire lead from radio antenna to radio loading coil. Tighten knurled nut.

w. **Connect Headlights.** Insert right headlight lead into headlight receptacle box. Make sure packing is properly placed around lead in conduit opening. Connect lead to proper terminal within box. Place cover on box, and install four bolts and nuts. Install one bolt, nut, and flat washer in clamp which holds conduit to hull. Insert left headlight lead through conduit opening into control panel. Connect lead to proper terminal within control panel. Install control panel cover (par. 138).

123. ADJUSTMENT OF BRAKE SHOES.

a. From within cab remove two pipe plugs at rear of transmission. Insert a 1 1/16-inch deep socket wrench (41-W-462-200) through the pipe plug hole, and engage brake adjusting nut. Turn adjusting nut in one-half turn steps. Clockwise rotation decreases the clearance between the brake shoe and drum. Correct adjustment will be obtained when the shoe just engages as steering lever is in vertical position. The cushioning effect of the differential oil being squeezed out between shoe and drum should be felt as the lever is pulled back from the forward position. Repeat the operation to adjust the second brake shoe. Install pipe plugs. CAUTION: *When turning the brake adjusting nut, make sure that the nut is seated on its cross pin. The end of the adjusting nut is cylindrical where it contacts the cross pin.*

124. FLANGE ADAPTER.

a. **Description.** A flange adapter is used to provide a means of connecting the interior final drive to the final drive coupling. The exterior final drive is then mounted on the coupling, and driving sprockets are mounted on the exterior final drive.

b. **Maintenance.** If the flange adapter is broken or otherwise damaged, install a new flange adapter.

c. **Removal** (fig. 153). Break tracks (par. 128). Remove driving sprockets, final drive assembly, and sprocket bell housing (par. 118). Remove the eight elastic stop nuts which secure the flange adapter to the studs mounted on the interior final drive. Lift off the flange adapter.

TM 9-775
124

LANDING VEHICLE TRACKED MK. I AND MK. II

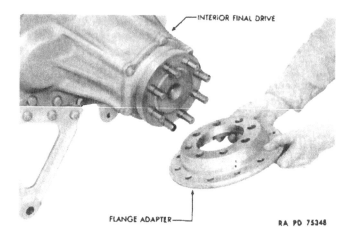

Figure 153 — Removing Flange Adapter

Figure 154 — Controlled Differential — One Brake Shoe Removed

POWER TRAIN

d. **Installation** (fig. 153). Place the flange adapter on the studs of the final drive, and install the eight retaining elastic stop nuts. Install sprocket bell housing, final drive assembly, driving sprockets (par. 118), and connect tracks (par. 129).

125. **CONTROLLED DIFFERENTIAL.**

a. **Description.** Each half of the controlled differential incorporates a brake drum and brake shoe. Pulling back on one of the steering levers contracts the brake shoe to which the lever is connected, thus slowing down that half of the differential. The vehicle is steered through the use of these brake shoes. When both brake shoes are contracted and locked in the contracted position by means of ratchet handles provided in the top of steering levers, the brake shoes act as parking brakes.

b. **Removal.**

(1) Remove transmission and differential from vehicle (par. 121).

(2) DRAIN OIL. Place oil receptacle under either or both magnetic drain plugs which are located on lower front side of transmission and differential case. Remove drain plugs. Remove oil level plugs located on front of case to facilitate drainage of oil.

(3) REMOVE CONTROLLED DIFFERENTIAL COVER. Remove 18 lock nuts and 13 flat washers (the 5 studs at front of cover have no washers) that secure cover to case. Pry cover up and off studs, using two small pry bars (personnel will use caution when using pry bars, in order not to damage cover). Remove cover gasket.

(4) REMOVE BRAKE SHOE ASSEMBLIES (fig. 154). Remove cotter pin from brake shoe pin. Remove cotter pin from link pin. Remove link pins (use small drift), and lift out brake shoe assembly by pulling it around the brake drum. Remove other brake shoe assembly in same manner.

(5) REMOVE BRAKE CAMSHAFT. Remove cap screw and lock washer in brake levers. Drive lever off shaft. Remove brake camshaft. Other brake camshaft is removed in same manner.

(6) REMOVE CONTROLLED DIFFERENTIAL ASSEMBLY (figs. 155 and 156). Remove safety wire at differential bearing support covers. Remove nuts from four studs. Remove bearing support caps using a slide hammer, or a ½-inch bolt, 6 inches long, and a bar. Lower hoist hook, and insert it through one of the holes in center of compensating case. Lift controlled differential out of transmission case, and lower carefully to floor or bench.

c. **Installation.**

(1) INSTALL CONTROLLED DIFERENTIAL. Install hoisting hook through one of the holes in center of compensating case. Lift differential and lower it into place in bearing bores. If necessary, tap bear-

TM 9-775

LANDING VEHICLE TRACKED MK. I AND MK. II

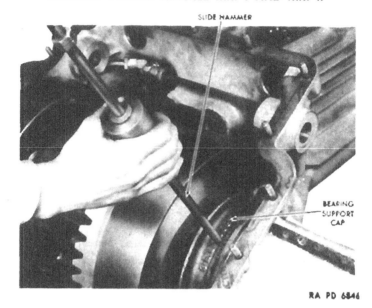

Figure 155 — Removal of Differential Bearing Support Cap

ings into place with a soft hammer. Facing the differential compartment, the bevel gear goes to right-hand side. Be sure bevel gear is in this position. (A wrong installation is possible, and will result in vehicle running backward.) Install bearing support covers. The machined side of cover must go against machined side-of case. The punch marks on the covers must be toward a similar marking on top of case. Place nuts on support studs and tighten nuts, then secure nuts with safety wire.

(2) INSTALL BRAKE CAMSHAFTS. Place the two brake camshafts into their respective holes in case so that camshaft pins are at the top of cam plates. The adjustable rod and spring should extend to the rear and rest on brake drum.

(3) INSTALL BRAKE SHOE ASSEMBLIES (fig. 154). Viewing one of the brake shoe assemblies (three shoes), note that open link pin holes at assembly ends are of different size. Stand facing differential chamber, and start the shoe having small link pin hole down around the forward side of brake drum, and up back side of drum. Aline cotter pin hole in connecting pin with cotter pin hole in brake shoe end, and install connecting pin. Install cotter pin from front. Loosen brake shoe adjusting nut to give necessary length to adjustable rod. Connect other end of brake shoe assembly, and install cotter pins on both ends of connecting pin.

Figure 156 — Removing Differential Assembly

TM 9-775
LANDING VEHICLE TRACKED MK. I AND MK. II

(4) ADJUST BAND CLEARANCE. With a feeler gage, adjust clearance between the band and drum to between 0.020 and 0.030 inch all the way around the band. To perform this adjustment, use only the brake band adjusting nut. If it is not possible to perform this adjustment because brake band contacts the drum at one or more places and is too far at other places, notify ordnance maintenance personnel. The linkage between steering lever and brake band may not be set properly.

(5) INSTALL CONTROLLED DIFFERENTIAL COVER. Install new differential cover gasket. Place cover on transmission case. Install 13 washers and 18 lock nuts (5 studs at front of cover have no washers) that secure differential cover to transmission case.

(6) FILL TRANSMISSION CASE WITH OIL. Screw the two magnetic drain plugs into tapped openings provided on lower front side of transmission case, and tighten plugs securely. Screw oil level plugs loosely into tapped holes provided in front of transmission. Unscrew transmission oil filler cap and fill transmission case up to oil level. NOTE: *If transmission and all oil lines are completely drained, the system will require approximately 30 quarts of oil.* Refer to Lubrication Guide (par. 27) for quantity and grade of oil. Tighten oil level plugs. Install filler cap.

(7) Install transmission and differential in vehicle (par. 122).

Section XXIX
TRACKS AND SUSPENSION

	Paragraph
Description	126
Track tension adjustment	127
Removal of tracks	128
Installation of tracks	129
Grousers and chain cross plates	130
Bogie assembly	131
Bogie wheel rubber bumpers	132
Rear idler wheels	133
Rear idler slide assembly	134
Rear idler slide bar	135
Rear idler adjusting screw	136
Return idler	137

126. DESCRIPTION.

a. **Bogie Assemblies.** Twenty-two single-wheeled, rubber-tired bogie assemblies, eleven on each side, support the landing vehicle tracked. These are mounted beneath pontoons, next to hull, and ride

TRACKS AND SUSPENSION

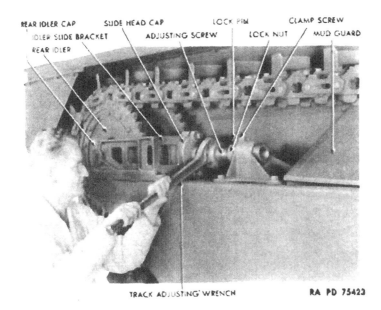

Figure 157 — Adjusting Track Tension

inside the endless track. Bogie assemblies are designed to provide vehicle support and maintain constant track tension while negotiating obstacles or irregular terrain. They also maintain track tension during water operation.

b. **Driving Sprockets.** Mounted at the front of the vehicle, and attached directly to the transmission final drives, are the driving sprockets. The driving sprocket draws the track over rear idler and return idlers, and lays it in the path of bogie assembly.

c. **Rear Idler.** Mounted at the rear of vehicle, one on each side, is an adjustable idler. A threaded adjusting screw provides a means of sliding rear idler toward front or rear of the vehicle, as necessary, in order to loosen or tighten track tension.

d. **Return Idlers.** Two rubber-tired return idlers are mounted on top of each pontoon, between rear idler and driving sprocket. These support track as it passes over top of pontoon.

127. **TRACK TENSION ADJUSTMENT.**

a. **General.** To determine track tension, inspect clearance between the slide head cap and idler slide bracket. If track tension is correct, cap will be flush against bracket. Incorrect track adjustment will be evidenced by an opening appearing between cap and bracket.

TM 9-775

LANDING VEHICLE TRACKED MK. I AND MK. II

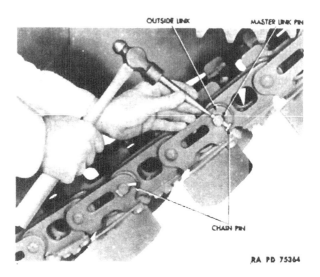

Figure 158 — Driving Out Chain Pin

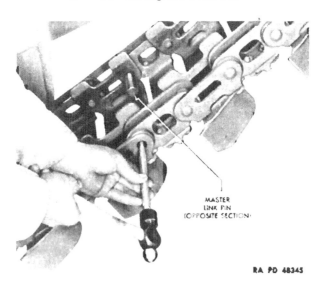

Figure 159 — Driving Out Master Link Pin

TRACKS AND SUSPENSION

b. **Adjusting Track Tension** (fig. 157). Loosen vertical clamp screw and horizontal lock pin which clamp lock nut on adjusting screw. Remove the two cap screws and toothed lock washers which secure mudguard above adjusting screw. Lift off mudguard. Using track adjusting wrench, turn adjusting screw to slide rear idler assembly toward rear to tighten track, or toward front to loosen track. After slide head cap is flush against idler slide bracket, turn screw three more notches. Track tension will then be correct. After proper track tension has been obtained, tighten clamp screw and lock pin. Install mudguard with its retaining cap screws and toothed lock washers.

128. REMOVAL OF TRACKS.

a. **General.** If possible, stop tank so that one of the three master link pins is opposite open area between front driving sprocket and skid mounted on the front of pontoon. This will facilitate work on the master link pin.

b. Loosen track tension (par. 127).

c. **Remove Chain Pin** (fig. 158). Tap chain pin straight from its bent condition, then drive out chain pin from master link pin. Repeat the operation to remove the chain pin securing the master link in opposite section of chain.

d. **Remove Master Link Pin** (fig. 159). Tap out the two master link pins. CAUTION: *Stand to one side while performing this operation. Even though track tension has been loosened, there is still enough tension in the track to cause it to snap free with considerable force when master link pins are driven out.*

e. **Remove Track.** After tracks have been disconnected, hook towing cable onto vehicle. Hook opposite end of cable to another vehicle or truck, and pull vehicle off the tracks.

129. INSTALLATION OF TRACKS.

a. **Roll Vehicle on Tracks.** Lay track in front of vehicle, extending it forward in a straight line from driving sprocket. Locate end of track under first bogie wheel as far as possible. Tow the vehicle forward on track by means of a prime mover or another vehicle, until forward end of track is directly under skid at the front of pontoon.

b. **Pull Track into Position.** Attach a cable to the rear end of track, and run cable over rear idler sprocket, along top of pontoon, and over front driving sprocket. Attach a prime mover or another vehicle to cable, and carefully pull track up over the rear idler, along top of pontoon, and over front driving sprocket. Make sure transmission gears are in neutral.

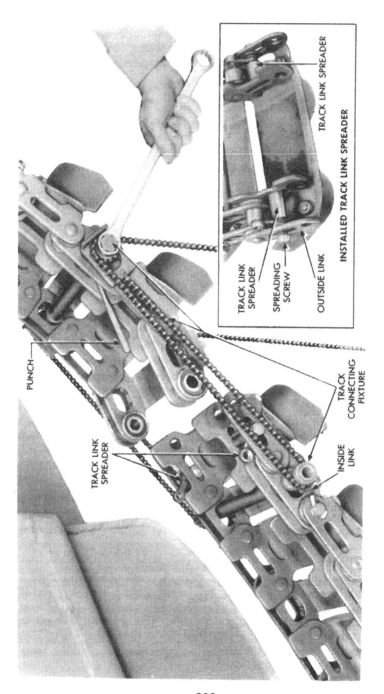

Figure 160 — Connecting Track

TRACKS AND SUSPENSION

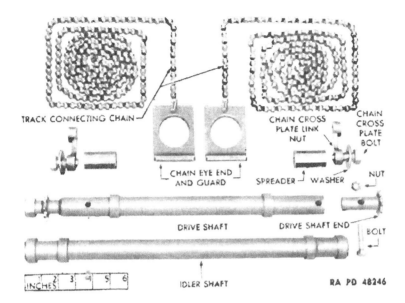

Figure 161 — Track Connecting Tools

c. **Install Track Connecting Fixture** (figs. 160 and 161). Work the two ends of track together as closely as possible. Using two track link spreaders and two chain cross plate link nuts and bolts, spread the two outside links apart as far as possible. Place a seal and a seal protector over ends of bushing which run through inside links (four seals and four seal protectors). Insert the idler shaft of track connecting fixture through inside links, just beneath the outside links to be connected. Remove drive shaft end from drive shaft, and insert shaft through inside links just above inside links to be connected. Install drive shaft end on drive shaft. Hook the two chain eye ends, one over each end of drive shaft (inside the sprocket), so that guards face outward. Run the chain down under idler shaft, back up and over the sprockets on the drive shaft.

d. **Connect Track** (fig. 160). With a box wrench, turn nut welded to end of the drive shaft to crank ends of track chain together. Insert a punch through the holes drilled through outer end of drive shaft to keep connecting chain from slipping when changing position of wrench. Carefully work inside links into outside links. Watch that seals and seal protectors are not damaged or pressed out of position. When inside links are lined up within outside links, tap the two master link pins through the links. Install a chain pin through each master link pin, and bend down end of chain pin (fig. 158). Remove track link spreaders, and remove the track connecting fixture from the track.

289

LANDING VEHICLE TRACKED MK. I AND MK. II

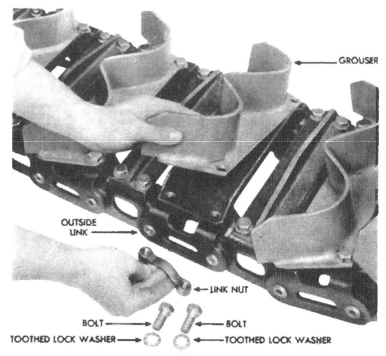

Figure 162 — Removing Grouser

e. **Tighten Track Tension.** Adjust tension to proper degree of tightness (par. 127).

130. GROUSERS AND CHAIN CROSS PLATES.

a. **Description.** Seventy-three grousers and seventy-three chain cross plates bolted to track chain, make up each track used on the landing vehicles tracked. Grousers are bolted on track chain from outside link to outside link. Chain cross plates are bolted on track chain from inside link to inside link.

b. **Maintenance.** Replace damaged or missing grousers or chain cross plates immediately. Unless absolutely necessary, never operate the vehicle when grousers or chain cross plates are broken or missing. The absence of a grouser throws a strain on the grousers on each side of the missing one, and serves to weaken track. When grousers are missing, the efficiency of the vehicle in water will naturally decrease, since grousers are the means of propelling the vehicle in water.

TM 9-775
130

TRACKS AND SUSPENSION

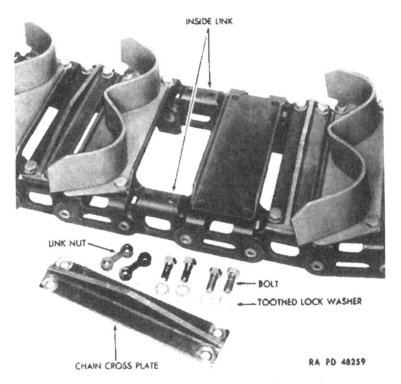

Figure 163 — Chain Cross Plate — Removed

c. **Removal of Grousers** (fig. 162). Remove the four bolts, toothed lock washers, and two link nuts which secure the grouser to the outside links of the track chain. Slide the link nut out from within outside chain link.

d. **Removal of Chain Cross Plate** (fig. 163). Remove the four bolts, toothed lock washers, and two link nuts which secure the chain cross plate to the track chain. Slide the two chain cross plate link nuts out from within inside chain links.

e. **Installation of Chain Cross Plate** (fig. 163). Place chain cross plate on track chain, from inside link to inside link. Lift chain cross plate link nuts up underneath inside links, and install the four bolts and toothed lock washers which secure cross plates to track chain.

f. **Installation of Grousers** (fig. 162). Place grousers on the track chain, from outside link to outside link. Lift the two grouser link nuts up underneath outside link, and install the four bolts and toothed lock washers which secure grouser to track chain.

LANDING VEHICLE TRACKED MK. I AND MK. II

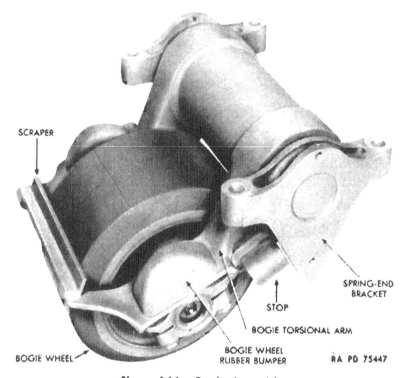

Figure 164 — Bogie Assembly

131. BOGIE ASSEMBLY.

a. Description (fig. 164). The bogie assembly used on landing vehicle tracked is new in design, and represents a distinct departure from the conventional bogie assembly used on light and medium tanks. No volute springs are employed. Instead, the torsional effect of a shaft "floating" in rubber is utilized to cushion and support the vehicle. A hollow shaft is placed inside another hollow shaft of larger diameter. Rubber is vulcanized between shafts, so that inner shaft "floats" in a cushion of rubber in the outer shaft. Welded on the ends of the outer shaft are bogie wheel arms between which is mounted the bogie wheel. Welded to the outer ends of the inner shaft are spring-end brackets, used to attach the bogie assembly to the pontoon. Thus, as the vehicle negotiates irregular terrain, the outer shaft (which is attached through the wheel arms to the bogie wheel) twists on the inner shaft (which is immovably bolted to the pontoon). The natural resistance of the rubber between shafts to twisting, acts to cushion upward and downward movement of the bogie wheel, and provides firm yet flexible support for vehicle.

TRACKS AND SUSPENSION

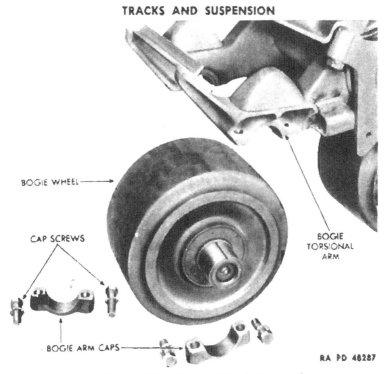

Figure 165 — Bogie Wheel Removed

b. **Removal of Bogie Wheel** (fig. 165). Remove the two cap screws and toothed lock washers which secure each of the two torsional bogie arm caps in position. Lift off caps and remove bogie wheel.

c. **Installation of Bogie Wheel** (fig. 165). Lift bogie wheel into position, place the two torsional bogie arm caps in position, and install the four retaining cap screws and toothed lock washers.

d. **Removal of Bogie Assembly** (figs. 166 and 167). Using a chisel and hammer, chisel off spot-weld which locks the four cap screws (two each side) securing bogie assembly to the pontoon. Remove the four cap screws (wrench 41-W-2964-275), and lift off bogie assembly.

e. **Installation of Bogie Assembly.** Block bogie assembly up under pontoon as close as possible to its correct position. Install the four retaining cap screws securing spring-end brackets of bogie assembly to pontoon. Spot-weld cap screws to spring-end brackets to secure screws in position.

132. BOGIE WHEEL RUBBER BUMPERS.

a. **Description.** Two rubber bumpers are used with each bogie assembly, one attached to each of the two torsional arms. Bumpers are

TM 9-775
132
LANDING VEHICLE TRACKED MK. I AND MK. II

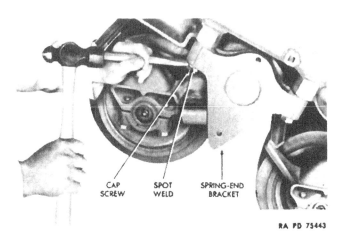

Figure 166 — Chiseling Spot-weld from Spring-end Bracket Retaining Screws

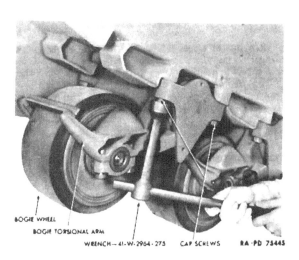

Figure 167 — Removing Bogie Assembly Cap Screws

TM 9-775
132—133

TRACKS AND SUSPENSION

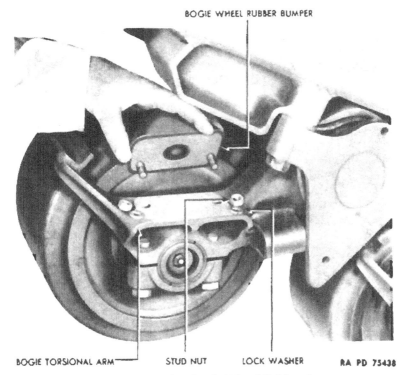

Figure 168 — Removing Bogie Wheel Rubber Bumper

designed and located to stop bogie wheels from bouncing against the under side of pontoon as vehicle is negotiating irregular terrain.

b. **Maintenance.** Examine rubber bumpers for torn or frayed spots. If rubber is cracked, or if chunks of rubber have been torn from bumper, install a new bumper.

c. **Removal** (fig. 168). Remove the two stud nuts which secure rubber bumper to torsional arm, and lift off bumper.

d. **Installation** (fig. 168). Place rubber bumpers in position, and install the two retaining stud nuts.

133. REAR IDLER WHEELS.

a. **Description.** An adjustable rear idler is located on each side of rear of the landing vehicle. Each rear idler is mounted on a slide assembly that is controlled by an adjusting screw. Track tension is obtained by turning adjusting screw, which slides rear idler backward or forward on slide bars, depending upon which way adjusting screw is turned.

TM 9-775

LANDING VEHICLE TRACKED MK. I AND MK. II

Figure 169 — Removing Rear Idler Sprocket

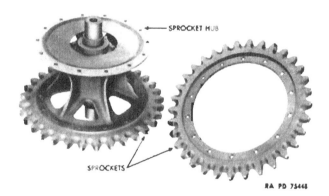

Figure 170 — Rear Idler Sprocket Removed from Sprocket Hub

TRACKS AND SUSPENSION

Figure 171 — Removing Rear Idler Shaft with Bearings

b. Maintenance. If rear idler sprocket teeth are chipped or broken off, or if the rear idler sprocket hub is damaged, replace either sprocket hub or sprocket.

c. Remove Rear Idler Sprocket Assembly. Break track (par. 128), and pull track off rear idler. Remove the two cap screws with toothed lock washers which secure each of the two rear idler sprocket caps to the rear idler slide assembly (fig. 157). Carefully lift the rear idler sprocket from its position on the rear idler slide assembly to the ground (fig. 169). CAUTION: *The rear idler is extremely heavy, awkward to handle, and will require two men to remove.*

d. Remove Rear Idler Sprocket from Sprocket Hub (fig. 170). Each of the two rear sprockets mounted on each rear idler sprocket hub is removed in same manner. Remove the 14 cap screws and toothed lock washers which secure rear idler sprocket to sprocket hub. Lift off rear idler sprocket.

e. Disassemble Sprocket Hub (fig. 172).

(1) REMOVE SPROCKET BEARING CAPS. Remove six cap screws and toothed lock washers which secure each of the two sprocket bearing caps to sprocket. Slide caps off the ends of shaft.

(2) REMOVE REAR IDLER SHAFT (fig. 171). Place sprocket in an arbor press. Using the special driver as a pilot (41-R-2384-40), press rear idler shaft with bearings and sleeves out of hub.

(3) REMOVE BEARINGS AND SLEEVES FROM SHAFT. Place rear idler shaft assembly in an arbor press with inner and outer races of bearing supported on steel plates. Steel plates should have approximately 1.66-inch radius cut out of side of plates to partially fit around circumference of shaft. Lower ram of press until it firmly contacts pilot. With one man holding lower end of shaft to prevent its dropping, press shaft out of upper sleeve and bearing. Sleeve and bearing on opposite end of shaft are removed in same manner.

f. Assemble Sprocket Hub.

(1) INSTALL FIRST BEARING ON SHAFT. Place rear idler shaft in vertical position on bed of arbor press. Place bearing over upper end of shaft, and aline bearing with bearing seat of shaft. Place special driver (41-R-2384-40) over end of shaft and against inner race of bearing. Lower ram of press and start bearing into bearing seat. Remove driver (41-2384-40). Place a film of white lead on inner surface of seal bearing sleeve; then place sleeve over upper end of shaft and against inner race of bearing. Place driver (41-R-2384-40) over end of shaft and against upper side of sleeve. Lower ram of arbor press, and press bearing and sleeve into position on shaft.

(2) INSTALL SHAFT IN HUB. Place rear idler sprocket hub on bed of arbor press. Place shaft into bore of hub with installed shaft bearing and sleeve in upper side of hub. Aline bearing with bearing

TRACKS AND SUSPENSION

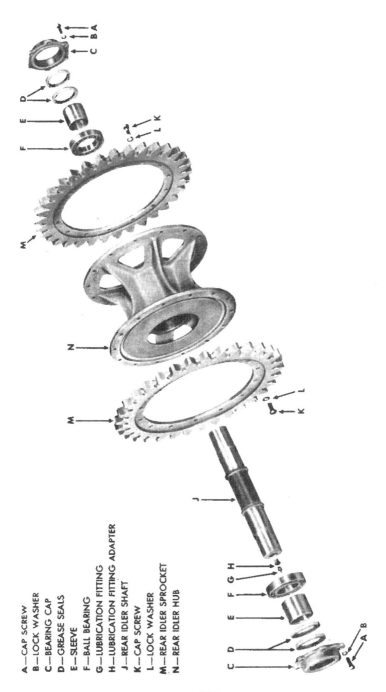

A—CAP SCREW
B—LOCK WASHER
C—BEARING CAP
D—GREASE SEALS
E—SLEEVE
F—BALL BEARING
G—LUBRICATION FITTING
H—LUBRICATION FITTING ADAPTER
J—REAR IDLER SHAFT
K—CAP SCREW
L—LOCK WASHER
M—REAR IDLER SPROCKET
N—REAR IDLER HUB

Figure 172 — Rear Idler — Disassembled

TM 9-775
133

LANDING VEHICLE TRACKED MK. I AND MK. II

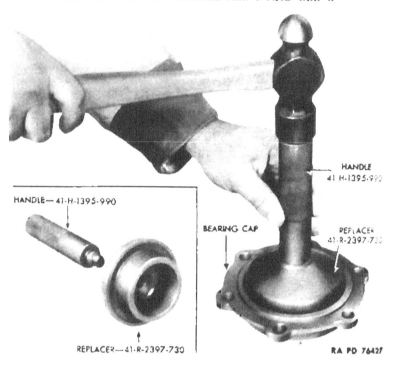

Figure 173 — Installing Rear Idler Bearing Cap Grease Seals

seat in hub. Place driver (41-R-2384-40) on upper side of sleeve. Lower press ram, and press shaft with its assembled bearing and sleeve into hub.

(3) INSTALL SECOND BEARING AND SLEEVE. Turn hub and shaft over on press bed with installed bearing end of shaft supported on press bed plate. Place second bearing over end of shaft. Install driver over end of shaft with driver alined with inner race of bearing. Lower ram and start bearing onto shaft. Remove driver. Place a film of white lead on second seal bearing sleeve. Place sleeve over end of shaft and against inner race of bearing. Aline driver (41-R-2384-40) with upper side of sleeve. Lower ram, and press second bearing and sleeve into position on shaft within hub.

(4) INSTALL GREASE SEALS IN BEARING CAPS. If at time of inspection, grease seals were removed from bearing cap, new grease seals must be installed. Place bearing cap on bench or wood block with mounting flange upward. Aline grease seal with opening provided in bearing cap. (Lip edge of grease seal is downward.) Use replacer (41-R-2397-730) with its assembled handle (41-H-1395-

TRACKS AND SUSPENSION

990) to drive seal into bearing cap (fig. 173). Second seal is installed in same manner. Bearing cap on opposite side of shaft is identical, and grease seal installation is the same.

(5) INSTALL REAR IDLER SHAFT. Place the rear idler sprocket hub in an arbor press. Aline bearings on shaft with openings provided in hub. Shaft must be vertical, and bearing correctly alined, to prevent their being cocked and thereby damaged. Place replacer (41-R-2384-40) over upper end of shaft. Lower ram of press to driver. Press shaft assembly into hub.

(6) INSTALL SPROCKET BEARING CAPS. Aline holes in bearing cap with tapped holes provided in hub. Install six cap screws and toothed lock washers to secure bearing cap to hub. The other bearing cap is installed in same manner.

g. Install Rear Idler Sprocket on Sprocket Hub. Each of the two rear idler sprockets is mounted in same manner. Place rear idler sprocket on sprocket hub so that the punched "o" mark on face of outer sprocket lines up with the punched "o" mark on face of the opposite sprocket. This will aline sprocket teeth correctly. Install the 14 retaining cap screws and toothed lock washers.

h. Install Rear Idler Sprocket Assembly. Using two men, lift rear idler wheel up into position on rear idler slide assembly. Place the two rear idler wheel caps in position, and install retaining cap screws and toothed lock washers. Connect track (par. 129).

134. REAR IDLER SLIDE ASSEMBLY.

a. Description. The rear idler slide assembly mounts rear idler wheel, and provides a method of adjusting track tension.

b. Maintenance. Inspect rear idler slide assembly frequently for evidence of inadequate lubrication. Lubricate slide bars and idler assembly as instructed in Lubrication Guide (par. 27). If rear idler slide assembly is damaged through accidental or combat use, install a new rear idler slide assembly.

c. Removal of Rear Idler Slide Assembly. Remove rear idler wheel (par. 133). Remove the four cap screws and toothed lock washers which secure adjusting screw head to adjusting screw (fig. 174). Slide rear idler slide assembly off slide bars, and lower to ground (fig. 175).

d. Installation of Rear Idler Slide Assembly. Lift rear idler slide assembly up into position, and slide it forward on slide bars. Measure distance on the follower screw between adjusting screw head and front face of cap attached to follower screw (fig. 175). Now measure distance on adjusting screw between rear face of screw (fig. 157) and lock nut on the screw. These two measurements must be

TM 9-775
134
LANDING VEHICLE TRACKED MK. I AND MK. II

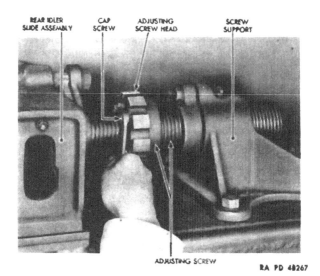

Figure 174 — Removing Adjusting Screw Head

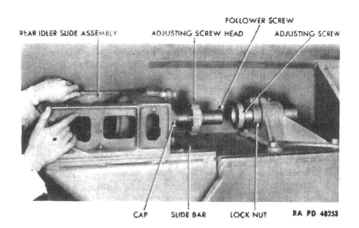

Figure 175 — Removing Rear Idler Slide Assembly

TRACKS AND SUSPENSION

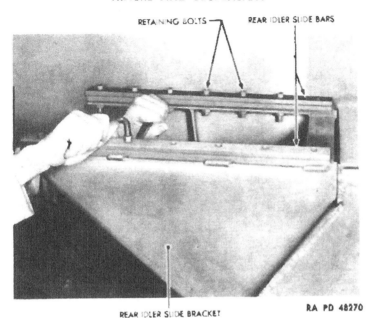

Figure 176 — Removing Rear Idler Slide Bar

approximately equal, so that when the rear idler slide assembly is installed, full adjustment will be realized from the adjusting screw. Install the four cap screws and toothed lock washers which secure screw head to adjusting screw (fig. 157). Install rear idler wheel (par. 131).

135. REAR IDLER SLIDE BAR.

a. **Description** (fig. 176). Two rear idler slide bars are used with each rear idler slide assembly These are mounted one on each side of the rear idler slide bracket.

b. **Maintenance.** If rear idler slide bars are broken or damaged so that they bind against rear idler slide assembly, install new rear idler slide bars.

c. **Removal** (fig. 176). Remove the six socket head set screws and nuts which secure each of the slide bars to rear idler bracket. Lift off the bars.

d. **Installation** (fig. 176). Place rear idler slide bars in position on rear idler bracket, and install the six socket-head set screws and nuts which secure each of the slide bars to bracket.

TM 9-775

LANDING VEHICLE TRACKED MK. I AND MK. II

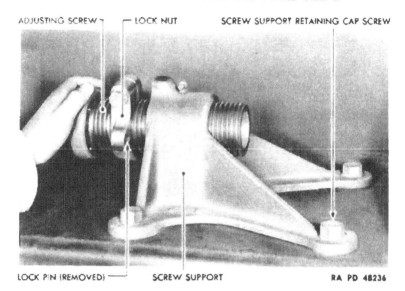

Figure 177 — Removing Adjusting Screw from Screw Support

136. REAR IDLER ADJUSTING SCREW.

a. Description. The rear idler adjusting screw (in mesh with the follower screw) is used to slide rear idler assembly back and forth on slide bars, and thus provide a method of loosening or tightening track tension.

b. Maintenance. If adjusting screw threads are chipped or broken, install new screw.

c. Removal. Remove track (par. 128). Remove rear idler wheel (par. 133). Remove rear idler slide assembly (par. 134). Remove lock nut pin (fig. 177), and unscrew adjusting screw from screw support. Remove the four cap screws and toothed lock washers which secure screw support to pontoon, and lift off screw support.

d. Installation. Place screw support in position, and install the four retaining cap screws and toothed lock washers (fig. 177). Screw adjusting screw into screw support. Rotate lock nut to a position where one thread can be seen between nut and screw support, and install lock pin (fig. 177). Install rear idler slide assembly (par. 134). Install the rear idler wheel (par. 133).

137. RETURN IDLER.

a. Description. Two return idlers are mounted on top of each pontoon, spaced equally between rear idler and driving sprocket. The function of the return idler is to provide support for track as it travels

TM 9-775
137

TRACKS AND SUSPENSION

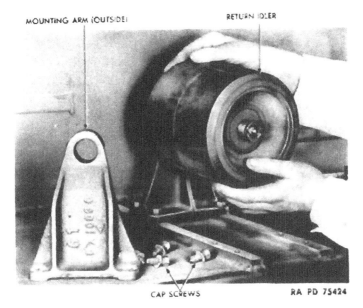

Figure 178 — Removing Return Idler

along the top of pontoon. The shaft on which return idler wheel revolves is held firmly by mounting arms bolted to welded brackets on pontoon.

b. Removal (fig. 178). Remove two cap screws and toothed lock washers which secure outside return idler mounting arm to bracket on top of pontoon. Lift the arm off the idler. Pull idler straight out to remove idler shaft from inside mounting arm. CAUTION: *Do not lose idler shaft dowel. The dowel is loosely slipped through inside seal sleeve into end of shaft, and falls out easily.*

c. Disassembly.

(1) REMOVE IDLER SHAFT (figs. 179 and 180). Remove the cotter pin which secures the idler shaft retaining nut. Remove grease fitting from opposite end of shaft. Hold opposite end of shaft with a drag link socket wrench, and remove nut. Pull the two idler shaft washers off ends of shaft, and lift out idler shaft dowel. Carefully pull idler shaft from idler wheel. Lift out inside and outside seal sleeve (one from each end of shaft).

(2) REMOVE IDLER SHAFT GREASE SEALS AND BEARINGS (fig. 181). Clean grease from the inside of return idler. Insert idler wheel bearing end grease seal puller (41-P-2900-18) into the idler wheel so that the small slide bar in the end of puller drops down into grease hole

305

TM 9-775

LANDING VEHICLE TRACKED MK. I AND MK. II

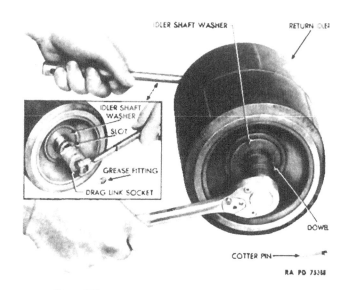

Figure 179 — Removing Return Idler Shaft Nut

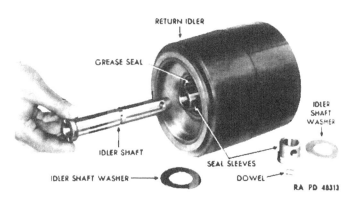

Figure 180 — Pulling Out Return Idler Shaft

TM 9-775

TRACKS AND SUSPENSION

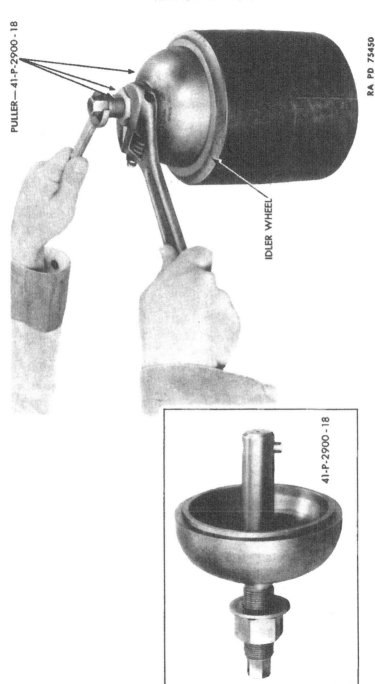

Figure 181 — Removing Idler Wheel Bearing and Grease Seal

TM 9-775
137
LANDING VEHICLE TRACKED MK. I AND MK. II

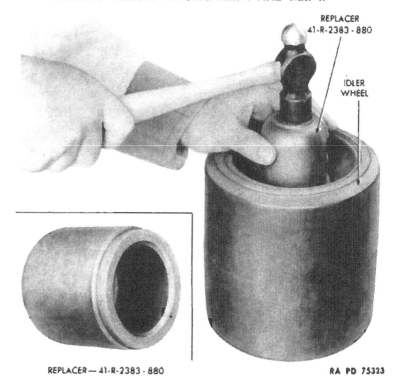

Figure 182 — Installing Idler Wheel Bearing and Grease Seal

in bearing spacer between the two idler wheel bearings. Pull out the bearing spacer along with one idler wheel bearing and grease seal. To remove the second idler wheel bearing and grease seal, insert a brass drift from the opposite end of the wheel, and tap out bearing and grease seal. Tap around complete circumference of the bearing to prevent cocking it.

d. Assembly.

(1) INSTALL BEARING AND GREASE SEALS (fig. 182). Using special idler wheel bearing and grease seal replacer (41-R-2383-880), simultaneously drive one bearing and grease seal into idler wheel. Turn wheel over and tap bearing spacer into idler wheel. Using bearing and grease seal replacer again, tap second bearing and grease seal into idler wheel.

(2) INSTALL IDLER SHAFT (fig. 180). Run a finger around the inside of each of the grease seals to open up seals slightly. Carefully rotate inside seal sleeve and outside seal sleeve into position within

308

PANELS AND INSTRUMENTS

seals. Now slide idler shaft through inside sleeve, through idler wheel, and out through outside sleeve. Slip an idler shaft washer over each end of shaft. Line up hole in idler shaft with hole in inside seal sleeve, and drop the idler shaft dowel in, to position. Using a drag link socket to hold shaft, install idler shaft retaining nut (fig. 179). Lock the nut with a cotter pin.

e. Installation (fig. 178). Lift the idler wheel into position. Slide idler wheel shaft into inside mounting arm so that idler shaft dowel slides through the slot in inside mounting arm. Place outside mounting arm over end of idler wheel shaft. Secure mounting arm to bracket welded on pontoon with two cap screws and toothed lock washers.

Section XXX
PANELS AND INSTRUMENTS

	Paragraph
Instrument panel	138
Regulator and relay panel	139
Control panel	140
Instruments and gages	141

138. INSTRUMENT PANEL.

a. Description. Located on the hull directly in front of the driver, the instrument panel contains a transmission oil pressure gage and engine oil pressure gage, tachometer, ammeter, and an oil temperature indicator. A brief description of the function of these instruments is given in section III.

b. Maintenance of Instrument Panel. Under ordinary circumstances it will not be necessary to replace the instrument panel as a unit. However, when damage occurs to the instrument panel case or cover, either by accident or through combat use, replace the complete instrument panel. Remove all operating gages and instruments from the panel, and install them in the new instrument panel.

c. Removal.

(1) REMOVE INSTRUMENT PANEL COVER. Remove the 11 cross-recessed screws which secure instrument panel cover to instrument panel. Tilt cover back and downward away from panel (fig. 188).

(2) DISCONNECT INSTRUMENTS. Disconnect all instruments except oil temperature indicator. Remove the two bolts, nuts, and washers securing oil temperature indicator to back cover of instrument panel. Place oil temperature indicator on cab floor out of way. NOTE: *On early models of the landing vehicle tracked, oil temperature line*

LANDING VEHICLE TRACKED MK. I AND MK. II

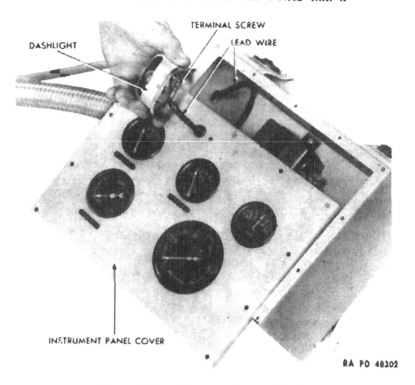

Figure 183 — Disconnecting Dashlight

was led through a hole in back of panel case. In order to remove line, so that panel can be removed, it is necessary to cut a slot with a hacksaw from opening in back of panel to edge of panel. The oil temperature indicator line can then be passed through this slot and free of panel.

(3) DISCONNECT DASHLIGHT (fig. 183). Working from front of instrument panel, remove the two bolts, nuts, and toothed lock washers which secure dashlight to instrument panel cover. Pull dashlight forward slightly, away from panel cover, and remove terminal screw which secures dashlight lead wire to dashlight.

(4) REMOVE INSTRUMENT PANEL (fig. 189). Loosen hexagonal nuts, and pull out the two conduits with enclosed lead wires from inside of panel. Push two oil pressure lines and tachometer cable out through their respective openings in back of instrument panel. Remove the cross-recessed screw securing ground strap to rear of instrument panel. Remove four cap screws (two each side) securing instrument panel brackets to hull. Remove the two bolts, nuts, and flat

PANELS AND INSTRUMENTS

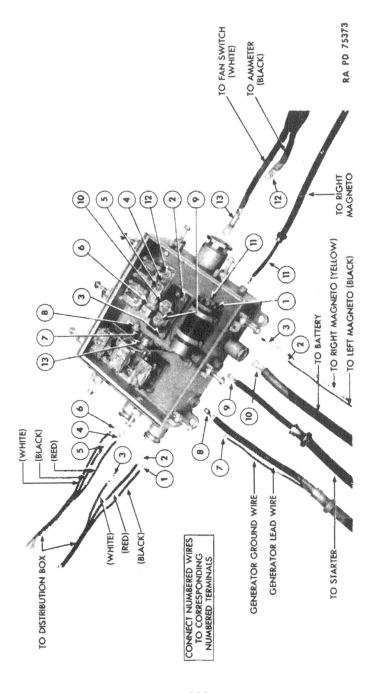

Figure 184 — Regulator and Relay Panel — Wires Disconnected

TM 9-775

LANDING VEHICLE TRACKED MK. I AND MK. II

washers securing instrument panel to two brackets extending out from hull to top of panel. Lift off instrument panel.

139. REGULATOR AND RELAY PANEL.

a. Description and Data. Located in battery compartment above battery is a regulator and relay panel. Access within regulator and relay panel is provided by a cover. Mounted in regulator and relay panel are generator regulator, booster coil, main battery relay, cranking motor relay, and condenser. The procedure for removing generator regulator is covered in paragraph 98. Procedure for removing booster coil is covered in paragraph 93. Function of regulator and relay panel is to provide watertight box in which to contain abovementioned electrical components.

b. Maintenance. In the event components within regulator and relay panel (such as generator regulator or booster coil) are defective, or have been damaged, they may be replaced as units. If entire regulator and relay panel is damaged or otherwise defective, the entire assembly may be replaced as a unit. No disassembly or repair of units within the box shall be attempted by using arm personnel.

c. Removal.

(1) GENERAL. In order to remove regulator and relay panel, all wires leading into panel must be disconnected at their terminals and pulled out of panel (fig. 184). As a guide to correct assembly of new panel to be installed, tag all wires as they are removed.

(2) Open hinged rear cover.

(3) Remove upper left rear baffle (par. 61).

(4) REMOVE REGULATOR AND RELAY PANEL COVER. Loosen the 10 hinged bolts which secure regulator and relay panel cover. Swing bolts out of way, and lift off cover (fig. 35).

(5) DISCONNECT REGULATOR AND RELAY PANEL (fig. 184). Disconnect all wires and cables leading into regulator and relay panel. Loosen nuts and remove knurled nuts which secure flexible conduit to regulator and relay panel. Pull conduits, with wiring, from regulator and relay panel.

(6) REMOVE REGULATOR AND RELAY PANEL. Remove six nuts, bolts, and toothed lock washers which secure regulator and relay panel to brackets on inside of hull. Lift out regulator and relay panel.

d. Installation.

(1) BOLT REGULATOR AND RELAY PANEL IN POSITION. Place regulator and relay panel in position and install six nuts, bolts, and toothed lock washers which secure panel to brackets on inside of hull.

(2) CONNECT REGULATOR AND RELAY PANEL (fig. 184). Insert lead wire (within conduits) into proper conduit opening in regulator

PANELS AND INSTRUMENTS

relay panel. Connect all wires to their proper terminals. To facilitate assembly, all wires and terminals should have been tagged at removal. Tighten hexagonal nuts and knurled nuts which secure flexible conduits within regulator and relay panel.

(3) INSTALL REGULATOR AND RELAY PANEL COVER. Place regulator and relay panel cover in position. Swing the 10 hinged bolts up into position in bracket, and tighten bolts.

(4) Install upper left rear baffle (par. 62).

(5) Close hinged rear cover.

140. CONTROL PANEL.

a. Description. Mounted on hull to left of driver, the control panel contains all switches used in the operation of landing vehicles tracked. These are the booster, cranking motor, blower, battery, magneto, dashlight, light, and fan switches. A detailed description of the switches and their function is given in paragraphs 6 and 7. On early models of landing vehicles tracked, the control panel includes six fuses located on a fuse block in center of control panel (fig. 185). All these fuses are 15-ampere fuses with the exception of the generator fuse, which is a 75-ampere fuse. On later models of landing vehicles tracked, the 75-ampere fuse has been discontinued and a circuit breaker, installed in lower left corner of control panel, has taken its place (fig. 186). The circuit breaker is reset by merely pressing in rubber push button in left side of control panel. Remove fuses with fuse extractor to prevent damage to fuses.

b. Maintenance. If, through accident or combat use, control panel is damaged, install new control panel.

c. Removal of Switches (fig. 185). Each of the eight switches in control panel are removed in same manner. Remove control panel cover. Disconnect wires attached to terminals of switches. Tag all wires to assure correct assembly. Remove the two screws and lock washers which secure switch within panel, and lift out switch.

d. Installation of Switches (fig. 185). All switches are installed in the same manner. Place the switch in position, and install the two retaining screws and lock washers. Connect the lead wires to terminals on switch. (These should have been tagged at removal.) Install control panel cover.

e. Removal of Control Panel.

(1) GENERAL. In order to remove control panel, all conduits and wires leading into panel must be disconnected at their terminal and pulled out of panel. As a guide to assist in assembly of new panel to be installed, tag all terminals and wires, as wires are disconnected.

TM 9-775
140

LANDING VEHICLE TRACKED MK. I AND MK. II

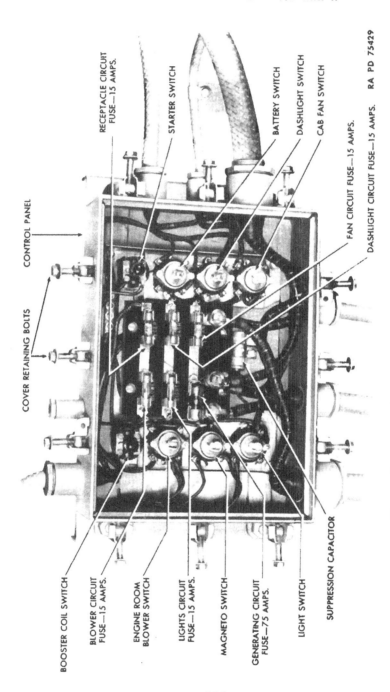

Figure 185 — Control Panel (With Fuses) — Cover Removed

314

TM 9-775
140

PANELS AND INSTRUMENTS

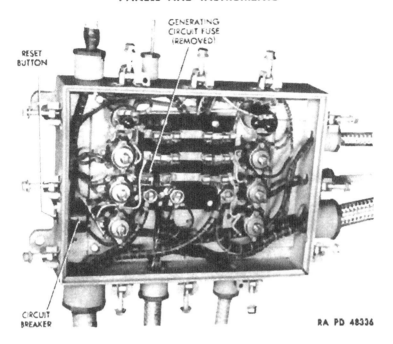

Figure 186 — Control Panel (With Circuit Breaker) — Cover Removed

(2) REMOVE THE CONTROL PANEL COVER (fig. 185). Loosen the 12 hinged bolts which secure control panel cover. Swing hinged bolts away from cover, and lift off cover.

(3) DISCONNECT CONTROL PANEL (fig. 187). Disconnect all wires leading into control panel. Loosen the hexagonal nuts and knurled nuts which secure conduit containing wires to control panel. Pull conduit, with wires, from control panel.

(4) REMOVE CONTROL PANEL. Remove the four stud nuts and toothed lock washers which secure control panel to side of hull. Lift off panel.

f. **Install Control Panel.**

(1) BOLT CONTROL PANEL IN POSITION. Place control panel in position, and install the four stud nuts and toothed lock washers which secure the control panel to side of hull.

(2) CONNECT CONTROL PANEL (fig. 185). Insert conduit, with wiring, into proper conduit openings in control panel. Connect wires to their proper terminals within control panel. Tighten hexagonal nuts and knurled nuts which secure conduits within control panel.

315

TM 9-775
140

LANDING VEHICLE TRACKED MK. I AND MK. II

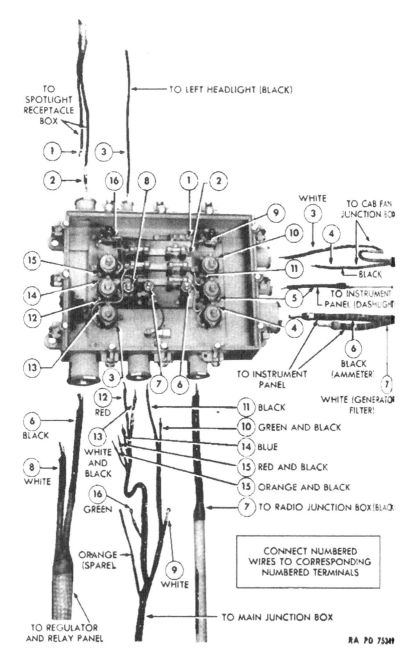

Figure 187 — Control Panel — Wires Disconnected

PANELS AND INSTRUMENTS

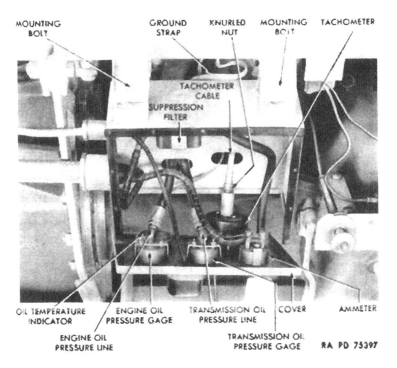

Figure 188 — Instrument Panel Cover in Position to Disconnect Instruments

(3) INSTALL CONTROL PANEL COVER. Place instrument panel cover in position on control panel. Swing the 12 hinged bolts up into position on brackets of control panel cover, and tighten bolts.

141. INSTRUMENTS AND GAGES.

a. Description. Five electrical and nonelectrical instruments are mounted on instrument panel used in landing vehicles tracked. A brief description of these instruments and their function is given in paragraphs 6 and 7. In subparagraphs c through j, below, procedure for removal and installation of these five instruments is given. In all cases, it will be necessary to remove instrument panel cover, and tilt panel cover face downward in order to provide access to instrument or gage to be moved.

b. Maintenance. Install new instruments and gages for all gages which become inoperative.

TM 9-775

LANDING VEHICLE TRACKED MK. I AND MK. II

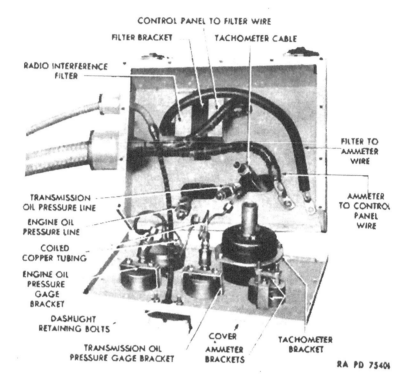

Figure 189 — Ammeter, Tachometer, and Oil Pressure Gages Disconnected (New-type Panel)

c. **Removal of Ammeter** (figs. 188 and 189). Disconnect the lead wire from the control panel. Disconnect the lead wire to radio interference filter. Remove two bracket retaining stud nuts, and lift off the two angle brackets. Pull ammeter out through the front of panel.

d. **Installation of Ammeter** (figs. 188 and 189). Place ammeter in position, and slip the two angle brackets over the two retaining studs. Install the two retaining stud nuts. Connect lead wire from radio interference filter, and then connect lead wire from the control panel.

e. **Removal of Engine or Transmission Oil Pressure Gage** (figs. 188 and 189). Disconnect oil pressure line (flexible tubing) at junction with copper tubing. Hold fitting on flexible tubing with one wrench while turning fitting on copper tubing with another wrench. This will prevent damage to either line. Remove the two bracket

PANELS AND INSTRUMENTS

retaining stud nuts. Lift off bracket, and pull oil pressure gage out through the front of instrument panel cover. NOTE: *On early model panels, the junction of the flexible tubing and copper tubing was located beneath hull in back of instrument panel. The lines were difficult to reach, to disconnect. On later model panels, a coil of copper tubing is connected to rear of gage, and flexible tubing is connected to coiled copper tubing (fig. 189).*

f. **Installation of Engine or Transmission Oil Pressure Gage** (figs. 188 and 189). Place the oil pressure gage in position in instrument panel cover. Slip retaining bracket over gage on retainnig studs, and install retaining stud nuts. Connect oil pressure line (flexible tubing) to fitting on copper tubing. Hold fitting on flexible tubing with one wrench while turning fitting on copper tubing with another wrench.

g. **Removal of Tachometer** (figs. 188 and 189). Unscrew knurled nut which secures tachometer cable in the back of the tachometer. Pull out tachometer cable. Loosen the two screws which clamp tachometer bracket in position. Rotate bracket so that slots in bracket will line up with flange on tachometer. Slide tachometer out front of the instrument panel.

h. **Installation of Tachometer** (figs. 188 and 189). Slide tachometer in the front of instrument panel. Place tachometer bracket on rear of tachometer, lining up slots in tachometer bracket with flanges on tachometer. Rotate bracket to lock it in position, then tighten the two bracket screws.

i. **Removal of Oil Temperature Indicator** (fig. 188). Unlike all other instruments, the oil temperature indicator line cannot be disconnected at gage. In order to install a new instrument, line must be disconnected at engine (par. 61). All blocks which secure line in place along control tunnel must be loosened, and line then pulled from engine, through control tunnel, and up to instrument panel. With line loose in instrument panel, remove the two retaining bolts, nuts, and washers which secure indicator to instrument panel cover, and remove indicator from the back of cover.

j. **Installation of Oil Temperature Indicator** (fig. 188). Place oil temperature indicator in position on back of instrument panel cover. Install the two retaining bolts, nuts, and washers. Run oil temperature indicator line through bottom of instrument panel, and feed line through all the blocks which hold it in place around transmission, through control tunnel, and in engine room. Install oil temperature indicator bulb at end of line in place in engine (par. 62). Tighten blocks which secure line along its passageway.

TM 9-775
142

LANDING VEHICLE TRACKED MK. I AND MK. II

Section XXXI

CONTROLS AND LINKAGE

	Paragraph
Description	142
Removal	143
Installation	144

142. DESCRIPTION.

   a. **Steering.** The vehicle steering and braking controls and linkage consist of the two steering levers, steering lever shafts, and steering lever mounting brackets, which are mounted in the driver's compartment ahead of the driver. The steering levers are connected to the transmission brake arms by the transmission control steering rods (fig. 190).

   b. **Clutch.** The clutch control mechanism consists of the clutch foot pedal, foot pedal shaft assembly, and the clutch control lever mounted in the driver's compartment. The clutch control lever is connected to the clutch by means of control rods and brackets, which are mounted in the control tunnel. The rods are held in place in the control tunnel by wooden guide blocks which are secured to the tunnel brackets. The control rod mechanism consists of the front clutch control rod with connecting yokes, which connect the clutch control lever on the clutch pedal shaft with the front bulkhead clutch control lever, mounted in a bracket on the forward bulkhead (rod-A, fig. 193). The intermediate clutch control rod connects the clutch control lever on the forward bulkhead with a turnbuckle mounted on the rear end of this intermediate rod (rod-B, fig. 193). The rear clutch control rod connects the turnbuckle in the center of the vehicle with the rocker arm assembly mounted on a bracket on the stern bulkhead (rod-C, fig. 193). The clutch operating rod connects the upper end of the rocker arm with the clutch throw-out yoke (rod-D, fig. 193).

   c. **Throttle.** The throttle control mechanism consists of the accelerator pedal, accelerator pedal shaft assembly, and the accelerator operating lever, mounted in the driver's compartment. The accelerator pedal assembly is connected to the throttle on the engine by means of control rods and brackets which are mounted in the control tunnel and on the engine support. The throttle control rod mechanism consists of the front throttle control rod (rod-E, fig. 193), which connects the accelerator operating lever with the throttle control lever mounted in a bracket on the forward bulkhead. The intermediate throttle control lever (rod-F, fig. 193) connects the throttle control lever on the forward bulkhead with a turnbuckle in the center of the

320

CONTROLS AND LINKAGE

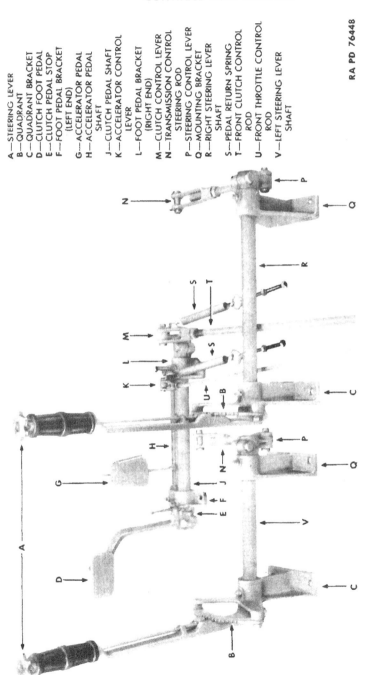

A—STEERING LEVER
B—QUADRANT
C—QUADRANT BRACKET
D—CLUTCH FOOT PEDAL
E—CLUTCH PEDAL STOP
F—FOOT PEDAL BRACKET (LEFT END)
G—ACCELERATOR PEDAL
H—ACCELERATOR PEDAL SHAFT
J—CLUTCH PEDAL SHAFT
K—ACCELERATOR CONTROL LEVER
L—FOOT PEDAL BRACKET (RIGHT END)
M—CLUTCH CONTROL LEVER
N—TRANSMISSION CONTROL STEERING ROD
P—STEERING CONTROL LEVER
Q—MOUNTING BRACKET
R—RIGHT STEERING LEVER SHAFT
S—PEDAL RETURN SPRING
T—FRONT CLUTCH CONTROL ROD
U—FRONT THROTTLE CONTROL ROD
V—LEFT STEERING LEVER SHAFT

RA PD 76448

Figure 190 — Controls and Levers

TM 9-775
LANDING VEHICLE TRACKED MK. I AND MK. II

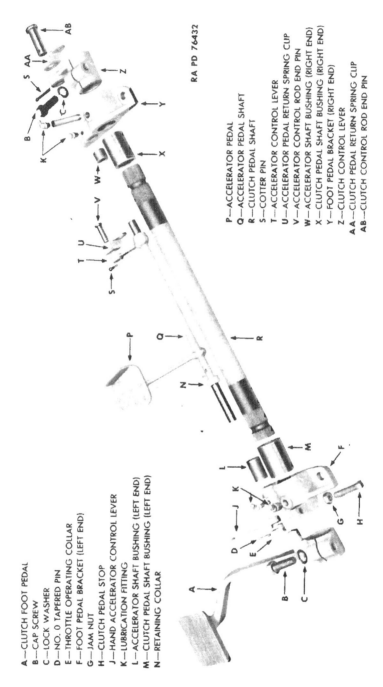

RA PD 76432

A—CLUTCH FOOT PEDAL
B—CAP SCREW
C—LOCK WASHER
D—NO. 0 TAPERED PIN
E—THROTTLE OPERATING COLLAR
F—FOOT PEDAL BRACKET (LEFT END)
G—JAM NUT
H—CLUTCH PEDAL STOP
J—HAND ACCELERATOR CONTROL LEVER
K—LUBRICATION FITTING
L—ACCELERATOR SHAFT BUSHING (LEFT END)
M—CLUTCH PEDAL SHAFT BUSHING (LEFT END)
N—RETAINING COLLAR
P—ACCELERATOR PEDAL
Q—ACCELERATOR PEDAL SHAFT
R—CLUTCH PEDAL SHAFT
S—COTTER PIN
T—ACCELERATOR CONTROL LEVER
U—ACCELERATOR PEDAL RETURN SPRING CLIP
V—ACCELERATOR CONTROL ROD END PIN
W—ACCELERATOR SHAFT BUSHING (RIGHT END)
X—CLUTCH PEDAL SHAFT BUSHING (RIGHT END)
Y—FOOT PEDAL BRACKET (RIGHT END)
Z—CLUTCH CONTROL LEVER
AA—CLUTCH PEDAL RETURN SPRING CLIP
AB—CLUTCH CONTROL ROD END PIN

Figure 191 — Clutch and Accelerator Shafts — Disassembled

CONTROLS AND LINKAGE

vehicle. The rear throttle control rod (rod-G, fig. 193) connects the turnbuckle with the rear throttle extension rod (rod-H, fig. 193). The rear throttle extension rod connects the rear throttle control rod with the throttle operating lever mounted on the engine mount in the rear of the engine. The throttle operating rod connects the throttle operating lever with the throttle arm on the carburetor.

143. REMOVAL.

a. Disassemble Clutch Pedal and Accelerator Pedal Shaft Assembly (fig. 191). Unscrew the four lubrication fittings from the shaft mounting brackets to prevent damage to the fittings. Remove cap screw and lock washer which secure clutch pedal to clutch pedal shaft. Mark relative position of clutch pedal to shaft with prick punch to facilitate assembly procedure. Tap clutch pedal off end of shaft. Remove cap screw and lock washer which secure clutch control lever to right end of shaft. With prick punch, mark the position of the lever on the shaft. Tap control lever off shaft. Pull the foot pedal mounting bracket (right end) and clutch pedal shaft off the accelerator pedal shaft and out of left end foot pedal bracket. Tap out tapered pin which secures hand throttle operating lever collar to accelerator pedal shaft. Remove collar from shaft. Slide hand throttle operating lever off end of accelerator shaft. Pull clutch foot pedal mounting bracket (left end) and its assembled shaft bushings from accelerator pedal shaft.

b. Disassemble Steering Lever Shaft Assembly (fig. 192). Unscrew the lubrication fittings from quadrant and steering lever shaft mounting brackets. Loosen jam nut from adjustable yoke on the transmission control steering rod, and unscrew the yoke from the control rod. Unscrew jam nut from control rod. Remove cotter pin and rod end pin which secure transmission control steering rod to steering control lever, and remove transmission control steering rod. Mark position of steering control lever on steering lever shaft with a prick punch. Remove cap screw and toothed lock washer which secure steering control lever to steering lever shaft. Tap steering control lever off steering lever shaft. Slide steering lever shaft bracket off end of shaft. Remove cap screw and toothed lock washer which secure steering lever on shaft, and tap steering lever off shaft. Slide quadrant mounting bracket assembly off end of steering lever shaft. Remove two cap screws and toothed lock washers which secure quadrant to quadrant bracket, and lift off quadrant. The other steering lever shaft assembly is disassembled in the same manner.

c. Disassemble Steering Lever (fig. 192). Loosen the jam nut which secures pawl rod to pawl. Lift pawl, pawl rod, and pawl spring from slot in steering lever. Unscrew pawl from pawl rod. Slide pawl spring and spring stop from pawl rod. Pull handle and pawl rod from

TM 9-775
143

LANDING VEHICLE TRACKED MK. I AND MK. II

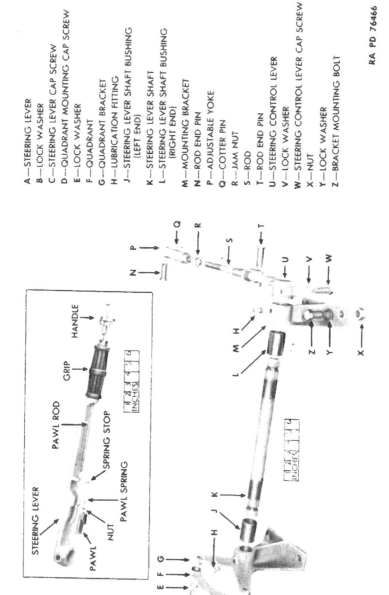

A—STEERING LEVER
B—LOCK WASHER
C—STEERING LEVER CAP SCREW
D—QUADRANT MOUNTING CAP SCREW
E—LOCK WASHER
F—QUADRANT
G—QUADRANT BRACKET
H—LUBRICATION FITTING
J—STEERING LEVER SHAFT BUSHING (LEFT END)
K—STEERING LEVER SHAFT
L—STEERING LEVER SHAFT BUSHING (RIGHT END)
M—MOUNTING BRACKET
N—ROD END PIN
P—ADJUSTABLE YOKE
Q—COTTER PIN
R—JAM NUT
S—ROD
T—ROD END PIN
U—STEERING CONTROL LEVER
V—LOCK WASHER
W—STEERING CONTROL LEVER CAP SCREW
X—NUT
Y—LOCK WASHER
Z—BRACKET MOUNTING BOLT

RA PD 76466

Figure 192 — Steering Lever and Shaft Assembly

324

CONTROLS AND LINKAGE

top of steering lever. The other steering lever is disassembled in the same manner.

d. Control Rods and Linkage (fig. 193). Remove control tunnel side panels making control rods available. Remove bolts, nuts, and toothed lock washers which secure wooden guide blocks to brackets in control tunnel. Separate guide blocks. Remove cotter pin and end pin from yoke end of rod to be removed. If rod has turnbuckle on one end, loosen the jam nut on the side of the turnbuckle toward the rod to be removed, and unscrew the rod from the turnbuckle so as not to disturb the adjustment of the turnbuckle on the rod remaining in the tunnel. CAUTION: *Be careful not to disturb the adjustment of the rest of the linkage until the rod has been replaced in exactly the same position as before it was removed.*

144. INSTALLATION.

a. Throttle and Clutch Control Rods (fig. 193). Adjust yoke or turnbuckle on the adjustable end of the rod to be replaced so that it may be installed in the vehicle without disturbing the linkage already in the vehicle. If the rod being installed fits into a turnbuckle at one end, screw this end into the turnbuckle before installing the rod end pin and cotter pin which secure the yoke end of the rod. Then fasten the yoke end. If there are yokes at both ends, fasten the adjustable end first. Place the wooden guide blocks around the control rods, and secure to the mounting brackets in the tunnel with bolts, nuts, and toothed lock washers.

b. Assemble Front Bulkhead Clutch and Throttle Control Levers (fig. 193). Aline the fulcrum end of forward bulkhead throttle control lever inside the fulcrum end of the forward bulkhead clutch control lever. Aline the fulcrum end of both levers with mounting holes provided in the forward bulkhead bracket. Install two cap screws and toothed lock washers which secure forward bulkhead clutch and throttle control levers in bracket.

c. Assemble Steering Levers (fig. 192). Slide pawl handle into position on pawl rod. Hold handle in position, and guide pawl rod into grip end of steering lever. Slide spring stop and pawl spring over lower end of pawl rod, and screw nut on end of pawl rod. Place pawl rod, pawl spring, and spring stop into slot provided in steering lever. Tighten nut to secure pawl rod to pawl. The other steering lever is assembled in the same manner.

d. Assemble Steering Lever Shaft Assembly (fig. 192). Slide quadrant bracket, with its assembled bushing, over left end of steering lever shaft. Slide steering lever shaft mounting bracket, with its assembled bushing, over right end of steering lever shaft. Aline prick punch mark on steering control lever with punch mark on shaft; then

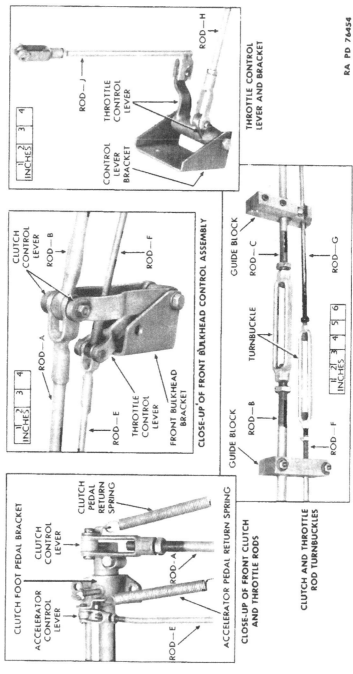

CONTROLS AND LINKAGE

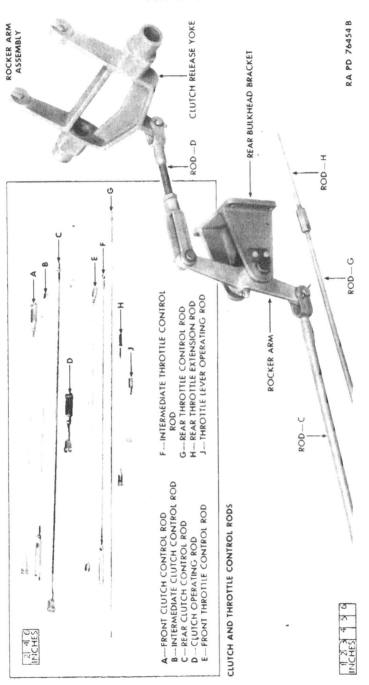

A—FRONT CLUTCH CONTROL ROD
B—INTERMEDIATE CLUTCH CONTROL ROD
C—REAR CLUTCH CONTROL ROD
D—CLUTCH OPERATING ROD
E—FRONT THROTTLE CONTROL ROD
F—INTERMEDIATE THROTTLE CONTROL ROD
G—REAR THROTTLE CONTROL ROD
H—REAR THROTTLE EXTENSION ROD
J—THROTTLE LEVER OPERATING ROD

Figure 193 — Clutch and Throttle Control Rods and Linkage

LANDING VEHICLE TRACKED MK. I AND MK. II

slide steering control lever, which is serrated on ends, over serrations on right end of steering lever shaft. Aline mounting screw hole in steering control lever with groove provided between serrations on end of shaft, and install cap screw and toothed lock washer to secure steering control lever in position on shaft. Position quadrant on quadrant bracket with quadrant teeth pointing toward rear. Aline holes in quadrant with tapped holes in quadrant bracket, and install two cap screws and toothed lock washers to secure quadrant to bracket. Aline prick punch mark on steering lever with punch mark on shaft and slide steering lever, which is serrated, over left end of steering lever shaft, which is also serrated. Aline cap screw mounting hole in steering lever with groove provided in left end of shaft, and install cap screw and toothed lock washer to secure steering lever on shaft. Screw jam nut and adjustable yoke onto the transmission control steering rod. Aline transmission control steering rod yoke with steering control lever, and install rod end pin and cotter pin to secure transmission control steering rod assembly to the steering control lever. Install lubrication fitting in each mounting bracket. The other steering lever shaft assembly is assembled in the same manner.

c. **Assemble Clutch Pedal and Accelerator Pedal Shaft Assembly** (fig. 191). Aline openings provided in foot pedal and accelerator pedal shafts mounting brackets with the ends of the shafts. Slide the mounting brackets into position on the shafts. Aline prick punch marks on clutch control lever and on clutch foot pedal with punch marks on clutch foot pedal shaft, then slide the clutch control lever and clutch foot pedal on serrated ends of shaft. Aline the cap screw mounting holes in both clutch control lever and in foot pedal with grooves provided in serrations, and install cap screw and toothed lock washer (one in each part) to hold the control lever and foot pedal, preventing lateral movement on the shaft. Slide hand throttle operating lever over left end of accelerator shaft, with jaw of lever facing outward. Place the hand throttle operating lever collar over the left end of the accelerator shaft, with jaw of collar facing toward hand throttle operating lever. Aline pin hole in collar with pin hole provided in shaft, and tap tapered pin into collar and shaft to secure hand throttle operating lever collar on accelerator shaft. Screw the four lubrication fittings (two in each mounting bracket) in position in the brackets.

Section XXXII

FIRE EXTINGUISHING SYSTEM

	Paragraph
Description	145
Removal of fixed fire extinguishers	146
Care of fire extinguishers	147
Handling of fire extinguishers	148
Installation of fixed fire extinguishers	149
Remote controls (fixed fire extinguisher)	150
Lines and nozzles	151

145. DESCRIPTION.

a. **Fixed Fire Extinguishers.** To combat fire in engine room, there is a fixed fire extinguisher system mounted just in back of right fuel tank. This system consists of two 10-pound carbon dioxide cylinders, four shielded nozzles, dual control cables, tubing, and connections.

b. **Portable Fire Extinguisher.** There is a portable fire extinguisher on floor of the driver's compartment, to the left of the driver. This unit consists of a 15-pound carbon dioxide cylinder, to which is attached a flexible hose, a shielded nozzle, and a control knob.

146. REMOVAL OF FIXED FIRE EXTINGUISHERS.

a. Open hinged rear cover.

b. **Remove Upper Right Rear Baffle.** Loosen clamp nuts which secure two clamps (upper and lower) holding upper right rear baffle in position (fig. 61). Rotate clamps to a vertical position. Pry baffle outward and remove.

c. **Disconnect Fixed Fire Extinguisher Release Head** (fig. 195). If fire extinguishers are being removed for inspection or weighing, but have not been discharged, it will not be necessary to remove release head cover to disconnect release cable. Unscrew release head union, and lay release head to one side (fig. 195). Disconnect tubing between release head and dual pull mechanism. Be careful not to pull head, since this will rotate the mechanism as if the release cable were pulled, and force release plunger down in bottom of head.

d. **Disconnect Pressure Connecting Tubing** (fig. 195). Disconnect the pressure connecting tubing at pressure connecting head on each of the two fire extinguisher cylinders. Lift off pressure connecting tubing.

e. **Disconnect Extinguisher Tubing.** Disconnect unions on copper tubing running to each cylinder (fig. 195).

LANDING VEHICLE TRACKED MK. I AND MK. II

Figure 194 — Fixed Fire Extinguishers

f. Remove Fixed Fire Extinguishers (fig. 195). Remove the six cap screws and toothed lock washers which secure the dual cylinder bracket. Remove upper and lower halves of bracket. Lift out the two fire extinguisher cylinders. CAUTION: *Do not drop cylinders or handle them roughly. The cylinders are loaded. They are dangerous, and may explode if dropped.*

147. CARE OF FIRE EXTINGUISHERS.

a. Fixed Fire Extinguishers. Weight of empty cylinder (34½ lb) and weight of the carbon dioxide (10 lb) are stamped on each cylinder head. Every 4 months, or more often if necessary, weigh each cylinder. If weight of carbon dioxide is less than 9 pounds, exchange unit for one fully charged. If red blow-off seal on valve head indicates the cylinder has been discharged due to high temperature, replace cylinder.

b. Portable Fire Extinguisher. Weight of empty cylinder (38½ lb) without hose and discharge horn, is stamped on each cylin-

FIRE EXTINGUISHING SYSTEM

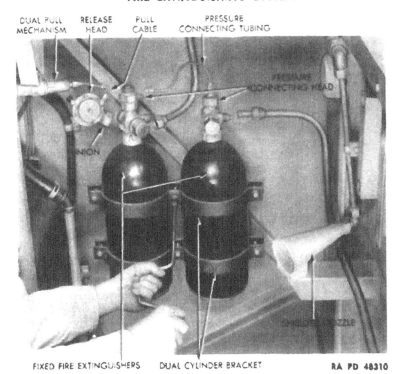

Figure 195 — Removing Fixed Fire Extinguishers

der head. Every 4 months, or more often if necessary, weigh each cylinder without hose and discharge horn. If weight is less than 13½ pounds plus container weight marked on cylinder, exchange unit for a fully charged one.

148. HANDLING OF FIRE EXTINGUISHERS.

a. A cylinder containing gas at high pressure is as dangerous as a loaded shell. The extinguisher cylinders (fixed and portable) never should be dropped, struck, handled roughly, or exposed to unnecessary heat.

149. INSTALLATION OF FIXED FIRE EXTINGUISHERS.

a. Place Extinguishers in Position (fig. 195). Lift the two fire extinguisher cylinders into position in vehicle. The cylinder to which the release head is attached is mounted in forward position. Place upper and lower halves of dual cylinder bracket in position, and loosely install the six cap screws and toothed lock washers which secure

TM 9-775

LANDING VEHICLE TRACKED MK. I AND MK. II

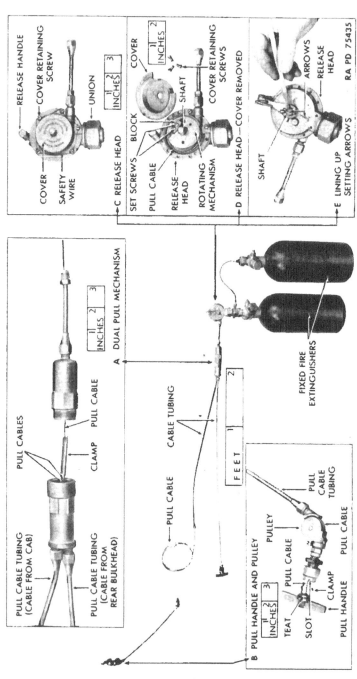

Figure 196 — Fixed Fire Extinguishers and Controls

FIRE EXTINGUISHING SYSTEM

bracket. (These will be tightened later after all connections have been made.)

b. Connect Extinguisher Tubing (fig. 195). Install and tighten unions which secure extinguisher tubing to cylinders.

c. Connect Pressure Connecting Tubing (fig. 195). Place pressure connecting tubing on the two cylinder heads and tighten unions, securing tubing.

d. Set Release Head. Remove the lacing wire and three cap screws which secure the release head cover to release head (fig. 196). Lift off cover. Within head is a rotating mechanism which encloses a small brass block equipped with two set screws (fig. 196). These clamp and pull cable at block. Rotate mechanism clockwise as far as it will go (with finger pressure), until arrow on front of rotating mechanism shaft lines up with arrow on front face of release head (fig. 196). Then turn counterclockwise slightly to ease strain on pull cable.

e. Install Release Head. Place release head cover on head (fig. 196). Install three retaining cap screws, and lace screws with locking wire. Place release head on fire extinguisher cylinder, and do not pull head so as to pull out on the cable connected within head. Tighten union, securing release head to fire extinguisher cylinder, then tighten union which secures tubing running between dual pull mechanism and release head (figs. 195 and 196).

f. Tighten Dual Cylinder Bracket Nuts (fig. 195). If all connections are securely fastened, tighten nuts, securing upper and lower halves of dual cylinder bracket.

g. Install Upper Right Rear Baffle. Place upper right rear baffle in position. Rotate the two securing clamps to a horizontal position, then tighten clamp nuts (fig. 61).

h. Close hinged rear cover.

150. REMOTE CONTROLS (FIXED FIRE EXTINGUISHER):

a. Description. A remote control pull handle is mounted on forward face of forward bulkhead behind assistant driver. To this handle is attached a pull cable (fig. 196) which is threaded through a tube, and extends back along the upper right side of the cargo compartment through the stern bulkhead, and is connected to the dual pull mechanism near the two fixed carbon dioxide cylinders. A remote control pull handle of the same type is mounted on forward face of stern bulkhead in upper right rear corner of the cargo compartment. To this handle is attached a similar, but necessarily shorter, pull cable that is also connected to dual pull mechanism. Inside dual pull mechanism, remote control cables are joined together, and a single

LANDING VEHICLE TRACKED MK. I AND MK. II

cable emerges from the dual pull mechanism, and is connected to release head (par. 146). Pulleys assist free movement of the remote control cables.

b. **Replacing Broken Cables** (Both Cables Replaced in Identical Manner). If a remote control cable breaks, remove upper rear right baffle to gain access to fire extinguishing system. Unscrew the two halves of dual pull mechanism. Remove end of broken cable from pull cable clamp within the dual pull mechanism. Pull out broken cable end. Pull out the other broken cable end by pulling out pull handle (cable end is welded to handle). To install new cable and handle assembly, push cable end through handle socket assembly mounted on bulkhead until end of cable approaches the dual pull mechanism. Within the dual pull mechanism, join cable end with the other remote control cable, using a new pull cable clamp. Pinch tube to make this connection. Push single cable end through the other end of the dual pull mechanism, and connect cable to release head (par. 149).

151. LINES AND NOZZLES.

a. As carbon dioxide is released, it passes through $\frac{1}{4}$-inch and $\frac{1}{2}$-inch copper tubing to four shielded nozzles mounted in engine room. A $\frac{1}{4}$-inch copper line extends from one fire extinguisher cylinder to the other cylinder. When replacing a damaged piece of tubing, use same size and materials. Cut the new piece the same length, and bend to same shape as tube to be replaced.

Section XXXIII

HULL

	Paragraph
Cab seats	152
Safety belts	153
Escape hatches	154
Cab escape windows	155
Bilge pump	156
Floor plates	157
Cab fan	158
Controls and levers	159
Draining water from pontoons	160

152. CAB SEATS.

a. **Description.** Two cab seats are used in the landing vehicles tracked; one for the driver, and one for the assistant driver. Seats are adjustable upward, downward, backward, and forward.

HULL

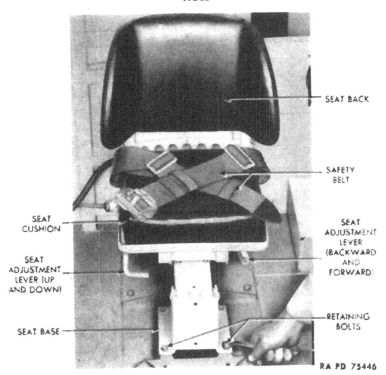

Figure 197 — Removing Cab Seat

b. **Maintenance.** Examine seat cushions for rips and tears. Replace torn or badly worn seat cushions with new cushions. Test the seats for adjustments. Seats should move upward, downward, backward, and forward easily, and lock securely in any position. If adjusting levers will not operate, or if the seat pedestal is damaged, install new seat.

c. **Removal** (fig. 197). Remove the four bolts, nuts, and lock washers which secure seat pedestal to brackets on floor. Lift out seat.

d. **Installation** (fig. 197). Place seat in position, and install the four retaining nuts, bolts, and lock washers.

153. SAFETY BELTS.

a. **Description.** Each of the seats in cab is equipped with a web-type safety belt. These are the quickly removable type used in airplanes.

b. **Maintenance.** Check action of safety belt buckle. In an emer-

TM 9-775
LANDING VEHICLE TRACKED MK. I AND MK. II

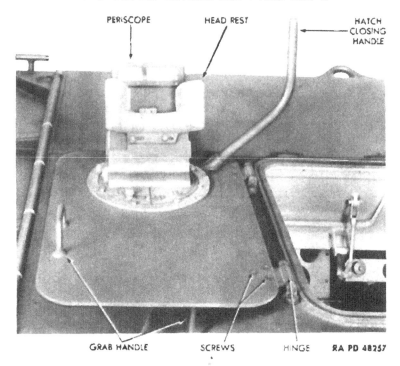

Figure 198 — Driver's Escape Hatch — Open Position

gency it may be necessary to release belt quickly. Examine webbing carefully. If webbing is torn or frayed, install new safety belt.

c. **Removal** (fig. 197). Loosen one belt strap in belt adjusting buckle, and remove end which goes through belt support bracket.

d. **Installation.** With adjustable buckle on strap about 8 inches from end of strap, place strap through belt support bracket, and back through both sides of the adjustable buckle. Make sure belt is right-side out. Install the other half of the safety belt exactly as the first half was installed.

154. ESCAPE HATCHES.

a. **Description** (figs. 3 and 198). Two escape hatches are used on the armored landing vehicles tracked (LVT (A) (1) and LVT (A) (2)). Cab escape windows are used in the unarmored cargo carrier. Escape hatches are each equipped with a periscope. A hatch-closing handle is located on the under side of the hatch to facilitate closing the hatch. A grab handle on the escape hatch serves as a stop to lock the escape hatch in its closed position.

TM 9-775
154—155

HULL

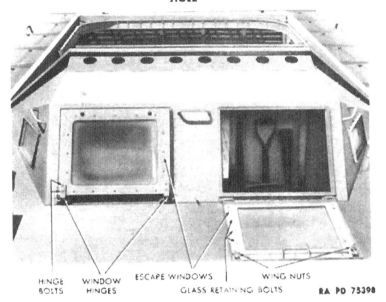

Figure 199 — Escape Windows — Unarmored Landing Vehicle Tracked (LVT (2))

b. Removal (fig. 198). Remove the six screws which secure the escape hatch to the two hatch hinges, and lift off hatch.

c. Installation (fig. 198). Place hatch in position, and install the six screws which secure the hatch to the two hatch hinges.

155. CAB ESCAPE WINDOWS.

a. Description. On the unarmored cargo carrier LVT (2), two cab escape windows are used instead of the escape hatches used on other models of the landing vehicles tracked. The windows hinge downward and, when open, rest in a horizontal position on the front hull. Two hinged wing nuts, one on each top corner of the window, are used to lock the window in the closed position (fig. 199). A grab handle, located at the top center of the window, facilitates opening or closing the window.

b. Maintenance. If the window frame is bent, warped, or damaged by combat use, replace complete window frame. If glass is broken, replace glass. Keep windows clean. Periodically inspect condition of rubber seal around glass. If seal is broken or in poor condition, permitting entrance of water, remove glass and install a new seal.

c. Removal of Window and Frame (fig. 199). With the window in a closed position, remove the two bolts and nuts which secure each of the two hinges to window frame. Lift off window and frame.

337

TM 9-775
155

LANDING VEHICLE TRACKED MK. I AND MK. II

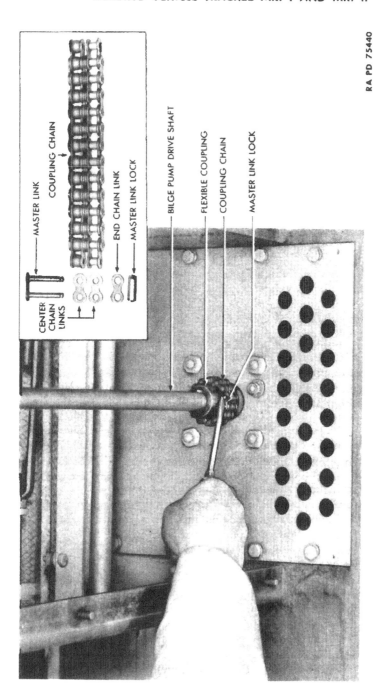

Figure 200 — Disconnecting Bilge Pump Drive Shaft — First Type

TM 9-775
155-156

HULL

Figure 201 — Lifting Out Bilge Pump Drive Shaft — Second Type

d. **Removal of Window from Frame** (fig. 199). Remove the 16 bolts and nuts which secure window within the window frame. Lift out window with retaining glass frame and gasket.

e. **Installation of Window in Window Frame** (fig. 199). Place the rubber gasket against window, and place window in position in window frame. Place the glass frame over glass, and install the 16 retaining bolts and nuts.

f. **Installation of Window and Frame** (fig. 199). Place window and frame on the two hinges, and install the four retaining bolts and nuts (two, each hinge).

156. **BILGE PUMP.**

a. **Description and Data.** The bilge pump is mounted beneath the floor plates in the center of the vehicle. Power to drive the bilge is transmitted through the bilge pump drive shaft from the power take-off. A direct drive is employed, insuring that the bilge pump turns at the same number of revolutions per minute as the engine. Vehicles

TM 9-775
LANDING VEHICLE TRACKED MK. I AND MK. II

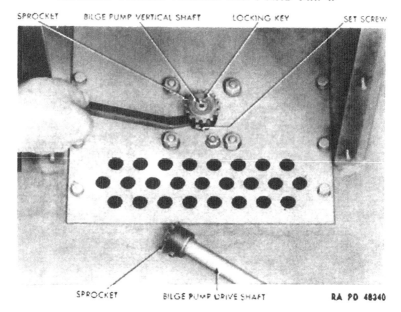

Figure 202 — Prying Off Coupling Chain Sprocket

numbered from 1 to 25 inclusive were equipped with a bilge pump drive shaft having a double sprocket at each end coupled by a chain (fig. 200). From vehicle number 26 and up, a drive shaft employing disk-type flexible couplings was used (fig. 201). Pumping capacity of the bilge pump at various revolutions per minute of the engine is:

Engine Revolutions Per Minute	Pump Capacity Gallons Per Minute
2,400	500
2,000	430
1,800	375

b. Maintenance. It is essential that service and lubrication of the bilge pump be performed frequently. Since the pump turns whenever the engine is running, there is possibility of wear and damage due to insufficient lubrication. If bilge pump is broken or damaged so that it will not operate to its full pumping capacity, install a new pump.

c. Removal.

(1) REMOVE POWER TAKE-OFF SUPPORT CASE COVER. Remove the 10 stud nuts and toothed lock washers which secure power take-off support case cover to support case. Lift off cover.

TM 9-775
156

HULL

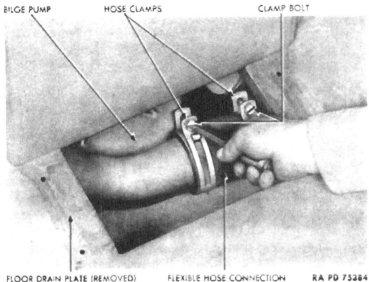

Figure 203 — Loosening Bilge Pump Flexible Hose Connection

(2) REMOVE BILGE PUMP DRIVE SHAFT — FIRST TYPE (fig. 200). Pry off lock which secures coupling chain master link at upper end of drive shaft. Pull master link out of chain, lift out the three chain links (two center, one end), and unwind chain. Repeat the operation to disconnect the sprockets at lower end. Lift out drive shaft.

(3) REMOVE BILGE PUMP DRIVE SHAFT — SECOND TYPE (fig. 201). Remove two cotter pins, nuts, bolts, and six flat washers which hold the disk-type coupling at the upper end of drive shaft. Repeat the operations to disconnect lower end, and remove drive shaft.

(4) REMOVE COUPLING SPROCKET — FIRST TYPE. Loosen the set screw which secures sprocket to bilge pump vertical shaft. Pry sprocket off shaft (fig. 202). Pull out locking key which secures sprocket on shaft.

(5) REMOVE POWER TAKE-OFF COUPLING YOKE — SECOND TYPE. Cut locking wire from set screw in power take-off coupling yoke, and loosen set screw. Pull coupling yoke and its two assembled disks straight downward from power take-off shaft. Remove coupling key from keyway.

(6) REMOVE BILGE PUMP COUPLING YOKE — SECOND TYPE. Repeat procedure in step (5) to remove coupling yoke from bilge pump shaft.

TM 9-775

LANDING VEHICLE TRACKED MK. I AND MK. II

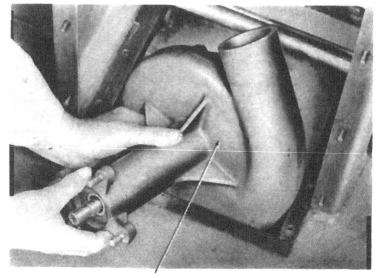

BILGE PUMP   RA PD 75401

Figure 204 — Removing Bilge Pump

(7) REMOVE BILGE PUMP COVER PLATE. Remove the eight cap screws and toothed lock washers which secure bilge pump cover plate to floor plates. Remove the four stud nuts which secure bilge pump cover plate to bilge pump. Lift out cover plate.

(8) REMOVE BILGE PUMP. Remove the four cap screws and toothed lock washers which secure the floor drain plate on the right side of control tunnel. Lift off drain plate. Reach down into bilge pump compartment, and loosen bolts securing the two hose clamps on flexible hose which connects bilge pump to bilge pump discharge channel (fig. 203). Work flexible hose back on discharge channel and off bilge pump. Now, manipulate bilge pump until it may be removed from its compartment (fig. 204).

d. Installation.

(1) PLACE BILGE PUMP IN POSITION. Manipulate bilge pump into position in its compartment on floor of vehicle. The bilge pump discharge opening must fit into flexible hose on bilge pump discharge channel (fig. 204). Work hose into position, and tighten bolts which secure the two hose clamps on flexible hose.

(2) INSTALL BILGE PUMP COVER PLATE. From the left side of control tunnel, place bilge pump cover plate on top of the bilge pump. With the assistance of another man to lift up on the bilge pump, install the four stud nuts which secure bilge pump cover plate to bilge

TM 9-775
156

HULL

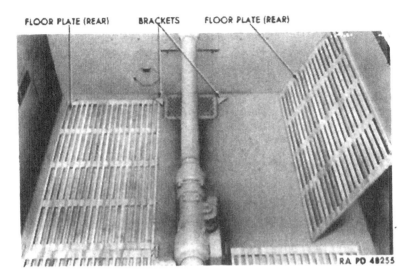

Figure 205 — Floor Plates — Unarmored Landing Vehicle Tracked (LVT (2))

pump. Line up holes, and install the eight cap screws and toothed lock washers which secure bilge pump cover plate to floor plate.

(3) INSTALL COUPLING SPROCKET — FIRST TYPE. Tap the coupling sprocket on bilge pump vertical shaft. Tap the locking key in the keyway forward between sprocket and shaft. Tighten set screw.

(4) INSTALL COUPLING YOKES — SECOND TYPE. Tap yoke into position on bilge pump shaft. Tighten set screw, and install locking wire. Tap yoke into position on power take-off vertical shaft. Tighten set screw, and install locking wire.

(5) INSTALL BILGE PUMP DRIVE SHAFT — FIRST TYPE (figs. 200 and 204). Place the bilge pump drive shaft on bilge pump vertical shaft. Line up the coupling sprocket on the lower part of bilge pump drive shaft with the coupling sprocket on bilge pump vertical shaft. Line up the coupling sprocket on the upper part of bilge pump drive shaft with the coupling sprocket on power take-off vertical shaft. Wind coupling chain around coupling sprocket. Insert the master link in the two ends of the chain, making sure that two center chain links and one end chain link are installed. Install master link lock. Repeat the operation to connect coupling sprocket at opposite end of bilge pump drive shaft.

(6) INSTALL BILGE PUMP DRIVE SHAFT — SECOND TYPE. Place drive shaft in position on disk-type couplings. Install flat washers,

TM 9-775
156-157

LANDING VEHICLE TRACKED MK. I AND MK. II

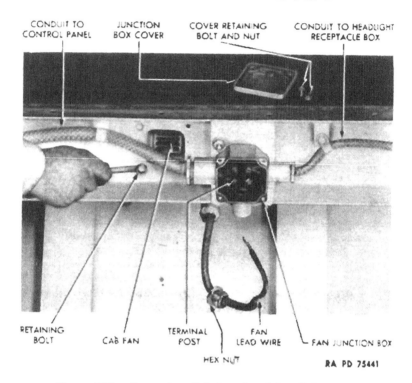

Figure 206 — Removing Cab Fan Retaining Bolts

bolts, and nuts which hold drive shaft yokes to disks. Install cotter pins.

(7) INSTALL POWER TAKE-OFF SUPPORT CASE COVER. Place power take-off support case cover in position, and install the 10 stud nuts and toothed lock washers which secure cover to support case.

157. FLOOR PLATES.

a. Description (fig. 205). Four wooden, lattice-type floor plate assemblies are used in the unarmored landing vehicle tracked, LVT (2). These plates are designed to raise the level of the cargo slightly above the steel floor. Their open design permits drainage of water from the cargo, and provides a way of keeping cargo away from water on the floor of the vehicle. Floor plates lie flat on the floor, and are held in position at one end by means of triangular brackets welded to the floor. If floor plates are damaged or broken so badly that cargo rests against the steel floor, lift out the damaged floor plate and install a new plate.

TM 9-775
158

HULL

CAB FAN   RA PD 48311

Figure 207 — Removing Cab Fan

158. CAB FAN.

a. Description. A cab fan, similar in construction to the engine room blowers, is mounted under the hull in the cab, between driver and assistant driver. The cab fan switch is located on the control panel. Function of the cab fan is to control circulation of air in the cab, and assist ventilation.

b. Maintenance. If cab fan will not operate after switch and wires have been inspected, and have been found in operating condition, install new fan.

c. Removal (figs. 206 and 207). Remove the four bolts and nuts which secure the cab fan junction box cover. Lift off cover. Unscrew the terminal stud nut which retains cab fan lead wire in junction box. Unscrew the conduit hexagonal nut on the lead, and pull conduit, with lead wire, out of junction box. Remove the four bolts, nuts, and toothed lock washers which hold fan in position, and remove fan.

d. Installation (fig. 206). Place cab fan in position, and install

TM 9-775
158

LANDING VEHICLE TRACKED MK. I AND MK. II

Figure 208 — Controls and Levers

HULL

the four retaining bolts, nuts, and toothed lock washers. Insert conduit, with lead wire, into fan junction box, and place lead wire on its proper terminal. Install terminal stud nut. Tighten conduit hexagonal nut. Place fan junction box cover in position, and install the four retaining bolts and nuts.

159. CONTROLS AND LEVERS.

a. **Description** (fig. 208). Controls and levers used to operate the vehicle consist mainly of steering levers and steering lever shafts, clutch pedal, clutch pedal shaft and control rods, accelerator pedal and accelerator pedal shaft, and accelerator control rod. Steering levers, clutch pedal, and accelerator pedal are mounted on the ends of their respective shafts in front of the driver. The clutch control rod and accelerator control rod are attached by means of adjustable rod ends to arms on the clutch pedal shaft and accelerator pedal shaft. These control rods extend back under the transmission through the control tunnel to the engine room. They then attach to other control rods which, in turn, are connected to the components which they operate on the engine.

b. **Maintenance.** Maintenance of controls and levers is concerned primarily with frequent and thorough inspection of all control rods for wear or damage. Adjustment of the steering levers is covered under the adjustment of steering brake shoes (par. 123). Adjustment of the clutch pedal is covered under the clutch assembly (par. 114). Frequently inspect shafts and control rods. Make sure all cotter pins, rod end pins, jam nuts, socket head set screws, and other devices used in the installation of the controls and levers are firmly secured.

c. **Removal of Steering Lever** (fig. 208). Loosen the one clamp screw which clamps the steering lever to the splined steering lever shaft. Pull steering lever off shaft.

d. **Installation of Steering Lever** (fig. 208). Tap the steering lever on the splined steering lever shaft, and tighten clamp screw which secures it in position.

e. **Removal of Steering Lever Shaft** (fig. 208). Remove steering lever. Loosen clamp screw, and remove steering rod arm from opposite end of steering lever shaft. Tap steering lever shaft out of steering lever shaft brackets bolted to floor.

f. **Installation of Steering Lever Shaft** (fig. 208). Tap the steering lever shaft into position through the two steering lever shaft brackets. Tap steering rod arms onto the right end (facing forward) of the shaft, and tighten clamp screw. Install steering lever.

g. **Removal of Clutch Pedal Shaft** (fig. 208). Loosen clamp screw and tap clutch pedal off end of clutch pedal shaft. Unhook clutch pedal return spring. Remove cotter pin and rod end pin which secure clutch control rod adjustable rod end to clutch pedal return spring arm. Loosen clamp screw, and tap arm off right-hand end of clutch pedal shaft. Tap clutch pedal shaft out of the two brackets bolted to floor.

h. **Installation of Clutch Pedal Shaft** (fig. 208). Tap clutch pedal shaft into the two brackets bolted to cab floor. Tap clutch pedal return spring arm on right end of clutch pedal shaft. Tighten clamp screw. Tap clutch pedal on left end of shaft (make sure clutch pedal fits on splines of shaft). Tighten clamp screw. Place clutch control rod over clutch pedal return spring, and install rod end pin and cotter pin. Hook up clutch pedal return spring.

i. **Removal of Accelerator and Shaft** (fig. 208). Remove clutch pedal shaft (subpar. g, above). Unhook accelerator pedal return spring. Remove cotter pin and rod end pin which secure hand throttle cable to arm on accelerator shaft. Remove cotter pin and rod end pin which secure accelerator control rod to arm on accelerator pedal shaft. Remove four cap screws and toothed lock washers which secure the two brackets in which the accelerator pedal shaft is mounted. Tap out locking pin from collar on each outer end of accelerator pedal shaft.

j. **Installation of Accelerator Pedal and Shaft** (fig. 208). Slide the floor brackets on ends of accelerator pedal shaft. Slip a collar over each outer end of accelerator pedal shaft, and tap in locking pins. Bolt the two brackets to floor with four cap screws and toothed lock washers (two screws in each bracket). Attach the adjustable rod end on hand throttle cable to arm on accelerator pedal shaft with a rod end pin and cotter pin. Place accelerator control rod on arm on accelerator pedal shaft, and install the retaining rod end pin and cotter pin. Hook up accelerator pedal return spring.

160. DRAINING WATER FROM PONTOONS.

a. **When to Drain.** Water in pontoons affects steering, and causes vehicle to track incorrectly (travel in circles) when operating in water. Drain pontoons whenever water is present.

b. **Draining.** Remove the six drain plugs from the bottom of each pontoon, and allow all water to drain.

c. **Installing Drain Plugs.** Clean threads and apply white lead base anti-seize compound. Tighten plugs to a snug fit, but avoid damage to threads.

TM 9-775
161–162

Section XXXIV

TURRET

	Paragraph
Description	161
Traversing control assembly	162
Oil pump motor	163
Gear box assembly	164
Hydraulic motor	165
Traversing oil pump	166
Oil pot (reservoir)	167
Turret doors	168
Turret seats	169
Removal of turret	170
Installation of turret	171

161. DESCRIPTION.

a. The turret (fig. 209) on model (LVT) (A) (1) is of welded armor-plate construction. The front half of this turret is a semicircular shape. The rear sides are flat, as is the back of the turret. The top forward section of turret is sloped downward toward the combination gun mount. The balance of the top of turret slopes slightly toward the rear. The armored combination gun mount covers the front side of the turret. Two hinged turret doors are mounted on top rear side of turret (fig. 17). The doors may be opened from the outside or inside when not locked. However, when locked from the inside, the doors cannot be opened except from within the turret. Three periscopes are mounted in top of turret. The two on right side are for the commander, and one on left side for the gunner. The turret may be traversed 360 degrees either by a hand crank or by a hydraulic turret traversing mechanism (fig. 18). This turret traversing mechanism is described in section IV. The turret is secured to the vehicle by means of six roller assemblies and six segments. The roller assemblies and segments are covered by a dust cover.

162. TRAVERSING CONTROL ASSEMBLY.

a. Description (figs. 211 and 212). A pistol-grip control handle is mounted on the switch box in front of and to the left of the gunner. This handle enables the gunner to control the speed and direction of the traverse. Another control handle is mounted at the top of turret in front of the commander, and is connected to the pistol-grip control by suitable linkage. The pistol-grip control handle also is connected to the traverse oil pump located under turret basket by suitable linkage.

b. Maintenance. All of the control linkage, or sections of the linkage, can be removed in event of wear or damage to linkage.

TM 9-775

LANDING VEHICLE TRACKED MK. I AND MK. II

Figure 209 — Turret Assembly

TURRET

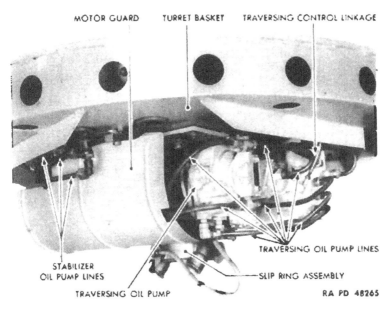

Figure 210 — Turret Basket — Under Side

c. **Removal.** Turn main battery switch to the "OFF" position. Disconnect linkage at traversing oil pump (fig. 210) by removing cotter pin from retaining screw, then remove retaining screw. Remove bolt, nut, and toothed lock washer that secure lower vertical control rod to control linkage crank at left of switch, and remove lower vertical control rod (switch to traversing oil pump rod). Remove bolt, nut, and toothed lock washer that secure lower end of upper vertical control rod to control linkage crank at left of switch. Remove bolt, nut, and toothed lock washer that secure upper end of this vertical control rod to L-shaped crank at top of turret, and remove upper vertical control rod. Remove cotter pin and flat washer that secure L-crank pin at upper left side of turret. Remove cotter pin and flat washer from commander's control handle pin, then lift the horizontal control rod, L-crank, and commander's control handle from turret. Cut safety wire from four socket head screws on top of switch box, and remove screws. Carefully lift lid and switches from switch box. Remove three cap screws and toothed lock washers which hold switch bracket to turret, and remove control handle and switch bracket. Disconnect conduit connector at top of switch box. Mark and disconnect the three wires from switches.

d. **Installation.** Place switch box mounting bracket and control handle assembly in position on turret, and secure with three cap

TM 9-775
LANDING VEHICLE TRACKED MK. I AND MK. II

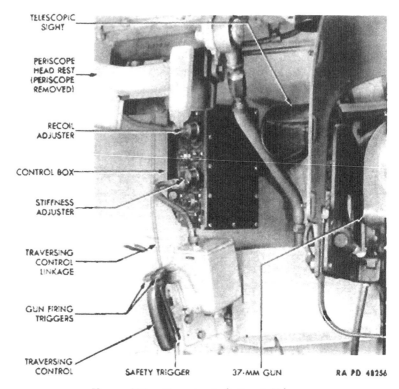

Figure 211 — Turret — Left Front Side

screws with toothed lock washers. Connect the three wires from the conduit (which have been previously marked) to switches, then connect conduit to lid of switch box, and tighten connector. Install three switch push rods in switch box. Insert switches in switch box. Tilt lid of box so that switch hinge is between side of box and ends of three push rods. CAUTION: *Be careful not to bend these push rods. Even a slight bend will cause them to bind, and thus prevent the guns from firing.* With lid of box properly seated, install the four socket-head screws that secure lid to switch box, and lace lid screws with safety wire. Screw lower vertical control rod into ball and socket joint on traversing oil pump (fig. 210). Secure upper end of this rod to crank at left of switch by installing bolt, nut, and toothed lock washer. Adjust tension of rod by means of retaining screw. Install horizontal control rod by inserting L-crank pin and commander's control handle pin in holes in brackets provided at top of turret, and secure crank and handle with flat washer and cotter pin. Connect upper vertical control rod to crank at left of switch and to

TM 9-775
162–163

Figure 212 — Turret — Right Front Side

L-crank at top of turret by installing a bolt, toothed lock washer, and nut at each end of upper vertical rod. Adjust linkage by means of retaining screws provided on threaded ends of rods.

163. OIL PUMP MOTOR.

a. Description. The oil pump motor (fig. 213) is mounted on the under side of the turret basket. It is held in place by two cap screws through the left-side motor mounting flange, and a cap screw through right-side mounting flange. This motor is rated 1.4 horsepower at 10.5 volts and 1,800 revolutions per minute. It is a totally enclosed, ball-bearing, shunt-wound, direct-current motor, serving two purposes: driving the traversing oil pump, and driving the stabilizer oil pump. The batteries mounted in the engine room of the vehicle supply power for driving the motor.

b. Maintenance. The oil pump motor is replaceable as a unit. No repair of its internal parts should be attempted in the field.

c. Removal. Turn main battery switch to the "OFF" position. Remove four cap screws and toothed lock washers that secure the motor guard to under side of turret basket, and remove motor guard

TM 9-775
163

LANDING VEHICLE TRACKED MK. I AND MK. II

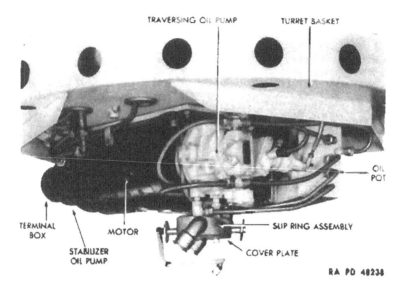

Figure 213 — Oil Pumps and Motor

(fig. 210). Remove all eight oil lines by unscrewing flange nuts. Make provision to catch oil lost when disconnecting oil lines. Plug lines with cloth. Disconnect the traversing oil pump control linkage at pump. Remove four cap screws and lock washers that secure traversing oil pump to motor. CAUTION: *Have one man hold traversing oil pump while another man removes the mounting cap screws.* Remove oil pump motor from mounting plate under turret by unscrewing three cap screws and toothed lock washers. Lower motor carefully from plate. CAUTION: *Motor should be held or blocked in position while removing motor from mounting plate.* Unscrew and remove four screws and lock washers which hold terminal box to stabilizer oil pump, and remove terminal box. Disconnect battery cable from motor terminal by removing nut and lock washer. Remove four cap screws and toothed lock washers that secure stabilizer pump mounting flange to oil pump motor (these are screws not safety wired). Tap flange lightly, and remove stabilizer oil pump.

d. Installation. Install stabilizer oil pump on end of motor (fig. 213), and secure with four socket-head screws. Connect battery cable to motor terminal with nut and lock washer. Position terminal box on stabilizer oil pump (fig. 213), and secure by installing four screws and lock washers. Install traversing oil pump on motor (fig. 213), securing it with four cap screws and lock washers. CAUTION: *Be sure to place the small driving coupling between the traversing*

TURRET

oil pump shaft and driving shaft. Remove plugs from oil lines, and connect oil lines securely by tightening flare nuts. Connect pump control linkage at pump. Lift motor to position on bottom of turret basket, and secure motor to mounting plate by installing three cap screws and toothed lock washers. Aline motor guard holes with holes provided on bottom of turret basket, and install four cap screws and toothed lock washers to secure guard to basket.

164. GEAR BOX ASSEMBLY.

a. **Description.** The gear box assembly (fig. 18) is mounted in the rear of the turret between the gunner's seat and commander's seat. A series of gears within the gear box is engaged with the turret gear by means of the clutch lever mounted on top of the gear box. The gears are driven by a hydraulic motor mounted on the side of the gear case nearest the gunner's seat. The hand traversing crank is mounted on the side of the gear box nearest the commander's seat.

b. **Maintenance.** The gear box assembly can be removed as a unit, or the hydraulic motor can first be removed from gear box (par. 161). No repairs to gear box assembly should be attempted in the field.

c. **Removal.** Turn main battery switch to the "OFF" position. Disconnect three oil lines at hydraulic motor by unscrewing flare nuts on oil lines (fig. 18). Use drain pan to catch oil. Plug oil lines with cloth. Remove gear box from turret.

d. **Installation.** Aline the two dowel pin holes in gear box assembly with dowel pins provided on rear of turret. Install three socket-head screws with lock washers to secure gear box to turret. Remove cloth plugs from oil lines, and connect lines to hydraulic motor with flare nuts. Tighten flare nuts. Fill with oil (par. 27).

165. HYDRAULIC MOTOR.

a. **Description.** The hydraulic motor is mounted on the gear box, and connected to the traversing oil pump by three oil pipes (fig. 18). This motor transmits fluid power into drive for operation of the turret.

b. **Maintenance.** The hydraulic motor is replaceable as a unit. Repairs should not be attempted in the field.

c. **Removal.** Turn main battery switch to the "OFF" position. Disconnect the three oil lines by unscrewing flare nuts (fig. 18). Use drain pan to catch oil. Plug lines with cloth. Remove safety wire from the eight socket-head screws (do not remove wire from screws) (fig. 18). Remove four of the eight socket-head screws. Remove alternate screws, starting with the one at extreme top of motor. Do not remove the other four screws. Lift hydraulic motor off gear box. Remove gasket.

d. *Installation*. Place hydraulic motor with new gasket in position on gear box (fig. 18). Aline holes. Install the four socket-head screws that hold hydraulic motor to gear box. Lace all eight socket-heads screws with safety wire. Remove cloth plugs from oil lines; then connect oil lines on hydraulic motor, and tighten flare nuts securely.

166. TRAVERSING OIL PUMP.

a. *Description*. The traversing oil pump, located on the left end of the oil pump motor (fig. 213), supplies power to the hydraulic motor for the turret drive.

b. *Maintenance*. The pump is replaceable as a unit. No repair of its internal parts should be attempted in the field.

c. *Removal*. Removal of pump is covered in paragraph 163.

d. *Installation*. Installation of traversing oil pump is covered in paragraph 163.

167. OIL POT (RESERVOIR).

a. *Description*. The oil pot, mounted to left of the slip ring under turret basket (fig. 213), holds approximately 1 gallon of special hydraulic oil (par. 27).

b. *Maintenance*. The oil pot is replaceable as a unit, and should be replaced in event of damage.

c. *Removal*. Turn main battery switch to the "OFF" position. Disconnect five oil lines to oil pot by unscrewing flare nuts (fig. 213). Catch oil in drain pan. Plug lines with cloth. With one man holding oil pot, remove three safety nuts that secure oil pot to bottom of turret basket, and remove oil pot from basket.

d. *Installation*. Have one man hold oil pot in position under turret basket, and install three bolts and safety nuts to secure oil pot to turret basket. Remove cloth plugs from oil lines, and connect oil lines to oil pot by tightening the five flare nuts. Fill oil pot with special hydraulic oil (par. 27) if necessary.

168. TURRET DOORS.

a. *Description*. Two turret doors are mounted on rear top side of turret (fig. 17). They are hinged to open upward and rearward, and can be locked in their open positions. The commander's door has a periscope assembly mounted in the door. For a detailed description of turret doors, refer to paragraph 10.

TURRET

b. **Maintenance.** The three periscope mounts, one in commander's door and two on top forward side of turret (fig. 17), can be removed. Both turret doors are also replaceable as units. Repair of these parts should not be attempted in the field.

c. **Removal.** Open turret doors. Remove periscope from periscope mount in commander's door. Remove periscope mount from commander's door by unscrewing eight cap screws and lock washers that secure periscope mount retaining ring to under side of door (fig. 17). Cut out the hinge pin safety plugs, and drive the hinge pin out of the hinge, to remove door.

d. **Installation.** Position turret doors on turret (fig. 17). Door with periscope mount hole is installed on right side of turret for commander's periscope. Aline hinge pin hole on door with hinge pin holes in bracket provided on rear of turret, and install hinge pin. Install a new safety plug on each end of hinge pin, and flatten safety plugs to secure pin in hinge. All hinge pins are installed in same manner. Place periscope mounting ring assembly in position in commander's door. Place periscope mount retaining ring in position on under side of door (fig. 17). Aline holes in retaining ring with tapped holes provided in door, and install eight cap screws and lock washers to secure periscope mounting assembly in turret door. Slide periscope in periscope mount, and secure in position by tightening knurled knob on periscope. Other two periscope mounts on forward side of turret top (fig. 17) are installed in similar manner.

169. **TURRET SEATS.**

a. **Description.** Two adjustable seats are mounted in the turret (fig. 18). The one in left rear of turret is the gunner's seat. The seat in right rear of turret is the commander's seat.

b. **Maintenance.** The seat cushions can be replaced as a unit in event of damage to them. The complete seat and bracket assembly can be removed from the turret and replaced as a unit.

c. **Removal.** Lift seat cushion from seat. Remove two cap screws and toothed lock washers that secure the conduit bracket to rear of commander's seat, and remove bracket (fig. 18). Remove four bolts and safety nuts that secure seat mounting bracket to floor of turret basket. Remove four cap screws and lock washers that secure seat mounting bracket to turret; then lift seat and bracket from turret. The gunner's seat assembly does not have a conduit bracket, but is removed in a similar manner.

d. **Installation.** Position commander's seat mounting bracket in turret (fig. 18). Aline holes in upper end of bracket with tapped holes provided in turret, and install four cap screws with lock washers to secure upper end of seat bracket assembly. Aline holes in

LANDING VEHICLE TRACKED MK. I AND MK. II

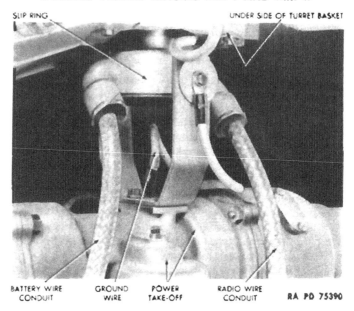

Figure 214 — Electrical Leads to Turret Basket

lower end of bracket with holes provided in turret basket floor, and install four bolts and safety nuts to secure lower end of seat bracket. Place master switch conduits in conduit bracket, and aline this bracket on rear of commander's seat; install two cap screws and toothed lock washers to secure conduit bracket to commander's seat assembly. With the exception of conduit bracket, gunner's seat assembly (fig. 18) is installed in same manner.

170. REMOVAL OF TURRET.

a. **Disconnect Electrical Connections.** Disconnect the two conduits and battery ground wire from slip ring assembly (fig. 214). Unscrew two knurled nuts from cover plate on bottom of slip ring assembly, and remove cover (fig. 213). Disconnect conduit wires. Tag all wires and terminals for ready identification when installing turret.

b. **Remove Roller Assemblies and Segments.** Remove three cap screws and lock washers that secure the short section of dust cover on side of turret (short section of dust cover may be installed on either right or left side of turret) (fig. 215). Remove lubrication fitting from segment (fig. 216). Remove four cap screws and toothed lock washers that secure segment to the inside of vehicle (fig. 215). It may be necessary to pry off the short section of dust cover and

TM 9-775
170

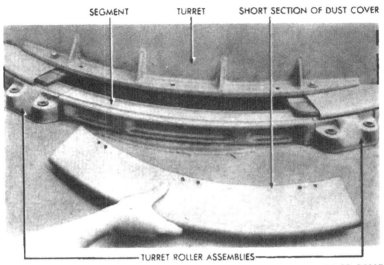

Figure 215 — Removing Short Section of Turret Dust Cover

the segment as a unit with a small pry bar. With this section of dust cover removed, the segments and roller assemblies are accessible. Traverse the turret until ends of segment to be removed will clear opening provided when short section of dust cover was removed (fig. 215). Lift segment from vehicle (fig. 217). Traverse turret until the adjacent roller assembly is accessible, and remove this roller assembly (fig. 218). Two men are required to remove the roller assemblies. One man on top of vehicle holds the socket-head screws which secure the roller assemblies from the outside (fig. 218), while a second man, inside the vehicle, removes nuts from socket-head screws. Remove nut and toothed lock washer from each of the two socket-head screws, and remove screws from roller assembly. From inside vehicle remove two cap screws and toothed lock washer that secure roller assembly from inside vehicle (fig. 217), and remove roller assembly. NOTE: *The roller assemblies and segments, and their relative positions on hull should be marked with a prick punch, so they can be reinstalled later in same locations.* Repeat above operations until all six roller assemblies and six segments are removed from the vehicle.

c. **Disconnect Turret Traversing Mechanism.** Disconnect turret traversing gear box from turret (fig. 18). Support turret traversing gear box in place with an hydraulic jack or wood blocks; and remove three socket-head screws and lock washers that secure traversing gear box to turret. Pull traversing gear mechanism inward just far enough

359

LANDING VEHICLE TRACKED MK. I AND MK. II

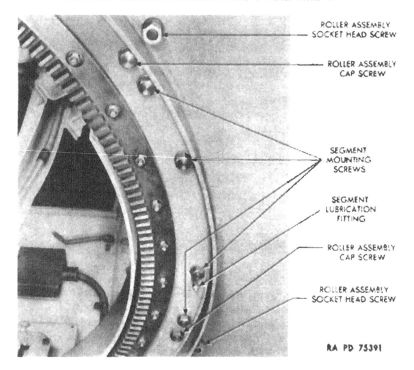

Figure 216 — Turret Segment Mounting Screws

to disengage pinion gear from turret ring gear (keep traversing mechanism blocked in this position). NOTE: *It is not necessary to disconnect the three oil lines from hydraulic motor (fig. 18) if traversing gear box is maintained in its relative installed position, as disconnecting these oil lines involves the extra operation of purging the system.* Traversing gear box may be secured in its disengaged position as follows: Obtain a piece of wood approximately 1¾ x 4 x 12 inches long. Drill three holes in wood to correspond with position of tapped holes in turret, and large enough in diameter to accommodate cap screws of same diameter as mounting socket-head screws. Aline holes in turret with holes in wood block, and install three cap screws through mounting flange of traversing gear mechanism, through wood block, and into turret, to secure box to turret (cap screws must be approximately 1¾ in. longer than socket-head mounting screws to allow for thickness of wood block).

d. **Lift Turret from Vehicle.** Move crane or A-frame over center of turret, with turret turned toward front of vehicle. Install triple chain sling on turret (two lifting eyes are provided on front of turret

TM 9-775
170–171

TURRET

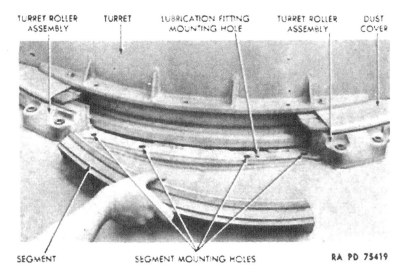

Figure 217 — Removing a Turret Segment

(fig. 209), and one lifting eye on rear of turret). The two short chains of the triple chain sling (6 in. shorter than rear chain) are installed on front lifting eyes. The long chain is hooked to the rear lifting eye. Lift turret straight upward and lower to convenient place on floor.

171. INSTALLATION OF TURRET.

a. **Position Turret on Vehicle.** Install triple chain sling on turret (par. 170 c). With hoist or A-frame, raise turret upward, and position directly over turret opening in vehicle. Lower turret to vehicle (two men will guide turret while it is being lowered to position on vehicle). CAUTION: *Be sure turret and turret ring assembly are correctly alined on all sides before lowering turret to vehicle, to prevent damage to turret parts.*

b. **Connect Turret Traversing Mechanism.** With traversing gear mechanism blocked in position (par. 170), remove three cap screws and wood block that hold traversing gear mechanism in disengaged position. Aline dowel pin holes in traversing gear mechanism with dowel pins provided on turret. Aline pinion gear with turret ring gear, and position traversing gear mechanism on wall of turret. Secure in position by installing three socket-head screws with lock washers. Remove hydraulic jack or wood blocks.

c. **Install Roller Assemblies and Segments.** Aline roller assembly with mounting holes in vehicle. Roller assemblies and their location have previously been marked with a prick punch (par. 170).

TM 9-775
LANDING VEHICLE TRACKED MK. I AND MK. II

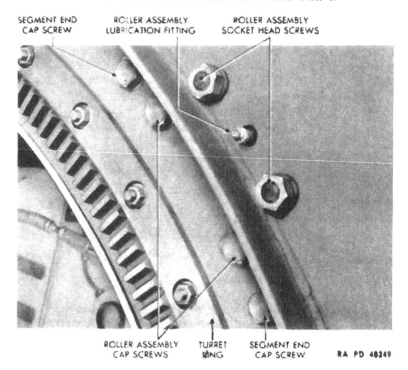

Figure 218 — Turret Roller Assembly Mounting Screws

Install two socket-head screws from outside vehicle (two men required). One man holds socket-head screws with socket-head set screw wrench while second man, inside tank, installs two nuts with toothed lock washers. Install two cap screws and toothed lock washers that secure roller assembly from inside vehicle. Turn turret until segment adjoining roller assembly can be positioned. Position segment adjoining installed roller assembly, and install four cap screws and lock washers to secure segment from inside vehicle. Repeat the above operations until all six roller assemblies and segments have been installed. Aline short section of dust cover on side of turret with ends of dust cover sections already installed (fig. 215), and secure cover by installing three cap screws with lock washers.

d. **Connect Electrical Connections** (fig. 214). Connect conduit wires to terminals in slip ring assembly (wires and terminals have previously been tagged for ready identification). Secure conduits to slip ring assembly by tightening hexagonal nut on conduit. Position cover plate on bottom of slip ring assembly, and secure by tightening two knurled nuts provided on cover. Connect battery ground wire to slip

TM 9-775
171-172

AUXILIARY GENERATOR

ring assembly mounting screw. Check turret installation by rotating turret with hand crank on turret traversing mechanism; then check turret hydraulic traversing mechanism

---

Section XXXV

AUXILIARY GENERATOR

	Paragraph
General	172
Maintenance	173
Removal	174
Installation	175

172. GENERAL.

a. A one-cylinder, gasoline-powered, air-cooled auxiliary generator (model DR 7812) is provided to charge the vehicle batteries, and for

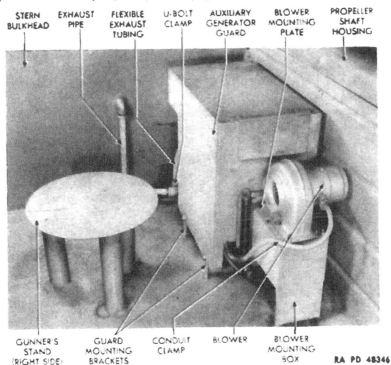

Figure 219 — Auxiliary Generator — Installed

TM 9-775
LANDING VEHICLE TRACKED MK. I AND MK. II

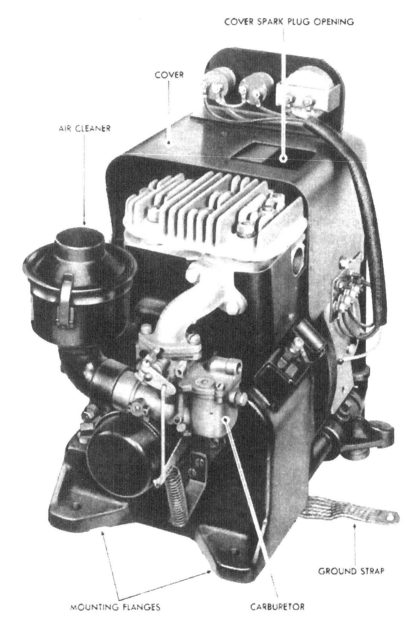

Figure 220 — Auxiliary Generator — Rear View

AUXILIARY GENERATOR

emergency operation of the electrically-controlled turret guns and the turret traversing mechanism. This auxiliary generator is mounted on the right side of the control tunnel in front of the stern bulkhead (fig. 219). A guard is provided to protect the generator from damage. A blower is mounted in front of the generator for air circulation, and to disperse any exhaust fumes which may possibly be present in the vicinity of the generator. A further detailed description of the auxiliary generator, including operation and controls, is contained in the section on auxiliary controls and operation (sec. V). For trouble shooting on the auxiliary generator, see section XIII.

173. MAINTENANCE.

a. General. For efficient operation of the auxiliary generator, make the following inspections and adjustments frequently.

b. Oil. Change oil periodically. Drain oil from crankcase by removing drain plug at base of filler pipe. Use engine oil SAE 10 in temperatures below 32° F. and engine oil SAE 20 in temperatures above 32° F. The crankcase has a capacity of 3 pints.

c. Spark Plug. Check spark plug gap with a feeler gage. Set gap at 0.028 to 0.033 inch.

d. Air Cleaner (fig. 224). Disengage clips and remove filter element. Wash element with gasoline. Clean out oil base and fill with clean engine oil SAE 20. Assemble after element has thoroughly drained.

e. Carburetor. Replace a defective carburetor as a complete assembly. If carburetor adjustment is necessary, notify ordnance maintenance personnel.

f. Mounting Bolts and Connections. Check mounting bolts and electrical connections to make sure they are tight.

174. REMOVAL.

a. Close Fuel Shut-off Valve. Open hinged rear cover, and remove upper right rear baffle. Turn fuel shut-off valve handle (at auxiliary generator fuel tank above fire extinguisher) clockwise to stop the flow of fuel to generator.

b. Remove Auxiliary Generator Guard. Remove two nuts and lock washers from each of two U-bolt clamps that secure the upper and lower ends of flexible exhaust tubing to exhaust pipe, and remove the U-bolt clamps. Remove exhaust tubing from exhaust pipe. Unscrew the vertical nipple from upper section of exhaust pipe that is mounted to auxiliary generator; unscrew the upper horizontal section of exhaust pipe from auxiliary generator exhaust port elbow. Remove four cap screws and toothed lock washers that secure auxiliary

TM 9-775
LANDING VEHICLE TRACKED MK. I AND MK. II

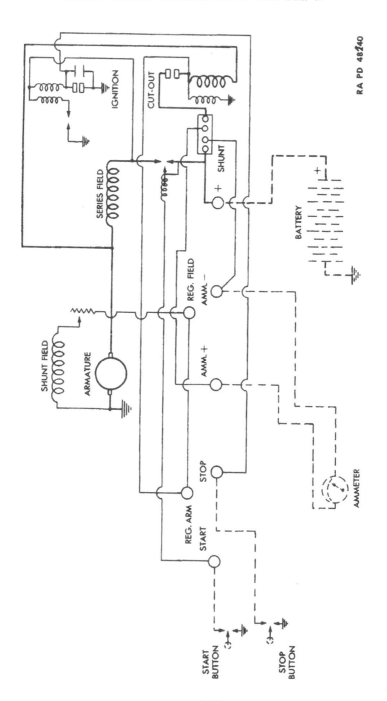

Figure 221 — Wiring Diagram — Auxiliary Generator

AUXILIARY GENERATOR

generator guard to mounting brackets on floor of cargo compartment (fig. 219), and lift guard from generator.

c. **Remove Blower** (fig. 222). Unscrew the hexagonal nut that secures electrical conduit to blower. Disconnect wires from blower motor (tag wires for ready identification when reinstalling blower). Unscrew thumb nut that secures conduit to blower mounting plate, and remove conduit from mounting plate. Unscrew six thumb screws that secure blower mounting plate to blower mounting plate box, and lift off blower and blower mounting plate from blower mounting box.

d. **Disconnect Fuel Line.** Disconnect auxiliary generator flexible fuel line to carburetor at the carburetor (fig. 222).

e. **Disconnect Electrical Lead** (fig. 222). Disconnect the electrical lead by removing elastic stop nut and lock washer that secure "BATT" terminal electrical lead to the battery terminal post on auxiliary generator, and remove electrical lead.

f. **Disconnect Ground Strap** (fig. 220). Remove nut, bolt, lock washer, and flat washer which secure ground strap to mounting bracket on floor of hull.

g. **Remove Mounting Stud Nuts.** Remove eight stud nuts (two on each stud) and four toothed lock washers which secure auxiliary generator to mounting bracket. Lift off auxiliary generator, allowing the four rubber mounting pads to remain on the mounting studs.

175. INSTALLATION.

a. **Install Mounting Stud Nuts.** Place auxiliary generator in position on the four rubber mounting pads, and install eight stud nuts (two on each stud) and four toothed lock washers which secure the auxiliary generator to the brackets in floor of hull.

b. **Connect Ground Strap.** Install nut, bolt, lock washer, and flat washer which secure ground strap to mounting bracket on floor of hull.

c. **Connect Electrical Lead.** Connect "BATT" terminal electrical lead at "BATT" terminal post on auxiliary generator, and secure with lock washer and elastic stop nut.

d. **Connect Fuel Line.** Connect flexible line to carburetor.

e. **Install Blower** (fig. 222). Position blower on blower mounting box. Aline holes in blower mounting plate, and install six thumb screws which secure blower and blower mounting plate to blower mounting box. Connect wires in electrical conduit to motor in blower. (Wires have previously been tagged for ready identification.) Secure conduit to blower by tightening hexagonal nut on conduit. Aline hole in clamp on conduit with hole provided in blower

TM 9-775

LANDING VEHICLE TRACKED MK. I AND MK. II

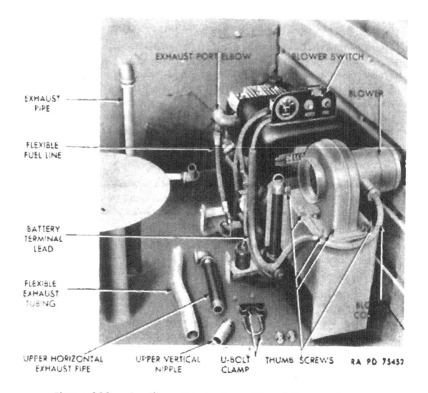

Figure 222 — Auxiliary Generator — Guard Removed

mounting plate, and install thumb screw to secure conduit to mounting plate.

f. **Install Auxiliary Generator Guard** (fig. 219). Place guard in position over auxiliary generator, and install four cap screws and toothed lock washers that secure guard to mounting brackets welded to floor of vehicle (fig. 219). Screw the upper horizontal section of exhaust pipe into exhaust port elbow on generator. Screw the vertical nipple into end of upper horizontal section of exhaust pipe. Place flexible exhaust tubing over upper and lower nipple of exhaust pipe. Secure flexible exhaust tubing by installing a U-bolt clamp on each end of flexible tubing (fig. 222). Tighten clamps in position by installing two nuts and lock washers on each U-bolt clamp.

g. **Open Fuel Shut-off Valve.** Turn auxiliary generator fuel shut-off valve (at auxiliary generator fuel tank above fire extinguisher) clockwise to start flow of fuel to the generator. Install upper right rear baffle, and close hinged rear cover.

PART THREE — ARMAMENT

Section XXXVI
INTRODUCTION

	Paragraph
General	176

176. GENERAL.

a. This part of the manual is devoted to instructions covering the care, operation, removal, and installation of component parts of the turret assembly used on the Landing Vehicle Tracked, Mk. I (LVT) (A) (1). Reference is made to existing manuals covering cleaning, lubrication, malfunctions, corrections, inspection, and adjustments of the 37-mm Gun, M6 and Machine Gun, cal. .30, M1919A5 used in the Combination Gun Mount M44 on turret of this vehicle. For operation of the turret traverse controls, refer to the section on turret controls and operation (sec. IV).

Section XXXVII
DESCRIPTION OF GUNS AND MOUNT

	Paragraph
Guns and gun mount	177

177. GUNS AND GUN MOUNT.

a. **Combination Gun Mount M44.** This gun mount is located in the turret, and mounts a 37-mm Gun, M6 and a Machine Gun, cal. .30, M1919A5, fixed, both of which move together as a unit. The gunner sits in the left seat of the turret basket to the left of the gun mount and behind the sighting telescope. The commander (loader) sits on the right side of the turret. A recoil guard is attached to the rear of the gun mount to provide protection for personnel.

b. **Traverse.** Traverse of 360 degrees is provided so that the guns may be fired in any direction. The turret assembly is traversed either manually or by the hydraulic mechanism (par. 11).

c. **Elevating.** Elevation or depression of the combination mount is obtained by means of the handwheel located on the left side of gun mount (fig. 223). Elevation or depression is obtained and controlled while the vehicle is in motion by means of a stabilizer (fig. 212). Turning of the elevating handwheel counterclockwise depresses the guns a maximum of 10 degrees, while turning handwheel clockwise elevates guns to a maximum of 20 degrees.

d. **Gun Positioning Control.** A stabilizer is attached to the Combination Gun Mount M44 to maintain the positioning of the guns while the vehicle is in motion, passing over uneven terrain, or in rough seas. For a detailed description of the construction and operation of the stabilizer used on Combination Gun Mount M44, see section XL.

e. **Firing.** The guns in the Combination Gun Mount M44 can be fired either manually or electrically. For a detailed description of the operation of the guns, refer to section XXXVIII.

Section XXXVIII
OPERATION OF GUNS

	Paragraph
Loading the guns	178
Traversing the guns	179
Elevating and depressing the guns	180
Firing the guns	181

178. LOADING THE GUNS.

a. **Loading 37-mm Gun.** Lower recoil guard to its horizontal position. Manually open the breech on the 37-mm gun by pulling downward on breech-operating handle (fig. 216). The two extractors rotate rearward by means of cam action, and lock the breechlock in its open position. The commander (loader) starts the cartridge into the breech, and then with the heel of his hand slams the cartridge into the breech. The flared end of the cartridge case engages the extractors, tripping them, and allowing breechlock to rise to its closed position. Gun now is ready to fire. CAUTION: *Do not push cartridge slowly into breech, as fingers will be injured when breechlock rises.*

b. **Loading the Cal. .30 Machine Gun.** See pertinent Field Manuals and Technical Manuals.

179. TRAVERSING THE GUNS.

a. The guns can not be traversed independently of turret, but are traversed as a unit with turret. The turret can be traversed 360 degrees, enabling commander to fire in any direction. For full description of operation of turret traversing mechanism, see paragraph 11.

REMOVAL AND INSTALLATION OF GUNS

180. ELEVATING AND DEPRESSING THE GUNS.

a. The guns may be elevated or depressed as a unit by means of a handwheel (par. 177 c). A stabilizer maintains the position of the guns while the vehicle is in motion. For a detailed description of the construction and operation of stabilizer, see section XL.

181. FIRING THE GUNS.

a. Manually Firing the Guns. The 37-mm gun is fired manually by means of a button located in the center of the elevating wheel (fig. 223). The cal. .30 machine gun is fired manually with a trigger on the gun itself.

b. Electrical Firing of Guns. To fire either of the guns with the electric controls, the gun firing switches located on control panel (fig. 16) must be turned to the "ON" position. Two firing buttons, one for each gun, are combined with a safety pistol grip in the control handle of the turret traverse mechanism (fig. 211). The pistol grip must be compressed to release the safety mechanism; when this is done, the left-hand button fires the 37-mm gun, and the right-hand button fires the cal. .30 machine gun.

Section XXXIX

REMOVAL AND INSTALLATION OF GUNS

	Paragraph
Turret machine gun, cal. .30, M1919A5	182
37-mm gun, M6	183

182. TURRET MACHINE GUN, CAL. .30, M1919A5.

a. Description. A Machine Gun, cal. .30, M1919A5 is mounted on the Combination Gun Mount M44. For information on care, preservation, operation, and malfunctions of this gun, refer to pertinent Field Manuals and Technical Manuals.

b. Removal. Pull out the two locking pins that secure the cal. .30 machine gun in gun mount. Pull the machine gun to the rear and out of combination gun mount.

c. Installation. Aline the muzzle of the machine gun with the opening provided on right side of combination gun mount, slide gun into this opening, and position gun in mount. Aline the two locking pin holes in machine gun with holes provided in gun mount, and install the two locking pins.

LANDING VEHICLE TRACKED MK. I AND MK. II

183. 37-MM GUN, M6.

a. *Description.* A 37-mm Gun, M6 is mounted on the Combination Gun Mount M44 (fig. 213). For information on care, preservation, malfunctions, adjustments, and operation of the 37-mm gun, refer to TM 9-250.

b. *Removal.* Remove the cal. .30 machine gun from combination gun mount (par. 182). Loosen the two socket-head set screws in spanner nut on muzzle of gun; unscrew the spanner nut that secures muzzle of gun to sleigh assembly. Remove four cap screws and toothed lock washers that secure recoil guard to gun mount, and remove recoil guard. Remove two nuts and toothed lock washers that secure the rear machine gun mounting bracket to the mount, and remove bracket. Pull down on breech opening handle to open breech. This makes the 37-mm gun mounting bolt accessible. Hold this bolt head with a screwdriver (bolt head is provided with a screwdriver slot) while removing safety nut from bolt, and remove bolt. Remove four cap screws and lock washers that secure the armor plate cover opening in rear of turret, and remove armor plate. Aline gun with opening in rear of turret, and slide gun (two men required) rearward and out of turret. CAUTION: *Be careful not to damage the spanner nut thread provided on gun tube.*

c. *Installation.* Slide gun (muzzle first) through opening provided in rear of turret (two men required). Aline gun with opening provided in the combination gun mount, and also aline bronze key, provided on right side of gun, with keyway in gun mount. Slide the gun forward into position on gun mount. CAUTION: *Be careful not to damage threads provided on gun tube.* Open the breech so that mounting bolt hole provided on rear of gun is accessible. Aline mounting bolt hole in gun with bolt hole provided in gun mount, and install mounting bolt and safety nut that secure breech end of gun to gun mount. Install spanner nut on gun tube to secure gun to sleigh mechanism, and tighten the two socket-head set screws in spanner nut. Aline holes in recoil guard with holes provided in breech end of gun mount, and install four cap screws and toothed lock washers to secure recoil to gun mount. Install machine gun rear mounting bracket. Install machine gun (par. 182).

Section XI

STABILIZER

	Paragraph
Description of stabilizer	184
Operation of stabilizer	185
Stabilizer trouble shooting	186
"Purging" or "bleeding" the system	187
Gyro control unit	188
Stabilizer oil pump	189
Piston and cylinder assembly	190
Recoil switch	191
Control box	192
Mounting shaft and bracket	193
Disengaging switch	194
Flexible shaft	195
Gyro control mounting shaft	196

184. DESCRIPTION OF STABILIZER.

a. Function. The purpose of the stabilizer is to maintain the position of the guns in the Combination Gun Mount M44 in a vertical plane, so that the gunner can aim and fire the guns accurately while the landing vehicle tracked is in motion. In order to accomplish this purpose, the stabilizer unit must be properly maintained. The using arms must not make repairs to individual assemblies. They should be replaced as complete units.

b. Location of Stabilizer Assembly. The motor which supplies power for the stabilizer oil pump is mounted on the under side of the turret basket (fig. 210). This motor supplies power both for the hydraulic turret traverse mechanism and for the stabilizer. The stabilizer oil pump is mounted at one end of the electric motor (fig. 210). The mounting of the gyro control unit, piston and cylinder assembly, and mounting bracket assembly is shown in figures 212 and 226. The master switch is located in rear of turret, and is mounted to the right of turret traverse mechanism (fig. 18). The master switch also controls the turret traverse mechanism.

185. OPERATION OF STABILIZER.

a. Starting the Unit. Set the stiffness control at zero (fig. 211). Take the hand elevating gears out of mesh by pulling worm gear locking pin (fig. 223) to the right. Push lock assembly upward (worm gear lowers from elevating segment gear), and insert locking pin in upper locking pin hole to secure worm gear in the disengaged position. Turn elevating handwheel (fig. 223) until gyro control unit is approximately in a vertical position. Start the oil pump motor by turning master switch (fig. 18) to the "ON" position. Start the gyro control by turning stabilizer switch in control box to the "ON" position (fig. 211). CAUTION: *In cold weather, the oil must be permitted to warm up to obtain full control from stabilizer equipment. In sub-zero weather, allow 1½ minutes running time for each degree of temperature below 0° F., or a total running time of 30 minutes at −20° F.*

b. **Control of Turret Guns.** It is important that the stabilizer be in operation only when vehicle is moving, and when control of guns is desired. When the stabilizer equipment is in operation, guns are elevated or depressed in the usual manner by turning handwheel (fig. 223). This action changes the angular relation between guns and gyro control unit, and guns automatically take up the new desired position. If the stabilizer equipment is operating correctly, it will keep guns very near their set angular position within the elevating range when the vehicle is pitching normally. When guns are aimed, stabilizer must be allowed to control the positioning of guns. The handwheel should not be turned after gun has reached its maximum limit of travel in elevation or depression (par. 177). CAUTION: *Continued turning of handwheel with the guns against either stop will only displace the gyro control unit from its vertical position and overload the battery.*

c. **Adjusting the Stiffness Adjuster.** The stiffness adjuster, located in the control box (fig. 211), provides a means for the gunner to control the operation and effectiveness of the stabilizer. After the oil has warmed up, the knob of the stiffness adjuster should be turned clockwise slowly to "stiffen" action of stabilizer until desired action is obtained. An indication of too stiff an adjustment is a vigorous vibration of the guns. Insufficient stiffness is indicated by the guns "hunting" (slowly elevating and depressing from their aimed or set position). When the guns start to vibrate or "hunt" as the stiffness control knob is turned, decrease (vibration) or increase ("hunt") the adjustment by turning knob in the opposite direction until vibration or "hunting" is eliminated. To check the operation of the gyro control further, press on the breech of the 37-mm gun suddenly and release. If gun starts to vibrate, the stiffness adjustment must be decreased slightly. If gun comes to rest almost immediately after a sharp sudden displacement, it is in proper adjustment. It may be necessary for the operator to change the stiffness adjustment from time to time as the viscosity of the oil changes when the vehicle is in use. CAUTION: *Never push the 37-mm gun up and down rapidly by hand, as this will cause air to be drawn into the system around the piston and rod.*

d. **Adjusting the Recoil Adjuster.** The recoil adjuster located in the control box (fig. 211) provides a means for the gunner to con-

STABILIZER

trol the recoil action of the 37-mm gun. The recoil adjustment must be made by trial and error method while the gun is being fired, until the desired smooth recoil operation is obtained. The recoil adjustment knob should be gradually turned to the right (clockwise) until a point is reached where the gun will keep its angular setting during recoil. In case of faulty operation of the stabilizer during recoil, after above adjustment has been made, check for looseness in the mounting of the recoil switch.

e. *Test for Effective Operation.* After the stabilizer is operating, it should be checked for effectiveness or accuracy before the vehicle is used in combat. Operate vehicle over average rough terrain or rough water at normal speed. Aim the 37-mm gun in the usual manner, using the horizon or a fixed landmark as the target. If the gun does not fluctuate above or below the horizon, the stabilizer is operating satisfactorily. However, if the gun fluctuates above or below the horizon, after stiffness adjustment procedure has been followed (subpar. c, above), see paragraph 186.

f. *Oil Level.* The level of the oil in the oil reservoir (fig. 212) should be checked daily and maintained two-thirds full of hydraulic oil.

186. STABILIZER TROUBLE SHOOTING.

a. Trouble Shooting.

(1) GUN VIBRATES.

Possible Cause	Possible Remedy
Stiffness rheostat knob turned too far clockwise.	Adjust stiffness rheostat (par. 185 c).
Excess friction in trunnion bearings.	Notify ordnance personnel.

(2) GUN FLUCTUATES OR "HUNTS."

Stiffness adjustment weak.	Increase stiffness adjustment (par. 185 c).
Insufficient oil supply.	Fill oil reservoir (par. 185 f).
Air in system.	Purge system (par. 187)
Loose electrical connections.	Tighten electrical connections.
Leaking oil lines.	Replace leaky oil lines.
Loose mounting between gyro control mounting bracket and mount.	Tighten all mounting bolts and screws.
Excessive oil leakage around piston rod.	Install new piston and cylinder assembly (par. 190 d).
Lost motion between gyro control and handwheel.	Eliminate end play in worm bracket.

LANDING VEHICLE TRACKED MK. I AND MK. II

(3) BREECH OF GUN DROPS DURING RECOIL.

Possible Cause	Possible Remedy
Recoil rheostat knob not adjusted properly.	Adjust recoil rheostat knob (par. 185 d).

(4) BREECH OF GUN RISES DURING RECOIL.

Possible Cause	Possible Remedy
Recoil rheostat not adjusted.	Adjust recoil rheostat knob (par. 185 d).

b. Diagnosis.

(1) If oil supply is correct (par. 27), all mounting bolts tight, system has been purged, and stiffness adjustment tried without success, the trouble is probably caused by excess friction in trunnion bearings. Notify ordnance maintenance personnel.

(2) End play in the worm bracket will cause lost motion between gyro control and handwheel. Remove cover from gyro control gear box. Check to see if there is end play in worm gear. If there is end play, notify ordnance personnel.

(3) Little trouble from leaks at the cylinder and piston rod should occur, as the oil return line (fig 223) relieves the oil pressure on the packing. If a leak occurs around piston rod, the oil return line should be checked to see that it is not pinched, or otherwise restricted. If the leak persists, replace piston and cylinder (par. 190).

(4) If faulty operation is obtained from the stabilizer during recoil at all positions of the knob, check for looseness in the mounting, and for correct adjustment of the recoil switch (par. 191).

187. "PURGING" OR "BLEEDING" THE SYSTEM.

a. General. It is very important for proper operation of the stabilizer that all air trapped in the system be removed. This operation of removing air from the system is known as "purging" or "bleeding" the system.

b. Checking for Air in the System. With manual elevating gears engaged, start and stop oil pump motor. If oil level in oil reservoir drops, there is air in the system. To remove any air which may be trapped in the system, turn master switch to the "OFF" position. Disengage manual elevating mechanism (par. 185 a), and work breech of gun slowly up and down for a period of from 5 to 10 minutes. Engage elevating gears, and repeat starting and stopping of oil pump motor. If oil level in oil reservoir still drops when engine is started, purge the system (subpar. d, below). CAUTION: *Never push the gun up and down rapidly by hand, as this may cause air to be drawn into system around piston and rod.*

c. Supplying Oil to the System. Oil used should be hydraulic oil or Navy Symbol OS 1113 or OS 2943. Heat oil to approximately

TM 9-775
187

STABILIZER

Figure 223 — Elevating Gear and Stabilizer Cylinder

200° F. Oil may be added by removing cap on oil reservoir (fig. 212), and having one man maintain reservoir two-thirds full during purging operation. A considerable amount of time can be saved if oil is supplied to the system with a small amount of pressure on the supply line. This pressure may be obtained by use of a filler can placed on top of turret with a supply line connected to the ⅜-inch flare nut connection below oil reservoir, or by use of any suitable device for supplying the oil at pressure not to exceed 15 pounds per square inch. If a pressure supply system is employed, a shut-off valve should be placed in supply line just above the flare nut connection at oil reservoir for convenience in controlling oil flowing into system.

d. **Procedure for "Purging" or "Bleeding" the System.** Turn master switch to "OFF" position. Loosen oil return line (fig. 223). Remove small hexagonal plug, and loosen (but do not remove) bleeder valves (fig. 223) on cylinder. Add oil to the system (subpar. c, above) until a flow, free of bubbles, is obtained from oil return line. Tighten this connection permanently. When a solid flow of oil (free of bubbles) is obtained from bleeder valves (fig. 223), tighten the two bleeder valves finger-tight. Remove the pipe plug on stabilizer oil pump, and "bleed" until free of bubbles. Replace and tighten plug. Loosen top bleeder

377

LANDING VEHICLE TRACKED MK. I AND MK. II

valve (fig. 223). Push breech of 37-mm gun slowly to extreme upward position, then tighten top bleeder valve. Loosen lower bleeder valve (fig. 223), then push breech of gun slowly to extreme downward position, and tighten this bleeder valve. Repeat this operation several times until all air is removed from system. Remove pipe plug on oil pump, and again drain until free of bubbles. Replace plug and tighten securely. Disconnect pressure supply and other connections. Maintain oil reservoir at two-thirds full. This is important. Work breech of gun up and down slowly until no more signs of air appear in the oil reservoir. Start oil pump motor, and run pump approximately 10 minutes. Loosen both bleeder valves again (fig. 223), and after a solid flow of oil is obtained, tighten both bleeder valves securely. Install hexagonal nuts, and then stop motor. Check and fill the oil reservoir $\frac{2}{3}$ full, and install cap.

188. GYRO CONTROL UNIT.

a. **Description.** The gyro control, which controls the operation of the gun stabilizer, mounts on the flange of the gyro control mounting shaft (fig. 224).

b. **Maintenance.** The gyro control unit should be handled with extreme care at all times. Severe shock by dropping or jarring the unit may damage the internal working parts and cause erratic operation of the system. This unit is sealed by two ordnance lead seals. CAUTION: *Under no condition are these seals to be broken without proper authority.* The gyro control is replaceable as a unit, and no repair of its internal parts should be attempted.

c. **Remove Connector** (fig. 224). Turn the master switch to the "OFF" position to shut off power to stabilizer. Disconnect (pull straight outward) and remove the multi-prong connector from gyro control base.

d. **Remove Gyro Control Unit.** Remove four nuts that secure gyro control to gyro control mounting shaft (fig. 224), and remove gyro control unit. CAUTION: *The lead seals on the gyro control unit must never be broken except by ordnance personnel.*

e. **Install Gyro Control Unit.** Position gyro control unit on mounting shaft (fig. 224), and install the four mounting nuts.

f. **Install Connector.** Connect multi-prong connector to gyro control base (fig. 224), making sure that index plugs line up. The complete stabilizer assembly should be checked for operation after replacement (par. 185 e).

189. STABILIZER OIL PUMP.

a. **Description.** The oil pump is a dual-gear type with magnetically controlled oil valves. It is mounted directly on the oil pump

TM 9-775
189–190

STABILIZER

motor (fig. 213). The shafts of the pump and motor mesh together, forming a shaft-to-shaft coupling. The motor and pump, located under the floor of the basket (fig. 213), are held in place by bolts through the motor mounting flange.

b. **Maintenance.** The oil pump is replaceable as a unit, and no repair of its internal parts should be attempted in the field.

c. **Disconnect Oil Lines.** Throw battery switch to the "OFF" position. Throw master switch (fig. 18) to the "OFF" position. Disconnect oil lines from the pump. NOTE: *Make provision to catch the oil lost when disconnecting lines.* Plug the oil lines with ¼-inch and ⅛-inch flare plugs. Cap the oil connections on the oil pump with flare nuts and dead-heads.

d. **Disconnect Wires.** Remove terminal box cover (fig. 213), and disconnect the three wires (green, yellow, and white) from the terminals of the oil pump. NOTE: *Tag terminals and wires for easy identification when installing.*

e. **Remove Oil Pump.** Remove the four socket-head set screws around the oil pump mounting bracket. NOTE: *These are the screws that are not connected by locking wire.* Tap the base plate of the oil pump with a light hammer while pulling on the pump. Carefully lower pump, and remove from under basket.

f. **Install Oil Pump.** Raise pump into position on motor (fig. 213) so that the two shafts mesh properly. Install the four socket-head set screws mounting the pump to the motor. Connect the wires (green, yellow, and white) to their proper terminals (which have been previously tagged for identification), and install terminal box cover (fig. 213). Remove the caps and plugs from the oil lines, and connect the lines. Add oil and bleed system (par. 187) to remove any air trapped in the lines during the removal or installation of the pump.

190. PISTON AND CYLINDER ASSEMBLY.

a. **Description.** The piston and cylinder assembly (fig. 226) consists of a cylinder, piston and rod, and piston rod end. This assembly is mounted to the right of the 37-mm gun, and is located between the 37-mm gun and cal. .30 machine gun. The lower end of the cylinder is mounted on a pivot pin on right end of hand elevating wheel assembly (fig. 223). The upper end of the cylinder assembly is mounted to the piston rod and pivot pin in the mounting bracket.

b. **Maintenance.** The piston and cylinder assembly is replaceable as a unit, and no repair of its internal parts should be attempted in the field.

c. **Removal.** Turn master battery switch to the "OFF" position. Remove cal. .30 machine gun from combination gun mount (par.

TM 9-775

LANDING VEHICLE TRACKED MK. I AND MK. II

Figure 224 — Recoil Switch Installed

182). Disconnect and remove oil lines from cylinder, and plug oil lines. (Provision should be made to catch the oil lost when disconnecting the lines.) Disconnect and remove the firing cable (fig. 223). Loosen the lock nut and socket-head set screw, and remove the collar on the cylinder pivot pin (fig. 223). Loosen the socket-head set screws holding the piston rod end pivot pin (upper end of cylinder); then slide the pivot pin back into the mounting bracket while holding the gun in position to free the piston and cylinder. Work the cylinder end off its pivot pin, and lay cylinder assembly carefully to one side.

d. *Installation.* When mounting the piston and cylinder, use a small amount of O.D. grease No. 0 on pivot pins. Slide the lower end of cylinder onto its pivot pin. Position gun so that piston rod end pivot pin (upper end of cylinder) will slide into piston rod end. Slide pivot pin out of mounting bracket and into the piston rod, then tighten socket-head set screws. Install pivot pin collar and firing cable on cylinder pivot pin, and tighten the set screw and lock nut. The piston rod and cylinder should ride freely on their pivot pins. Remove plugs and reconnect oil lines. Install oil and bleed system

STABILIZER

(par. 187) to remove any air trapped in lines during removal and installation. Install cal. .30 machine gun in combination gun mount (par. 182) Test stabilizer operation (par. 185).

191. RECOIL SWITCH.

a. *Description* (fig. 224). The recoil switch, a normally closed switch held open by the 37-mm gun breech slide in battery position, is closed by the recoil of the gun. The switch bracket is mounted on the machine gun side of the 37-mm gun so the plunger will be released by the recoil mechanism when the gun is fired. The plunger must not protrude from the switch case more than $1/16$ inch when the gun is in battery position.

b. *Maintenance.* The recoil switch is replaceable as a unit, and no repair of its internal parts should be attempted in the field. The switch plunger is adjustable (subpar. d, below).

c. *Removal.* Turn master switch to the "OFF" position. Remove the two screws holding switch to its mounting bracket. Remove cover plate from switch. Disconnect electrical wiring at switch, melting solder when necessary. Unscrew shielded conduit fitting from recoil switch, and remove switch.

d. *Installation.* Connect electrical wiring to switch (solder where necessary). Install cover plate on switch. Position switch on mounting bracket, and install two mounting screws, but do not tighten. Adjust the switch so that its plunger does not protrude more than $1/16$ inch from the switch box (when the gun is in battery), by shifting the switch on its mounting bracket, and tightening the mounting bolts while the switch is held firmly in its correct position. In some cases, it may be necessary to elongate the switch mounting holes to permit shifting it sufficiently to obtain correct adjustment.

192. CONTROL BOX.

a. *Description.* The control box (fig. 211) includes two adjustable rheostats with control knobs, a master switch pilot light, a stabilizer pilot light, stabilizer switch, 37-mm gun switch, cal. .30 machine gun switch, and a dome light switch. One rheostat controls the sensitivity or stiffness of the gun, and the other maintains the position of the gun during recoil. The control box is installed near the top of the turret on left-front side of gunner.

b. *Maintenance.* The control box is replaceable as a unit, and no repair of its internal parts should be attempted in the field.

c. *Removal.* Throw master battery switch to the "OFF" position. Remove 12 screws and toothed lock washers, remove the right side cover from control box, and disconnect the three external wires. Re-

LANDING VEHICLE TRACKED MK. I AND MK. II

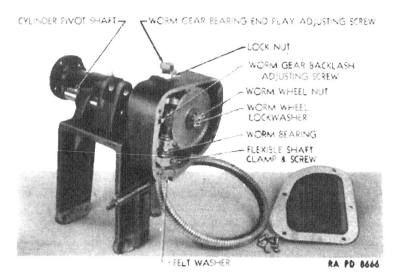

Figure 225 — Mounting Bracket Assembly

move screws holding control box to turret wall, and remove control box.

d. **Installation.** Position control box on turret wall, and install mounting screws. Connect wires to terminals on control box. Install right side cover plate on box with 12 screws and toothed lock washers.

193. MOUNTING SHAFT AND BRACKET.

a. **Description.** The gyro control mounting bracket (fig. 225) is bolted to the gun mount, and provides a mounting for the gyro control. The pivot pin connects the piston and cylinder to the mounting bracket. The worm and worm wheel are driven by a flexible shaft, which extends from the hand elevating wheel (fig. 223) to the gyro control gear box (fig. 225). The mounting bracket fastens to the gun mount in the same manner as a traveling lock bracket. Four mounting bolts are used. Three of the bolts are standard 1-inch long, $3/8$-24; but the fourth bolt is slightly shorter ($7/8$ inch). The shorter bolt is used on the breech end and cylinder side of bracket.

b. **Maintenance.** The mounting bracket and gyro control gear box is replaceable as a unit, or its component parts may be replaced.

c. **Removal.** Throw the master battery switch to the "OFF" position. Remove the direct vision telescope located on left side of 37-mm gun, and remove the rear telescope mounting bracket (par.

STABILIZER

196). Remove the gyro control unit (par. 188). Loosen the two piston rod pivot pin set screws. Slide the piston rod pivot pin out through the piston rod end. Remove gyro control gear box cover plate, loosen the flexible shaft clamp screw on the flexible shaft (fig. 225), and remove flexible shaft. Remove the four bolts holding mounting bracket assembly to gun mount. Lift mounting bracket up and toward breech of the gun to remove.

d. Installation. Lift assembly up over breech of gun, lower into position on gun mount, and install four mounting bolts (subpar. a, above). Install flexible shaft on elevating mechanism, and tighten clamp screw. Position cover plate on gyro control gear box, and install cover mounting screws. Aline piston rod with holes in mounting bracket, then slide piston rod pivot pin through piston rod and into mounting bracket. CAUTION: *Flats on pivot pin must face the set screws. Tighten the two piston rod pivot pin set screws. Install gyro control unit (par. 188). Install rear telescope mounting bracket, and install telescope.*

194. DISENGAGING SWITCH.

a. Description. The disengaging finger and the switch mounting bracket are mounted on the elevating gear housing and gun cradle support respectively. The disengaging switch is held open when the hand elevating gears are in mesh. It automatically returns to its normally closed position when the hand elevating gears are out of mesh. The stabilizer is rendered inoperative when this switch is open (elevating gears in mesh).

b. Maintenance. The disengaging switch is replaceable as a unit, and no repair of its internal parts should be attempted in the field. The plunger must not protrude more than $\frac{1}{16}$ inch from switch when hand elevating gears are in mesh. Plunger protrusion can be adjusted by shifting switch and/or switch mounting bracket. It may be necessary to enlarge mounting holes with a rat-tail file to give necessary adjustment. When correct adjustment is reached, hold switch firmly in its correct position and tighten mounting screws.

c. Removal (fig. 226). Turn motor switch to "OFF" position Remove the screws holding switch mounting bracket to gun mount. Remove two bolts, nuts, and flat washers that secure switch to its mounting bracket. Remove two stud nuts and flat washers that secure cover plate to switch, and remove cover plate. Disconnect electrical wiring to switch (tag wires for ready identification). Unscrew the knurled nut that secures shielded conduit to switch, and remove switch.

d. Installation (fig. 226). Connect switch to conduit, and connect wires to terminals in switch (wires have been tagged for identi-

LANDING VEHICLE TRACKED MK. I AND MK. II

Figure 226 — Disengaging Switch Installed

fication). Position cover plate on switch, and secure with two nuts and flat washers. Position switch on mounting bracket, and install two bolts, nuts, and flat washers to secure switch to mounting bracket. With the hand elevating gears in mesh, check the adjustment of the new disengaging switch (subpar. b, above).

195. FLEXIBLE SHAFT.

a. Description. The flexible shaft is a means of transmitting the action of the elevating handwheel mechanism to the gyro control gear box mechanism (fig. 223). A goose neck flexible shaft supports bracket mounts on the hand elevating wheel mechanism. A clamp holds the end of the flexible shaft casing. The flexible shaft meshes with an adapter which is a part of the hand elevating gear. An oil seal retains the grease, and prevents dirt and foreign matter from entering around the flexible shaft connections. The other end of the flexible shaft is connected to the worm gear in the gyro control gear box by means of a flexible shaft clamp screw (fig. 225). When the gyro control unit is operating, it is necessary only to turn the elevating handwheel in the normal manner. The flexible shaft turns the worm gear in the gyro control gear box, and this changes the relationship between the gyro control and the 37-mm gun. Since the gyro control unit tends to maintain itself in a vertical position, the

TM 9-775
195

STABILIZER

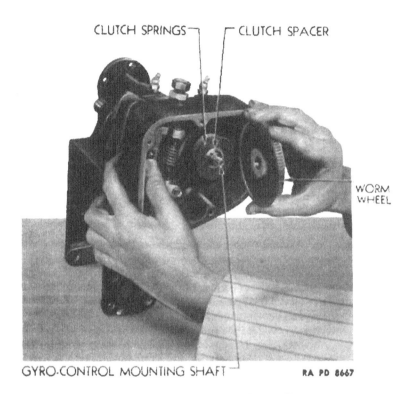

Figure 227 — Worm Wheel Installation

gun moves to a new position automatically, and restores the gyro control to its vertical position.

b. **Maintenance.** The flexible shaft is replaceable as a unit, and no repair of its parts should be attempted in the field.

c. **Removal.** Remove the direct vision telescope and its rear mounting bracket (par. 196). Remove the cover from the gyro control gear box. Loosen the flexible shaft clamp on worm bracket, and remove flexible shaft from gear box (fig. 225). Loosen the flexible shaft clamp on support bracket, and remove flexible shaft.

d. **Installation.** Position the flexible shaft in support bracket, making sure the flexible shaft coupling is in mesh; then tighten clamp screw. Position other end of flexible shaft in worm bracket, making sure that coupling is in mesh; tighten clamp screw (fig. 225). Position cover on gyro control gear box, and install cover mounting screws. Position telescope rear mounting bracket on gun mount, and install

LANDING VEHICLE TRACKED MK. I AND MK. II

three cap screws and toothed lock washers to secure bracket to mount. Place telescope in telescope bracket, and tighten telescope bracket thumb screw.

196. GYRO CONTROL MOUNTING SHAFT.

a. **Description.** The gyro control mounting shaft is located in the mounting bracket assembly (fig. 227). This shaft provides a mounting for the worm wheel, which is turned by the worm gear in the gyro control gear box. The turning of the worm gear by means of the elevating handwheel and flexible shaft actuates the worm wheel and the mounting shaft, and changes the relation between the gyro unit and gun.

b. **Maintenance.** The gyro control mounting shaft, together with its components, such as worm wheel, clutch spacer, clutch springs, and oil seals, is replaceable in event of wear or damage to this assembly.

c. **Removal.** Loosen thumbscrew on telescope clamp and remove telescope. Remove three cap screws and toothed lock washers that secure telescope rear mounting bracket to gun mount, and remove bracket. Turn master battery switch to the "OFF" position. Remove gyro control unit (par. 188). Remove five cap screws and toothed lock washers that secure gyro control gear box cover plate. Remove cover plate. Remove worm wheel lock nut and lock washer, and pull worm wheel from the shaft. Remove clutch spacer and two clutch springs (fig. 227). Pull the gyro control mounting shaft from bracket assembly. It may be necessary to drive the two oil seals from mounting bracket assembly.

d. **Installation.** Install two new oil seals in mounting bracket assembly. Install gyro control mounting shaft in bracket assembly, being careful not to damage oil seals. Install clutch spacer and two clutch springs on shaft (fig. 227). Install worm wheel on shaft, and secure with spanner nut and lock washer (fig. 227). Position cover plate on gyro control gear box, and install five cap screws and lock washers. Install gyro control unit (par. 188). Aline telescope rear mounting bracket holes with holes provided in gun mount, and install three cap screws and toothed lock washers to secure bracket to mount. Install telescope to mounting brackets, and secure with bracket thumbscrew.

TM 9-775
197–198

Section XLI

SHIPMENT AND TEMPORARY STORAGE

	Paragraph
General instructions	197
Preparation for temporary storage or domestic shipment	198
Loading and blocking for rail shipment	199

197. GENERAL INSTRUCTIONS.

a. Preparation for domestic shipment of the vehicle is the same as preparation for temporary storage or bivouac. Preparation for shipment by rail includes instructions for loading and unloading the vehicle, blocking necessary to secure the vehicle on freight cars, number of vehicles per freight car, clearance, weight, and other information necessary to properly prepare the vehicle for rail shipment. For more detailed information, and for preparation for indefinite storage, refer to AR 850-18.

198. PREPARATION FOR TEMPORARY STORAGE OR DOMESTIC SHIPMENT.

a. Vehicles to be prepared for temporary storage or domestic shipment are those ready for immediate service, but not used for less than 30 days. If vehicles are to be indefinitely stored after shipment by rail, they will be prepared for such storage at their destination.

b. If the vehicles are to be temporarily stored or bivouacked, take the following precautions:

(1) LUBRICATION. Lubricate the vehicle completely (par. 27).

(2) BATTERY. Check battery and terminals for corrosion and if necessary, clean and thoroughly service battery (par. 100).

(3) ROAD TEST. The preparation for limited storage will include a road test of at least 5 miles, after the battery, and lubrication services have been made, to check on general condition of the vehicle. Correct any defects noted in the vehicle operation, before the vehicle is stored, or note on a tag attached to the steering wheel, stating the repairs needed or describing the condition present. A written report of these items will then be made to the officer in charge.

(4) FUEL IN TANKS. It is not necessary to remove the fuel from the tanks for shipment within the United States, nor to label the tanks under Interstate Commerce Commission Regulations. Leave fuel in the tanks except when storing in locations where fire ordnances or other local regulations require removal of all gasoline before storage.

(5) EXTERIOR OF VEHICLE. Remove rust appearing on any part of the vehicle with flint paper. Repaint painted surfaces whenever necessary to protect wood or metal. Coat exposed polished metal surfaces susceptible to rust, such as winch cables, chains, and in the case of

TM 9-775
198–199
LANDING VEHICLE TRACKED MK. I AND MK. II

track-laying vehicles, metal tracks, with medium grade preventive lubricating oil. Close firmly all cab doors, windows, and windshields. Vehicles equipped with open-type cabs with collapsible tops will have the tops raised, all curtains in place, and the windshield closed. Make sure paulins and window curtains are in place and firmly secured. Leave rubber mats, such as floor mats, where provided, in an unrolled position on the floor; not rolled or curled up. Equipment such as pioneer and truck tools, tire chains, and fire extinguishers will remain in place in the vehicle.

(6) INSPECTION. Make a systematic inspection just before shipment or temporary storage to insure all above steps have been covered, and that the vehicle is ready for operation on call. Make a list of all missing or damaged items and attach it to the steering wheel. Refer to "Before-operation Service" (par. 22).

(7) ENGINE. To prepare the engine for storage, remove the air cleaners from the carburetor. Start the engine, and set the throttle to run the engines at a fast idle; pour 1 pint of medium grade preservative lubricating oil, Ordnance Department Specification AXS-674, of the latest issue in effect, into the carburetor throat, being careful not to choke the engine. Turn off the ignition switch as quickly as possible after the oil has been poured into the carburetor. With the engine switch off, open the throttle wide, and turn the engine five complete revolutions by means of the cranking motor. If the engine cannot be turned by the cranking motor with the switch off, turn it by hand, or disconnect the high-tension lead and ground it before turning the engine by means of the cranking motor. Then reinstall the air cleaner.

(8) BRAKES. Release brakes and chock the tracks.

c. Inspections in Limited Storage.

(1) Vehicles in limited storage will be inspected weekly for condition of battery. If water is added when freezing weather is anticipated, recharge the battery with a portable charger, or remove the battery for charging. Do not attempt to charge the battery by running the engine.

199. LOADING AND BLOCKING FOR RAIL SHIPMENT.

a. Preparation. In addition to the preparation described in paragraph 198, when ordnance vehicles are prepared for domestic shipment, the following preparations and precautions will be taken.

(1) EXTERIOR. Cover the body of the vehicle with a canvas cover supplied as an accessory.

(2) BATTERY. Disconnect the battery to prevent its discharge by vandalism or accident. This may be accomplished by disconnecting the positive lead, taping the end of the lead, and tying it back away from the battery.

SHIPMENT AND TEMPORARY STORAGE

(3) BRAKES. The brakes must be applied and the transmission placed in low gear after the vehicle has been placed in position with a brake wheel clearance of at least 6 inches (A, figure 232). The vehicles will be located on the car in such a manner as to prevent the car from carrying an unbalanced load.

(4) All cars containing ordnance vehicles must be placarded "DO NOT HUMP."

(5) Ordnance vehicles may be shipped on flat cars, end-door box cars, side-door cars, or drop-end gondola cars, whichever type car is the most convenient.

b. Facilities for Loading. Whenever possible, load and unload vehicles from open cars under their own power, using permanent end ramps and spanning platforms. Movement from one flat car to another along the length of the train is made possible by cross-over plates or spanning platforms. If no permanent end ramp is available, an improvised ramp can be made from railroad ties. Vehicles may be loaded in gondola cars without drop ends, by using a crane. In case of shipment in side-door box cars, use a dolly-type jack to warp the vehicles into position within the car.

c. Securing Vehicles. In securing or blocking a vehicle, three motions, lengthwise, sidewise, and bouncing, must be prevented. There are two approved methods of blocking tracked landing vehicles on freight cars, as described below.

(1) Method 1 (fig. 232). Locate four blocks "B," one to the front and one to the rear of each track. Nail the heel of each block to the car floor, using five 40-penny nails to each block. That portion of the block under the track will be toenailed to the car floor with two 40-penny nails to each block. Locate two cleats "C," one on each side of the vehicle on the outside of both tracks. These cleats may be located on the inside of the tracks if conditions warrant. Nail each cleat to the car floor with three 40-penny nails. In addition to blocking of tracks it is necessary to arrest all up and down movement of vehicle on bogie wheels. Block under hull to prevent movement at points opposite the third bogie from the front and also the third bogie from the rear at each side of the vehicle. Pass four strands, two wrappings of No. 8 gage, black annealed wire "H" through the towing eye at the front center of the vehicle and through a stake pocket on the railroad car, located slightly ahead of the vehicle. Repeat this operation, but fasten the wire to a stake pocket on the other side of the railroad car. Perform the same operation at the rear of the vehicle except that the wire "H" is passed through the splash guard bumper guard, located at each rear corner of the vehicle. Tighten all wrappings enough to remove slack. When a box car is used, this strapping must be applied in a similar fashion, and attached to the floor by the use of blocking or

TM 9-775

LANDING VEHICLE TRACKED MK. I AND MK. II

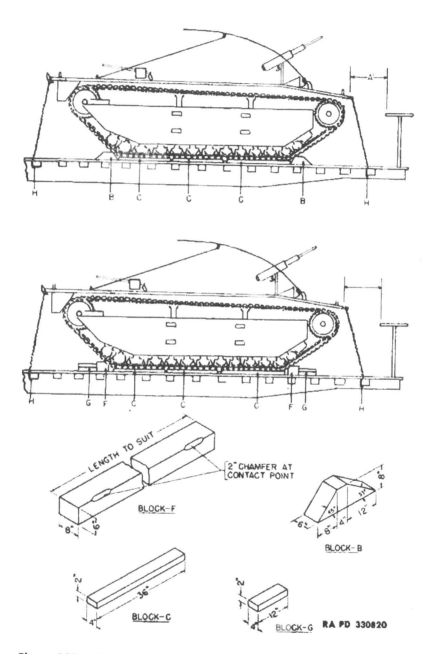

Figure 228 — Blocking Requirements for Securing Tracked Vehicle on Railroad Cars

SHIPMENT AND TEMPORARY STORAGE

anchor plates. This strapping is not repaired when gondola cars are used.

(2) Method 2 (fig. 232). Locate two blocks "F," one to the front and one to the rear of the tracks. These blocks are to be at least 8 inches wider than the overall width of the vehicle at the car floor. Locate eight cleats "G" against blocks "F" to the front and to the rear of each track. Nail the lower cleat to the car floor with three 40-penny nails, and the top cleat to the cleat below with three 40-penny nails. Locate two cleats "C" on each side of the vehicle on the outside of each track. These cleats may be placed on the inside of the tracks if conditions warrant. Nail each cleat to the car floor with three 40-penny nails. In addition to blocking of tracks it is necessary to arrest all up and down movement of vehicle on bogie wheels. Block under hull to prevent movement at points opposite the third bogie from the front and also the third bogie from the rear at each side of the vehicle. Pass four strands, two wrappings of No. 8 gage, black annealed wire "H" through the towing eye at the front center of the vehicle and through a stake pocket on the railroad car, located slightly ahead of the vehicle. Repeat this operation; but fasten the wire to a stake pocket on the other side of the railroad car. Perform the same operation at the rear of the vehicle except that the wire "H" is passed through the splash guard bumper guard, located at each rear corner of the vehicle. Tighten all wrappings enough to remove slack. Locate four strands, two wrappings, of No. 8 gage, black annealed wire "H" on each axle of each inside bogie wheel of the vehicle. Pass wire over the axle to the nearest stake pocket, and tighten enough to remove slack. When a box car is used, this strapping must be applied in a similar fashion, and attached to the floor by the use of blocking or anchor plates. This strapping is not required when gondola cars are used.

d. **Shipping Data.**

	LVT (2)	LVT (A) (2)	LVT (A) (1)
Length, over-all	26 ft 1 in.	26 ft 1 in.	26 ft 1 in.
Width, over-all	10 ft 10 in.	10 ft 10 in.	10 ft 10 in.
Height	8 ft 2 in.	8 ft 3 in.	8 ft 5 in.
Shipping weight	24,400 lb	27,600 lb	31,200 lb
Approximate floor area	283 sq ft	283 sq ft	283 sq ft
Approximate volume	2308 cu ft	2332 cu ft	2379 cu ft
Bearing pressure (lb per sq ft)	87	98	111

TM 9-775

LANDING VEHICLE TRACKED MK. I AND MK. II

REFERENCES

PUBLICATIONS INDEXES.

The following publications indexes should be consulted frequently for latest changes to revisions to the publications given in this list of references and for the new publications relating to materiel covered in this manual:

Introduction to ordnance catalog (explains SNL system) .... ASF Cat. ORD-1 IOC

Ordnance publications for supply index (index to SNL's) .... ASF Cat. ORD-2 OPSI

Index to ordnance publications (lists FM's, TM's, TC's and TB's of interest to Ordnance personnel, FSMWO's BSD, OPSR's, S of SR's OSSC's, and OFSB's. Includes Alphabetical List of Ordnance Major Items, and Publications Pertaining Thereto) OFSB 1-1

List of publications for training (lists MR's, MTP's, T/BA's, T/A's, FM's, TM's, and TR's, concerning training) .... FM 21-6

List of training films, film strips, and film bulletins (lists TF's, FS's, and FB's by serial number and subject) .... FM 21-7

Military training aids (lists graphic training aids, models, devices, and displays) .... FM 21-8

STANDARD NOMENCLATURE LISTS.

Vehicular.

Vehicle, landing, tracked (unarmored), Mk. II LVT (2) (formerly, tractor, amphibian T34) .... SNL G-167

Vehicle, landing, tracked (armored), Army type, Mk. II, LVT (A) (2) (formerly, tractor, amphibian, armored, T35) .... SNL G-168

Ammunition and armament.

Ammunition, fixed and semifixed, all types, for pack, light, and medium field artillery .... SNL R-1

REFERENCES

Ammunition, rifle, carbine, and automatic gun	SNL T-1
Gun, 37-mm, M5 and M6, and recoil mechanisms (tank)	SNL A-45
Gun, machine, cal. .30, Browning, M1919A4, M1919A5—fixed and flexible, and M1916A6—fixed, and ground mounts	SNL A-6
Mounts, small arms, for motor vehicles	SNL A-55
Periscopes, telescopes for periscopes, and direct sighting telescopes for use in tanks	SNL F-235
Stabilizers all types	SNL C-56

Maintenance.

Cleaning, preserving and lubrication materials, recoil fluids, special oils, and miscellaneous related items	SNL K-1
Interchangeability chart of ordnance maintenance tools.	
Interchangeability chart of organizational tools for ordnance vehicles	SNL G-19
Soldering, brazing and welding materials, gases and related items	SNL K-2
Tools, maintenance, for repair of automatic guns and antiaircraft materiel, automatic and semiautomatic cannon and mortars—Individual items and parts	SNL A-35
Tools, maintenance, for repair of pack, light, and medium field artillery; and armament of these calibers for airplane and combat vehicles	SNL C-18
Tool sets—Motor transport	SNL N-19

EXPLANATORY PUBLICATIONS.

Fundamental Principles.

Automotive electricity	TM 10-580
Basic Maintenance Manual	TM 38-250
Electrical fundamentals	TM 1-455
Military motor vehicles	AR 850-15

TM 9-775

LANDING VEHICLE TRACKED MK. I AND MK. II

Motor vehicle inspections and preventive maintenance services .... TM 9-2810

Precautions in handling gasoline .... AR 850-20

Standard Military Motor Vehicles .... TM 9-2800

Ammunition and Armament.

Ammunition for combat vehicles .... TM 9-1910

Ammunition, general .... TM 9-1900

Auxiliary fire-control instruments .... TM 9-575

37-mm gun materiel (tank), M5 and M6 .... TM 9-250

37-mm gun, tank, M6 (mounted in combat vehicles) FM 23-81

Browning machine gun, cal. .30, HB, M1919A4 (mounted in combat vehicles) .... FM 23-50

Maintenance and Repair.

Cleaning, preserving, lubricating and welding materials and similar items issued by the Ordnance Department .... TM 9-850

Cold weather lubrication and service of combat vehicles and automotive materiel .... OFSB 6-11

Ordnance maintenance: 37-mm gun materiel (tank) M5 and M6 .... TM 9-1250

Ordnance Maintenance: Browning machine gun, cal. .30, all types; U. S. machine gun, cal. .22, and trainer, cal. .22 .... TM 9-1205

Ordnance Maintenance: Carburetors (Stromberg) .. TM 9-1826B

Ordnance Maintenance: Clutch, power, take-off, bilge pump, tracks and suspension, and hull for tracked landing vehicles: Mk. II (armored) LVT (2) (A), Mk. II (unarmored) LVT (2), and Mk. I (armored) with turret LVT (2) (A) .... TM 9-1775

Ordnance Maintenance: Continental engine, Model W670-9A .... TM 9-1726

Ordnance Maintenance: Electrical equipment (Delco-Remy) .... TM 9-1825A

TM 9-775

REFERENCES

Ordnance Maintenance: Fire extinguishers	TM 9-1799
Ordnance Maintenance: Fuel and lubrication systems for light tanks M3 and M3A1	TM 9-1726
Ordnance Maintenance: Fuel pump	TM 9-1828A
Ordnance Maintenance: Generating, starting and ignition systems for light tank M3 and Modifications	TM 9-1726A
Ordnance Maintenance: Hull and turret for light tanks M5, M5A1 and 75-mm howitzer motor carriage M8	TM 9-1727C
Ordnance Maintenance: Hydraulic traversing mechanism for light tank M5 (Oilgear)	TM 9-1727K
Ordnance Maintenance: Power train for light tank M3 and M3A1	TM 9-1728
Tune-up and adjustment	TM 10-530

Protection of Materiel.

Camouflage	FM 5-20
Chemical decontamination, materials and equipment	TM 3-220
Decontamination of armored force vehicles	FM 17-59
Defense against chemical attack	FM 21-40
Explosives and demolitions	FM 5-25

Storage and Shipment.

Ordnance storage and shipment chart, group G — Major items	OSSC G
Registration of motor vehicles	AR 850-10
Rules governing the loading of mechanized and motorized army equipment, also major caliber guns, for the United States Army and Navy, on open top equipment published by Operations and Maintenance Department Association of American Railroads.	
Storage of motor vehicle equipment	AR 850-18

TM 9-775

LANDING VEHICLE TRACKED MK. I AND MK. II

INDEX

A

	Page No.
Accelerator pedal, description and functioning	23
Accelerator pedal shaft assembly	
assembly	328
disassembly	323
installation	348
removal	348
Accelerator pedal linkage, adjustment	165–166
Accessories, cold weather	43
Adjustment:	
accelerator pedal	165–166
brake shoes	279
breaker point gap	213
engine	134
recoil adjuster	374–375
stiffness adjuster	374
track tension	285–287
valves and valve push rods	158–160
After-operation and weekly service	51–56
Air cleaners	
after-operation and weekly service	53
at-halt service	51
description	184
lubrication	57
maintenance	184–186, 365
run-in test procedure	82
second echelon preventive maintenance	96
Air intake tubes	
installation	144
removal	141
Ammeter	
description and functioning	27
removal and installation	318–319
run-in test procedure	84
second echelon preventive maintenance	91
trouble shooting	118–119
Ammunition data	22
Angle support and upper front baffle	
installation	149
removal	134

	Page No.
Armament	
data	22
description	9
second echelon preventive maintenance	104–105
Assembly	
return idler	308–309
sprocket hub	298–301
At-halt service	50–51
Auxiliary engine generator filter, removal and installation	242
Auxiliary equipment controls and operation	
auxiliary generator	39–40
fixed fire extinguishers	40
portable fire extinguishers	40–41
Auxiliary generator	
description	39
installation	367–368
maintenance	365
operation	39–40
removal	365–367
second echelon preventive maintenance	103–104
trouble shooting	119–120
Auxiliary generator crankcase, lubrication	57–60
Auxiliary generator guard	
installation	368
removal	365–367

B

Baffles and cowling	
description	133
installation	156–158
maintenance	152
removal	152–155
Batteries and cables	
description and data	222
installation	224–225
maintenance	222–224
removal	224

INDEX

B — Cont'd

	Page No.
Battery and lighting system	
after-operation and weekly service	53
batteries and cables	222–225
circuit breaker	235
description	222
fan junction box	229–230
flexible conduits and wiring	235–236
headlight receptacle box	231–232
headlights	225–227
main junction box	232–235
radio junction box	237–238
run-in test procedure	82
solenoids	236
speaker junction box	238–240
spotlight receptacle box	229
taillights	227–229
trouble shooting	116–120
Battery and wiring, care in extreme cold	42
Battery switch, description and functioning	30
Bilge pump	
description and data	339–342
installation	342–344
maintenance	340
removal	340–342
Blower switch (engine room), description and functioning	30
Bogie assembly	
description	292
removal and installation	293
Bogie wheel rubber bumpers	
description	293–295
removal and installation	295
Booster coil, removal and installation	217
Booster switch, description and functioning	29
Brake shoe assemblies, installation	282
Brake shoes, adjustment	279
Breaker point gap, adjustment	213
Breech mechanism, lubrication	60

C

	Page No.
Cab, description	9–13
Cab escape windows	
description and maintenance	337
installation of window and frame	339
removal	
window and frame	337
window from frame	339
Cab fan	
description and maintenance	345
installation	345–347
removal	345
Cab seats	
description	334
maintenance	335
Carburetor	
description	164
inspection and cleaning	164
installation	165
maintenance	365
removal	164–165
second echelon preventive maintenance	96
Cargo compartment, description	13
Chain cross plates	
description and maintenance	290
removal and installation	291
Channel support	
installation	183
removal	173
Circuit breaker, removal and installation	235
Cleaning carburetor	164
Clutch	
description	320
organizational special tools and equipment	108
run-in test procedures	85
second echelon preventive maintenance	91
trouble shooting	125
Clutch assembly	
description	242
installation	247
maintenance	242–245
removal	245–247

LANDING VEHICLE TRACKED MK. I AND MK. II

C — Cont'd

	Page No.
Clutch pedal	
assembly	328
description and functioning	23
disassembly	323
installation	277
removal	272
Clutch pedal shaft	
installation	348
removal	347–348
Compass, description and functioning	29
Control box	
description	381
removal	381–382
Control panel	
description and maintenance	313
installation	315–317
switches	313
removal	313–315
switches	313
Control panel condenser, removal and installation	241
Control rods and linkage, removal	325
Controlled differential	
description	281
installation	281–284
removal	281
Controls and levers, description and maintenance	347
Controls and linkage	
description	320–323
installation	325–328
removal	323–325
Cooling system, description	195
Cranking motor	
organizational special tools and equipment	108
removal and installation	218
trouble shooting	112–114

D

	Page No.
Data	
ammunition	22
armament	22
engine	132–133
fuel tanks	171
vehicle specifications	21
Dashlight switch, description and functioning	30

	Page No.
Description:	
air cleaners	184
auxiliary generator	39
battery and lighting system	222
bogie assembly	292
carburetor	164
controls and linkage	320–323
engine	129–132
engine and transmission oil lines	203
engine exhaust system	188
fire extinguishers	329
fuel filter	167
fuel pump	166
fuel system	161–163
fuel tanks	170–171
generator regulator	220
generating system	219
gun and mount	369–370
ignition system	207
intake and exhaust system	188
lubrication system	195–197
magnetos	207
oil coolers	199–201
oil filter	197
oil pressure and scavenger pump	204
oil reservoir	201
oil screens	205
power train	261–264
priming pump	186
radio interference suppression system	240–241
stabilizer	373
starting system	217
tracks and suspension	284–285
turret	349
Differences among models of tracked landing vehicles	17–21
Differential (See Transmission and differential)	
Disassembly	
return idler	305–308
sprocket hub	298
Disengaging switch	
description and maintenance	383
installation	383–384
removal	383
Dragging clutch, trouble shooting	125

INDEX

D — Cont'd

Driving controls and operation
  instruments and controls 23-30
  use of in vehicular operation 30-34
  towing the vehicle 34-35
Driving sprockets
  description 285
  removal and installation 261

E

Electrical lines felt gasket
  installation 145
  removal 140
Engine
  baffles and cowling 133
  description and data 129-133
  engine support beam 133
  maintenance and adjustment of
    the vehicle 134
  organizational special tools and
    equipment 107
  removal and installation 134-161
  trouble shooting 111-112
  valves and valve push rods 134
Engine and transmission oil lines,
  description and maintenance 203
Engine exhaust system
  description 188
  installation 190
  removal 188-190
Engine mounting bolts and cap
  screws
  installation 144
  removal 141
Engine oil pressure gage
  description and functioning 23
Engine oil temperature gage
  description and functioning 27
  run-in test procedure 84
  second echelon preventive main-
    tenance 91
Engine oiling system 121-122
Engine room blower
  installation 192-193
  removal 190-192
Engine sling, installation 141, 143

Engine support beam
  description 133
  installation 150-152
  removal (engine out of vehicle) .. 150
Escape hatch
  description 336
  removal and installation 337
Exhaust system (See Intake and
  exhaust systems)

F

Facilities for loading 389
Fan junction box
  description 229
  removal and installation 230
Fan shroud
  installation 157
  removal 153
Fan switch, description and
  functioning 30
Final drive
  driving sprockets and sprocket
    hubs 261
  final drive assembly 254-259
  trouble shooting 126-127
Fire extinguisher
  after-operation and weekly
    service 53
  before-operation service 46
  fixed 40
  portable
    description and operation 40
    precautions 40-41
    run-in test procedures 81
Fire extinguishing system
  care of 330-331
  description 329
  handling of 331
  installation 331-333
  lines and nozzles 334
  remote controls 333-334
  removal of fixed fire
    extinguishers 329-330
Firing mechanism, lubrication 60

TM 9-775

LANDING VEHICLE TRACKED MK. I AND MK. II

F — Cont'd

First echelon preventive maintenance
- after-operation service 51–56
- at-halt service 50–51
- before-operation service 45–48
- during operation 48–50

Flange adapter
- description 279
- installation 281
- removal 279

Flexible conduits and wiring, installation of new conduits and wiring 236

Flexible shaft
- description 384–385
- installation 385–386
- removal 385

Flexible tubing and connections, description 188

Floor plates, description 344

Flywheel and fan assembly, description 195

Fuel filters
- after-operation and weekly service 54
- description and maintenance .. 167
- installation 169–170
- removal 167–169

Fuel flow system, description .. 161–163

Fuel lines, description 187

Fuel pump
- description and maintenance .. 166
- installation 166–167
- removal 166

Fuel tank filler pipe assembly
- installation 179–183
- removal 175

Fuel tanks
- data 171
- description 170–171
- installation 178–184
- removal 171–178

Fuel shut-off valve, description 163, 187

Fuel shut-off valve handles, description and functioning 30

Fuel system
- accelerator pedal linkage 165–166
- air cleaners 184–186
- carburetor 164–165
- description 161–163
- fuel filter 167–170
- fuel pump 166–167
- fuel tanks 170–184
- lines, valves, and fittings 187
- priming pump 186–187
- trouble shooting 121

G

Gas exhaust duct bottom
- installation 193
- removal 192

Gear box assembly, removal and installation 355

Gearshift lever, description and functioning 23

Generating system
- description 219
- generator 219–220
- generator regulator 220–221
- trouble shooting 116–120

Generator
- care in extreme cold 42
- installation 219–220
- maintenance 219
- removal 219
- second echelon preventive maintenance 95
- (See Auxiliary generator)

Generator regulator
- description and data 220
- installation 221
- removal 220–221

Generator tools, organizational special tools and equipment 108

Grousers
- description 290
- maintenance 290
- removal and installation 291

INDEX

G — Cont'd

Gun, machine, cal. .30, M1919A5, fixed
description 369
operation 370–371
removal and installation 371

Gun, 37-mm, M6
description 372
removal and installation 372

Gun equipment
ammunition 74
cal. .30 machine gun 74–75
cal. .50 machine gun 75

Gun spare parts
cal. .30 machine gun 76–77
cal. .50 machine gun 77–80

Gun tools 74

Gyro control mounting shaft, removal and installation 386

Gyro control unit, description and maintenance 378

H

Hand throttle, description and functioning 29

Headlight receptacle box
description 231
installation 231–232
removal 231

Headlights
description 225
installation 225–226
removal 225
bilge pump 339–344

Hull,
cab escape windows 337–339
cab fan 345–347
cab seats 334–335
controls and levers 347–348
description 9
draining water from pontoons 348
escape hatches 336–337
floor plates 344
safety belts 335–336

Hull and pontoons, trouble shooting 128–129

Hydraulic motor
description 355
installation 356
removal 355–356

Hydraulic traversing mechanism, trouble shooting 127–128

I

Ignition system
booster coil 216–217
description 207
magnetos 207–215
spark plugs 215–216
trouble shooting 114–116

Inspection
carburetor 164
in limited storage 388

Instrument panel
description and maintenance 309
removal 309–312

Instrument panel filter, removal and installation 241

Instruments and controls 23–30

Instruments and gages 317

Intake and exhaust systems
description 188
engine exhaust system 188–190
engine room blower
installation 192–193
removal 190–192
intake pipe gland nut 193

Intake and exhaust valves, organizational special tools and equipment 108

Intake pipe gland nut 193

Intercylinder baffling, removal 153–155

L

Light switch, description and functioning 30

Lighting system, trouble shooting 122–124

Lines and nozzles, description 334

Loading and blocking for rail shipment 388–391

Lower front baffle, installation 184

Lubrication
detailed instruction 57–64
lubrication guide 56–57
magnetos 207

401

LANDING VEHICLE TRACKED MK. I AND MK. II

L – Cont'd

	Page No.
Lubrication system	
description	195–197
engine and transmission oil lines	203
oil coolers	199–201
oil filter	197–199
oil pressure and scavenger pump	204–205
oil reservoir	201–203
oil screens	205–206

M

Magneto guard	
installation	149
removal	136
Magneto switch, description and functioning	30
Magnetos	
description and data	207
installation and timing of right magneto (engine removed)	211–214
maintenance	207
organizational special tools and equipment	108
removal	207–211
timing magnetos (engine in vehicle)	214–215
Main junction box	
description	232
installation	235
removal	234
Maintenance	
air cleaners	184–186
auxiliary generator	365
baffles and cowling	152
engine	134
engine and transmission oil lines	203
fuel filter	167
fuel pump	166
magnetos	207
oil coolers	201
oil filter	197
oil screens	205–206
priming pump	186

	Page No.
radio interference suppression system	241–242
spark plugs	215
valves and valve push rods	158–160
Manifolds, second echelon preventive maintenance	96
Mount, gun, combination, M44, description	369
Mounting shaft and bracket	
description and maintenance	382
installation	383
removal	382–383
Muffler and flexible tubing	
installation	149
removal	134
Mufflers, description	188
MWO and major unit assembly, replacement record	107

N

Nautical terms	5–8

O

Oil cooler transmission, second echelon preventive maintenance	98
Oil coolers	
description	199–201
installation and removal	201
maintenance	201
Oil filter	
description and maintenance	197
installation	198–199
removal	198
Oil pot (reservoir), removal and installation	356
Oil pressure and scavenger pump	
description	204
installation	205
removal	204
Oil pressure gage, second echelon preventive maintenance	90
Oil pressure pump screen, removal and installation	206
Oil pump motor	
description	353
installation	354–355
removal	353–354

INDEX

O — Cont'd

	Page No.
Oil reservoir	
description	201
installation	203
removal	202–203
Oil screens	
description	205
maintenance	205–206
Oil temperature indicator, installation and removal	319
Oilcan points	60–61
Ordnance personnel, lubrication by	61–64
Organizational tools and equipment	107–110
Operation	
auxiliary generator	39–40
guns	
elevating, depressing, and firing	371
loading and traversing	370
stabilizer	373–374
turret controls	37–38
under unusual conditions	
desert operation	43
in extreme cold	41–43
in extreme heat	43
in mud	44
preventive maintenance	41
submersion of engine in salt water or fresh water	43–44

P

	Page No.
Panels and instruments	
control panel	313–317
instrument panel	309–312
instrument and gages	317–319
regulator and relay panel	312–313
Periscopes	
description	35–36
operation	38
second echelon preventive maintenance	101
Piston and cylinder assembly	
description and maintenance	379
installation	380–381
removal	379–380
Pontoons, description	9

	Page No.
Power take-off	
description	252
installation	254
removal	252–254
Power train	
adjustment of brake shoes	279
controlled differential	281–284
description	261–264
flange adapter	279–281
organizational special tools and equipment	108
transmission and differential	
installation	274–279
lubrication system	264
removal	264–274
Primer lines, description	163
Priming pump	
description and maintenance	29, 186
installation	187
removal	186
Propeller shaft	
at-halt service	51
description	247
installation	250–251
maintenance	247
removal	248–250
run-in test procedures	83
trouble shooting	126
"Purging" or "bleeding" stabilizer system	376–378

R

	Page No.
Radio interference suppression system	
description	240–241
maintenance	241–242
trouble shooting	120–121
Radio junction box	
description and maintenance	237
installation	238
removal	237–238
Radio junction box condenser, removal and installation	242
Rear baffle, upper left	
installation	145
removal	140
Rear idler adjusting screw, removal and installation	304

LANDING VEHICLE TRACKED MK. I AND MK. II

R — Cont'd

	Page No.
Rear idler slide assembly	
description and maintenance	301
installation	301–303
removal	301
Rear idler slide bar, removal and installation	303
Rear idler wheels	
description	295
maintenance	298
rear idler sprocket	
installation on sprocket hub	301
removal from sprocket hub	298
sprocket hub	
assembly	298–301
disassembly	298
Recoil adjuster, adjustment	374–375
Recoil switch	
description and maintenance	381
removal and installation	381
Regulator and relay panel	
installation	312–313
removal	312
Regulator relay condenser, removal and installation	241
Remote controls	
description	330–334
replacing broken cables	334
Reports and records of lubrication	64
Return idler	
assembly	308–309
description	304–305
disassembly	305–308
removal	305
Roller assemblies and segments	
installation	361–362
removal	358–359
Run-in test, new vehicle	
correction of deficiencies	81
test procedures	81–87

S

	Page No.
Safety belts	
description	335
maintenance	335–336
removal and installation	336
Scavenger oil screen, removal and installation	206
Sealed beam lamp-unit	
installation	226–227
removal	226
Second echelon preventive maintenance	87–106
MWO and major unit assembly replacement record	107
Shipment and temporary storage	
loading and blocking for rail shipment	388–391
preparation for temporary storage or domestic shipment	387–388
Shipping data	391
Solenoids	236
Spark plug baffles	
installation	157
removal	152–153
Spark plugs	
installation	216
maintenance	215–365
removal	215–216
Speaker junction box	
description and maintenance	238
installation	238–240
removal	238
Specific gravity of battery	223–224
Spotlight receptacle box, removal and installation	229
Sprocket hub	
assembly	298–301
disassembly	298
Stabilizer	
control box	381–382
description	373
disengaging switch	383–384
flexible shaft	384–386
gyro control mounting shaft	386
gyro control unit	378
mounting shaft and bracket	382–383
operation	373–374
piston and cylinder assembly	379–381
"purging" or "bleeding" the system	376–378
recoil switch	381
stabilizer oil pump	378–379
trouble shooting	375–376
Starter switch, description and functioning	29–30

TM 9-775

INDEX

S — Cont'd

	Page No.
Starting system	
cranking motor	217–218
description	217
trouble shooting	112–114
Steering lever shaft, removal and installation	347
Steering lever shaft assembly	
assembly	325–328
disassembly	323
Steering levers	
description and functioning	23
disassembly	323–325
Stern cover	
installation	149
removal	134
Submersion of engine in salt or fresh water	43–44

T

Taillights	
description and maintenance	227
installation	228–229
removal	227–228
Throttle, description	320–323
Throttle and clutch control rods, installation	325
Throttle control levers, installation	325
Tools and equipment stowage on the vehicle	
care of tools and equipment	80
gun equipment	74–75
gun spare parts	75–80
gun tools	74
second echelon preventive maintenance	105–106
vehicle equipment	72–73
vehicle spare parts	73–74
vehicle tools	64–72
Track connecting fixture, installation	289
Track tension, adjustment	285–287
Tracks, organizational special tools and equipment	108–110
Tracks and suspension	
bogie assembly	292–293
bogie wheel rubber bumpers	293–295
description	284–285
grousers and chain cross plates	290–291

	Page No.
installation of tracks	287–290
return idler	304–309
rear idler adjusting screw	304
rear idler slide assembly	301–303
rear idler slide bar	303
rear idler wheels	295–301
removal of tracks	287
track tension adjustment	285–287
trouble shooting	127
Transmission (and differential)	
installation	274–279
lubrication system	264
removal	264–274
second echelon preventive maintenance	91
trouble shooting	126–127
Transmission oil pressure gage	
description and functioning	27
installation	319
removal	318–319
Traversing control assembly	
description and maintenance	349
installation	351–353
removal	351
Traversing mechanism, second echelon preventive maintenance	92
Traversing oil pump, description and maintenance	356
Trouble shooting	
batteries and generating system	116–120
clutch	125
engine	111–112
engine oiling system	121–122
fuel system	121
hull and pontoons	128–129
hydraulic traversing mechanism	127–128
ignition system	114–116
lighting system	122–124
propeller shaft	126
radio interference suppression	120–121
stabilizer	375–376
starting system	112–114
tracks and suspension	127
transmission and final drive assembly	126–127

405

LANDING VEHICLE TRACKED MK. I AND MK. II

T — Cont'd	Page No.
Turret	
description	13, 349
installation	361–363
gear box assembly	355
hydraulic motor	355–356
oil pot (reservoir)	356
oil pump motor	353–355
removal	358–361
traversing control assembly	349–353
traversing oil pump	356
turret doors	356–357
turret seats	357–358
Turret controls and operation	
description	35–37
operation	38
Turret doors	
description	356
removal and installation	357
Turret seats	
description and maintenance	357
installation	357–358
removal	357
Turret traverse controls	
description	36–37
operation	38

U	Page No.
Universal joints	
description	247
installation	250–251
lubrication	60
maintenance	247
removal	248–250

V	
Valves and valve push rods	
description	134
installation	161
maintenance and adjustment	158–160
removal	160–161
Vehicle equipment	72–73
Vehicle spare parts	73–74
Vehicle specifications	21
Vehicle tools	64–72
Vehicular operation, use of instruments and controls in	
before-operation service	30
from land to water	33
from water to land	34
on land	32–33
on water	33–34
starting engine	30–32
stopping engine	34

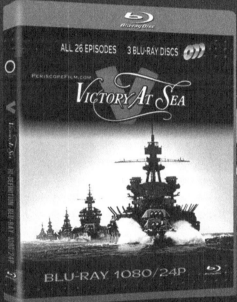

IN HIGH DEFINITION
NOW AVAILABLE!

COMPLETE LINE OF WWII AIRCRAFT FLIGHT MANUALS

WWW.PERISCOPEFILM.COM

©2013 Periscope Film LLC
All Rights Reserved
ISBN#978-1-937684-36-5
www.PeriscopeFilm.com

CPSIA information can be obtained at www.ICGtesting.com
Printed in the USA
LVOW05s1748261114

415840LV00035B/2058/P